Flexures

Flexures

Elements of Elastic Mechanisms

Stuart T. Smith
University of North Carolina, USA

CRC PRESS

Boca Raton London New York Washington, D.C.

Library of Congress Cataloging-in-Publication Data

Catalog record is available from the Library of Congress

International Standard Book Number 90-5699-261-9
Printed in the United States of America 4 5 6 7 8 9 0
Printed on acid-free paper

Contents

Preface

Flexures represent a broad range of compliant elements for connecting rigid bodies. Flexure mechanisms are produced through the successive connection of numerous rigid bodies, or 'links', and flexure elements, or 'joints', in such a way that, upon application of an appropriate force, there will be defined motions of one link relative to the others. Consequently, the goal of a flexure mechanism is to maintain a precise geometric relation between links while simultaneously providing sufficient compliance to accommodate relative motion in specific directions. Precise geometric relations can only be maintained through suitably applied constraints while relative motions require freedom. Because the required constraints and freedoms are in different directions, such a requirement is, in many cases, readily achieved. Such directionality can often be approximated using elastic elements, the geometry of which results in high and low compliance in the required directions. Though providing neither perfect constraint nor freedom, relative stiffness in different directions can vary by many orders of magnitude thereby producing a close approximation to the ideal. It is the design of flexure elements and their integration to form useful mechanisms that occupies this current text.

Being on the design of elastic flexure mechanisms, it is hoped that, by highlighting current issues and deficiencies in knowledge (possibly the authors), this has broken the ground for further developments. In many applications, and within limited ranges, flexures provide an inexpensive solution to problems requiring ultra-precise motions and/or forces. In some cases, the results are nothing short of spectacular. As a caveat, during the synthesis stage of design, there is always a danger of premature 'fine focus'. Too close attention to this book in isolation could induce 'near sight'. As has been the author's experience, flexures are not always the best solution to a fine motion or linear force problem. It is the objective of this book to provide ideas to enable 'good' flexure design and, possibly more importantly, an analytic framework to determine their suitability for a particular application. If the results of such analysis push the designer towards uneasy compromises, the chances are that alternatives may be more suitable. If this happens, it is probably best to look back to the possible solutions that were considered during initial brainstorming or reconsider the design requirements. Because of their simplicity, it is sometimes difficult to drop the idea of using a flexure in favor of more complicated alternatives. Some solace can be derived from the knowledge that the degree of difficulty sometimes represents a competitive edge in the marketplace.

This book provides the relevant mathematical tools and formulae for a broad range of flexure elements and their subsequent combination to form useful mechanisms. It is hoped that it will be of use for designers of precision mechanical instruments be they engineers, physicists or any other discipline in which fine instruments are required. Though this book is intended to be of use to professionals concerned with the design of precise mechanical instruments and machines, it could also be used to compliment graduate courses on precision mechanical design and will certainly provide extensive background material on flexure design to graduate researchers. It is the queries of the latter that created the initial motivation for starting this work and I hope that this will address some of the common, recurrent issues.

For the assessment, optimization and simplification of designs, symmetry has played probably the greatest part in the author's experience. Symmetric designs often tend to nullify possible errors, simplify analysis, de-couple vibration modes, reduce cost of manufacture and are inherently thermally and dynamically balanced. In addition to the above advantages, symmetry is 'easy on the eye' and can often be spotted at a glance. However, after a flexure has distorted, the symmetry is perturbed. In reality, vibration modes and forces originally considered to be independent are always weakly coupled and the engineer must live with the consequences.

This is drawn from collaborative work with many colleagues and students from around the world over many years. The support and faith of immediate coworkers at both the University of Warwick, UK, and the University of North Carolina at Charlotte, USA, has made this possible. In the production of this manuscript it became necessary to divert time from other duties. Bob Hocken without mention added a number of my more onerous burdens to his already substantial workload. Special mention goes to Shane Woody for working through many sections of the book, verifying some of the newly developed theories and removing a number of errors and misprints. Vivek Badami, Eric Coley, Jami Dale and Ashok Muralidhur were most helpful in providing case study examples. Dr's Harish Cherukuri and Jimmie Miller both carefully read the manuscript and, in many places, removed ambiguity and inelegant prose while adding considerable insight to the theoretical foundations upon which most of this book is built. Thanks also go to John Lovingood for both reading the proofs and providing the majority of the figures.

These notes represent the ideas and opinions of the author and, at the present time, much is left to say on this topic, possibly a second volume. Although every effort has been made to eliminate errors, experience dictates their insidious and inevitable presence in a work of this length. Extensive references are included to enable the reader to verify this work using articles that have been subject to the scrutiny of peer review. When detected, the author would be most grateful for guidance on errors or omissions.

1 Introduction

Then he took up the bow; with his right hand,
he tried the string; it sang as clearly as
a swallow's note[1].

1.0 Origins

Elastic deflections have been utilized in fine instrument mechanisms and precise machines for certainly more than three hundred years (a lot longer if the bow is considered a precision machine). While Galileo (1564-1642) sowed the seeds of scientific investigation of the static response of a built-in beam to applied forces, it was left to Hooke (1635-1703) and Marriott (1620 - 1684) to recognize the basis of linear elasticity. Subsequently, Augustine Cauchy (1789-1857), Barre Saint-Venant (1797-1886) and William Thompson (1824-1907 later Lord Kelvin) played a major part in the establishment of a rigorous framework upon which present theoretical investigations are founded, for a more complete historical discussion see Love, 1927, and Timoshenko, 1953. Possibly, an epocryphal moment marking first uses of precision flexures in instrumentation began with John Harrison's (1693-1776) development of a clock of unsurpassed precision for determination of longitude, see for example Sobel, 1995. With *Principia* as a guide, Harrison strove for most of his eighty-three years to produce a mechanism of temporal accuracy matching, as near as possible, the precision provided by the new Newtonian description of the universe (at that time with unprecedented success). Another period of note for which flexures can be identified as 'instrumental' to scientific progress, was the development of galvanometric instruments for the precise determination of the unit of electrical current (designers involved in the development of these early precision instruments include Helmholtz, Joule, Kelvin, Maxwell and Weber). Today flexure mechanisms in various forms find use in applications at the limits of precision and can also be found in commonplace consumer products. Examples at the extremes of precision include fine positioning stages for a wide variety of mechanical measuring instruments, probe microscopes, mass balances, step-and-repeat cameras and x-ray

---

[1] *The Odyssey* of Homer, trans. Allen Mandelbaum, 1990, University of California Press, Berkeley and Los Angeles, Ca, Book XXI, *Greek [401-428]*

interferometers. Commercially, flexures are ubiquitous and can be found in computer disk drives, compact disk players, coordinate measuring machines, optical scanners, optical interferometers, dial indicators, Fabry-Perot etalons, electronic lithography and almost any precision manufacturing machine, for a condensed review of various mechanisms see Geary, 1954, 1960, Sydenham, 1984. Possibly the main reason for their success is because flexures are easy to manufacture and provide smooth, friction free and wear free motion. In marked contrast, friction induced errors often account for a major portion of the precision motion errors of conventional mechanisms.

A review of the emergence of this specialized topic within the field of instrumentation must await the attention of a more qualified science historian. However, readers may derive some sense of the history of flexures within a scientific context from some of the case studies in this text and also the book of R. V. Jones (1911-1997) and references therein.

1.1 Objective

A flexure is usually considered to be a mechanism consisting of a series of rigid bodies connected by compliant elements that is designed to produce a geometrically well-defined motion upon application of a force. The objective of this book is to cover design of flexure mechanisms for precise control of motion or forces. In particular, techniques for analysis of flexure elements are discussed in some detail with some issues pertaining to manufacture and implementation covered in later chapters. Information on the design of the more conventional types of spring, including material properties and selection, can be found in more detail in the books of Carlson, 1978, and Wahl, 1964. These references provide a comprehensive discussion of coil and other, more traditional, spring designs and could be considered complimentary to this book. Before proceeding, it is worth reviewing the merits and limitations of flexure mechanisms for application in precision instrumentation and machinery.

1.2 Advantages of flexures

1. They are simple and inexpensive to manufacture and assemble
2. Unless fatigue cracks develop, the flexures undergo no irreversible deformations and are, therefore, wear-free.
3. Complete mechanisms can be produced from a single monolith.
4. Mechanical leverage is easily implemented.
5. Displacements are smooth and continuous. Even for applications requiring displacements of atomic resolution, flexures have been shown to readily produce predictable and repeatable motions at this level.

6. Failure mechanisms such as fatigue and yield are well understood.
7. They can be designed to be insensitive to thermal variations and mechanical disturbances (vibrations). Symmetric designs can be inherently compensated and balanced.
8. There will be a linear relationship between applied force and displacement for small distortions. For elastic distortions, this linear relationship is independent of manufacturing tolerance. However, the direction of motion will be less well defined as these tolerances are relaxed.

1.3 Disadvantages of flexures

1. Accurate prediction of force-displacement characteristics requires accurate knowledge of the elastic modulus and geometry/dimensions. Even tight manufacturing tolerances can produce relatively large uncertainty between predicted and actual performance.
2. At significant stresses there will be some hysteresis in the stress-strain characteristics of most materials.
3. Flexures are restricted in the length of translation for a given size and stiffness.
4. Out of plane stiffness values are relatively low and drive direction stiffness tends to be relatively high in comparison to other bearing systems.
5. They cannot tolerate large loads.
6. Accidental overload can be catastrophic or, at least, significantly reduce fatigue life.
7. At large loads there may be more than one state corresponding to equilibrium, possibly leading to instabilities such as buckling or 'tin-canning'.

1.4 Goals of flexure design

The goals of any flexure design are twofold:

1. To provide a precise displacement upon application of a specific applied force.
2. To provide a precisely known force upon application of a specific applied displacement.

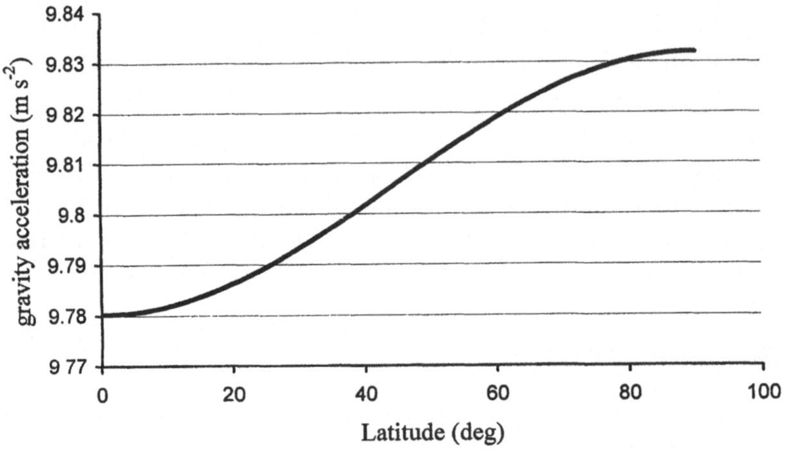

Figure 1.1: Local gravity acceleration at sea level, 0 degrees corresponds to either the 'true' north, or south, poles while 90 degrees is at the equator.

The former of these goals is surprisingly difficult. Were it not for the convenient constant effects of gravitation and a somewhat tenuous, artifact based, mass standard, things would be nigh impossible for the flexure designer and many other scientific investigations. To this day, the simplest method for deriving a 'known' force is to utilize the weight of a calibrated mass. In fact, if the latitude, ϕ, and height above sea level, h, are known, assuming smooth topography, the local gravity acceleration in the direction of the Earth's center of mass can be computed with uncertainty of less than one part in 10,000 from the equation, Howard and Fu, 1997,

$$g = 9.78024\left(1 + 0.0052884\sin^2\phi - 0.0000059\sin^2 2\phi\right) - 0.000003086h \qquad (1.1)$$

At sea level, the local gravity constant as a function of latitude is shown in figure 1.1. Irrespective of the value of gravity acceleration, at a fixed point relative to the Earth this will have a value that can be assumed constant for extremely long periods of time. Ignoring growth of contaminant films, wear and other 'decay' processes masses derived from solids can also maintain a relatively constant value over protracted periods of time. However, for most applications, a constant force would lead to a rather dull and unproductive machine, the gravity wound' clock being an exception.

Alternative methods for the production of forces requires known values for electromagnetic interactions that often include the requirement that electrical, mechanical and thermal properties of system elements be known. Solving Maxwell's equations for the system and computing Lorentz forces between elements will enable precise computation of dynamically varying forces. In many cases, symmetry, simple geometric construction and for slowly varying forces, the equations describing the system performance reduce to simple form.

The latter goal of flexure designs requires that the mechanism be driven by a prescribed displacement. Of course, a given displacement requires a specific force. However, for this latter system, it is possible to use a displacement transducer (direct methods such as capacitive, inductive or optical gaging or inferred displacements from strain gages, pressure transducers, etc.) so that the motions can be determined and feedback systems may be utilized to obtain prescribed displacements. With closed loop control, the limiting precision of achievable motions is changed from the performance of the drive to that of the sensors and control systems. Because many drive actuators have undesirable characteristics such as hysteresis, creep, thermal susceptibility and non-linear stiffness, systems incorporating closed loop control of displacement are often necessary to realize the highest attainable precision.

In some cases, the drive train may be sufficiently stiff so that less direct means of inferring displacement are sufficient. Examples might include a stepper motor drive of a feedscrew in which pulses to the stepper motor may be used to infer displacement, high torque motor drives with end stop monitoring and assuming constant velocity during motion, hydraulic drives in which pressure differences may be used to infer forces. If such a system is sufficiently reliable and performs to within the requirements, open-loop designs often represent an economic solution.

In general, for any flexure system

> *The relative stiffness of the drive and flexure represents a measure of achievable precision.*

More precisely this can be stated as

> *The stiffness of the components of the mechanism in which force and measurement loops coincide is a measure of the ability to maintain precision while doing work.*

Determination of flexure element and related component compliances and the subsequent distortions, of which some may be undesirable, occupies a significant fraction of this book. In many cases flexures are used to translate a 'rigid' body in a specific direction (i.e. linear motion or rotation about a fixed axis), deviations of motion from this path are referred to as parasitic errors throughout. For more general discussions the reader can consult, Rivin, 1994, Smith and Chetwynd, 1992, and Slocum, 1992.

1.5 Design considerations

While there are no systematic rules for the creative process required for the conception of a means for producing a mechanical effect to satisfy a need, often referred to as 'synthesis', a number of approaches and important considerations can guide the designer to the best of the available solutions. In this brief section, some general guidelines for the assessment of a design, and, in particular, a flexure design, will be outlined. In the following brief discussion, a background appreciation of the design process will be assumed. If the reader is not familiar with some of the terms used in this section, more detailed discussions can be found in Smith and Chetwynd, 1992 and Ertas and Jones, 1996.

Synthesis is the creative process that seeks the best solution for a specific need. Important stages of this are

1. Recognizing a need
2. Establishing design requirements
3. Conceptualization and creativity
4. Assessment of feasibility

Although brainstorming should be an open-minded process, unless guided by detailed and, where possible, quantitative *requirements*, it is unlikely that the solutions will satisfy all of the needs and it is also likely that time will be spent considering unnecessary issues. Establishing quantitative requirements is an essential first step in the design process. It is also important that these requirements are *independent* of a design. A frequent trap with any design is to think of the solution and tailor the requirements to suit. Such an approach will invariably result in poor design. As an example, the specification for the range of forces to be measured in a particular application might be expressed by either of the two following requirements

a. Forces ranging from 0.1 to 1,000 N must be measured
b. The cantilever of the force balance should be designed for applied forces ranging from 0.1 to 1,000 N

While both of the above specify the required range, it is obvious which requirement is independent of design. In many cases such a distinction is less obvious. The temptation of the second requirement is that it represents less work for the designer. Consequently, it might, incorrectly, be perceived that this is a lower cost path! While the first is open to any solution, the latter can be visualized, modeled and assessed for feasibility immediately. If the first stage of synthesis involves analysis, it is unlikely that a design process has occurred. Often, quantitative performance measures can be classified using the following terminology

- Range
- Resolution
- Precision, a measure of machine performance not effected by calibration

- Accuracy, dependent entirely on calibration
- Errors, often split into random and systematic
- Traceability
- Repeatability/stability
- Linearity
- Relevant standards and legal requirements

Analysis is the assessment of possible solutions using quantitative models. This is used to determine feasibility and optimize parameters of the model. If the model accurately represents the design, optimization of one implies the other. It is rare for analysis to correct, or even compensate, for poor synthesis.

Conceptualization and creativity relate to the skills, imagination and experience of the designer. The ability to apply knowledge, 'common sense' and experience is the key to good design. Some concepts can help to broaden these qualities. In particular, paradigm shifts are often the result of lateral thinking and an excellent discussion of this is to be found in the books of Edward de Bono, 1971, 1974. Although a discussion of the creative process is beyond the scope of this book, some general approaches are worthy of mention.

Inversion can transform a feasible but impractical design to an elegant solution. The conventional arch and the suspension bridge represent a classic example of inversion. The former traditionally utilizes concrete as the material of construction. Because of this, bridges were designed to produce predominantly compressive forces to favorably exploit the materials properties. However, the costs of such constructions become prohibitive for longer spans. Turning the bridge upside down results in a design that produces predominantly tensile forces thereby making steel wire a possible material of construction.

Reduction is also a powerful method for simplification. Often, designs tend to evolve through successive integration of individual solutions. Generally, each element of the solution is somehow connected to the other in the hope that, when combined, the complete system will satisfy all requirements. During this process it is common that elements have a 'base' and these are simply connected. Often the two connected components could be produced from a single block of material. Sometimes, a complex fabrication might be replaced by a mechanism produced from a single monolith, as will be shown for many successive flexure designs. Although it is rare that a complex design can be reduced to a single component, it is often surprising how many components can be eliminated by intervening ideas between the initial concept to the final product.

Probably one of the most important skills to be acquired by a designer is the ability to sketch. An idea has little chance of being accepted unless it is effectively communicated. In the same way that analysis should play little part in the initial conception of a design, detailed drawings can also constrain the solution prematurely. While the ability to sketch is not a prerequisite for good ideas, it is most frustrating to listen to a lengthy explanation of an idea only to realize that either a simple sketch could have portrayed the concept in a fraction of the time or, after much discussion, that a number of people have interpreted the description erroneously. The latter of these scenarios is time consuming and, sometimes, expensive. The key benefits of rapid sketching are, Jones, 1932,

- Sketches enhance the communication of ideas.
- Easily drawn sketches are readily modified or discarded, thereby preventing premature focus on less than optimal designs.
- Clear, well-proportioned sketches can be rapidly converted to engineering drawings.

Almost by definition, brainstorming is likely to produce many solutions to a given problem. In practice there will not be time to undertake a detailed feasibility study of them all. Often, the best strategy is to cut out the least feasible using the concept of Ockham's razor. This states that

When you have two competing theories producing exactly the same prediction, the one that is simpler is the better.

Another complimentary proposition also applies

A simple solution almost always exists.

Having selected a set of possible designs, analysis can be readily used to assess feasibility based on known physical laws. Such analysis tools represent the bulk of the material presented in this book. There are, however, more qualitative concepts that may be used to critically assess a design.

As mentioned in the preface to this book, symmetry confers many favorable characteristics to mechanism performance. Some of the benefits of a symmetric mechanism often include

a. Balancing of stresses can null distortion errors. Stresses about a line of symmetry are often of equal and opposite value resulting in no net displacement of points on the line.

b. Symmetric mechanisms are inherently balanced providing immunity to inertial forces and heat inputs. This is true only for disturbances of similar symmetry. For example, thermal expansion effects in a mechanism will balance if the heat source is placed at the line of symmetry. In practice this is likely to be the case if the heat source is symmetric.

c. Symmetry often simplifies analysis. Not only can this lead to boundary conditions such as zero slope and displacement that are easily modeled, symmetry often nulls some effects such that they no longer need to be considered.

d. Forces and vibration modes become decoupled. Under such conditions, each plane of symmetry can be analyzed independently. However, in practice, it is very difficult to prevent vibration in one plane of symmetry from coupling into other planes. To choose an example from Rayleigh, 1894, try setting a guitar string in vibration so that the string oscillates in a single plane. In some applications, it is also possible to use broken symmetry to alter or suppress modes of vibration to the designer's advantage.

e. Components are often easier to manufacture.

f. Symmetry can be spotted at a glance. Once recognized it is often most informative to consider the consequences thereof.

Another important design concept is that of force and measurement loops intrinsic to any instrument or machine (this discussion is adapted from notes prepared by Derek Chetwynd). By definition, a flexure mechanism produces defined displacements (be they linear, angular or combinations thereof) upon application of applied forces. For most applications it is necessary, at some stage, to measure the relative displacement of defined points on the flexure. Such measurements might occur during calibration or be continuously monitored by some integrated displacement sensor. Simplistically, most instruments and machines and almost all flexures require application of forces and measurement of relative displacements. As a consequence

Every *force* in a static system

- must pass from body to body in a closed path (Newton's third law)
- causes strain in those 'rigid' bodies

Every meaningful measurement requires a datum. This requires

- an arbitrary origin locked to a component of the mechanism
- a 'rigid' link from this origin to the gage probe

Both of these involve a structural loop. In some cases components of the measurement loop may consist of an optical path either in air or along a fiber. In either case the input and output points sources must be anchored to a frame and it is this frame plus any changes in the optical path length that constitutes the structural components of the loop. By definition

- *Force loops* carry loads and hold the system together and must necessarily distort.
- *Measurement loops* carry datum information around the system and must be rigid.

Errors may be introduced whenever components of the force and measurement loops coincide so, where possible, **keep them apart**.

In general, optimal designs often conform to the following rules for loops

- Make loop elements stiff.
- Keep forces (including self-weight) small.
- Keep bending moments in the 'rigid' bodies (very) small.
- Keep loading (thermal or mechanical) as symmetric as possible.
- Maintain stable temperatures.
- Avoid temperature gradients.
- Put heat sources at points of symmetry.
- Keep fundamental resonance frequencies as high as possible.
- Keep loops small. This infers high frequency response, small bending moments, low self-weight and fast thermal response.

Other considerations can have profound influence on the attainable precision of a flexure mechanism. Of these, compensation and both cosine and Abbe errors represent two powerful concepts for assessment of design potential.

A number of compensation strategies are commonly employed. These being

- Cancellation. This is achieved by adding equal and opposite effects and often occurs as a consequence of symmetry. One method is to use differential measurement for the elimination of common mode errors due to random disturbances (vibrations, electrical interference etc.) or systematic perturbations (inertial forces, thermal changes etc.). Balancing is also a classic example of a cancellation strategy.
- Correction. Repeatable errors can be removed by providing a look-up table to determine the true value from a measurement. Where appropriate, fitting the input/output relationship of a system to a reasonable mathematical model can be used to enhance the accuracy of an instrument or machine.

Often, measuring an artifact can be used to assess instrument errors. However, such measurements contain two systematic errors, those of the instrument and those of the artifact. Instrument errors are fixed to its frame while those of the latter move with the artifact. In many cases, using the instrument to measure the artifact in different orientations and comparing results can enable separation of the two errors. A review of error separation techniques can be found in Evans *et al.*, 1996.

For linear displacement measurements, cosine and Abbe errors often represent the dominant limit to instrument accuracy. These are caused by

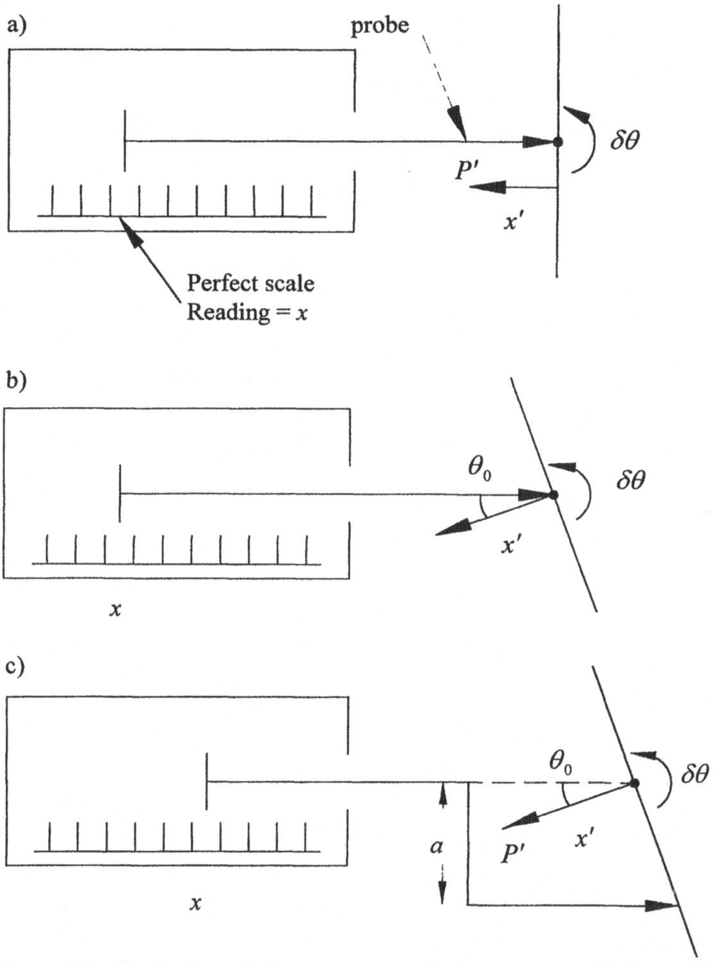

Figure 1.2: Illustration of alignment errors that can occur with linear displacement measurement, a) perfect measurement, b) probe misaligned but with axis initially passing through point to be measured, c) misalignment error with Abbe offset

differences between the line of action of the measuring probe and the motion of the point to be measured. Conceptually, the models of figures 1.2 (a-c) can represent these. Figure 1.2 (a) represents a perfect configuration. In this, motion of a point P on a surface is to be measured. It will be imagined that this undergoes a linear translation to point P' accompanied by a parasitic (i.e. undesirable and, often, unknown), small rotation about this point of magnitude $\delta\theta$. Clearly, if the probe is both *parallel* with the axis of motion and the axis of measurement *passes though* the point P then the perfect relationship between measurement, x, and true motion, x', is achieved i.e.

$$x = x' \tag{1.2}$$

In this case, parasitic rotations have no influence on the measurement. Such a situation does not occur in practice. Figure 1.2(b) represents a less ideal scenario where the line of action of the measurement is at an angle, θ_o, to that of the motion but its axis still, initially, passes through the point P. In this case, the relationship between the scale reading and true displacement is

$$\begin{aligned} x &= x'\cos\theta_o + x'\sin\theta_o \sin\delta\theta \\ &\approx x'\left(1 - \frac{\theta_o^2}{2} + \theta_o\delta\theta\right) \end{aligned} \tag{1.3}$$

For small angles, the error due to misalignment is of second order and in many cases is relatively small. Equation (1.3) implies that the parasitic rotation might compensate the cosine term. However, in practice, the sign of $\delta\theta$ is often arbitrary. Also, the magnitude of the parasitic rotation in comparison to initial misalignment is usually small. Consequently, the first term of equation (1.3) dominates and therefore this is often referred to as a 'cosine error'. More serious, and more common, in addition to a misalignment there will also be an offset a between the point to be measured and the line of action of the probe as shown in figure 1.2(c). In this case, the relationship between the scale reading and true displacement is

$$\begin{aligned} x &= x'(\cos\theta_o + \sin\theta_o \sin\delta\theta) - a\sin\delta\theta \\ &\approx x' - a\delta\theta \end{aligned} \tag{1.4}$$

From equation (1.4) it can be seen that the offset a results in an error that is directly proportional to the parasitic rotation. In many designs this represents the dominant error. Ernst Abbe realized the importance of this and it is to him that the alignment principle is named. Simply stated

> *When measuring displacement of a specific point it is necessary that the axis of measurement should be both parallel to and pass through the motion of the point*

When the errors are repeatable and free from hysteresis, this can often be removed, or at least reduced, by arranging for the calibration of the probe to satisfy the alignment principle.

While most of the above applies to design in general, some problems specific to flexures appear omnipresent. By definition, analysis requires a process of abstracting a real world component to enable quantitative mathematical modeling. In the process of reducing the complexity of a complete model to something that can be solved within a reasonable time it is necessary to make some assumptions. Often, the resulting model will still represent a reasonable correspondence between predicted and actual behavior. However, there are two commonly applied assumptions that should be used with utmost care. These are

1. Flexure elements are connected by infinitely rigid bodies, and
2. Flexures can be assembled with perfect fixtures that neither distort the flexure elements nor introduce significant stresses.

In many designs requiring high precision motion such assumptions may lead to significant discrepancy between modeled and actual behavior. The following caveats should always be attached to the above assumptions

1. Rigid bodies are not rigid, especially when levers are present.
2. For small motions, fabricated mechanisms often show some hysteresis due to tin canning, plastic deformations or insufficient fasteners.

Tin canning is often caused by curvature of leaf type flexures. Causes of this can be either pre-existing bending of 'flat' plates as supplied by the manufacturer (often due to rolling) or bending of the plates by the clamps. Microscopic plastic deformation can, once again, occur as a consequence of the clamps. Either, surfaces in the clamped region will experience plastic deformation of surface asperities (microscopic plastic deformation) or there might be some initial permanent bending of leaf springs (macroscopic plastic deformation). Additionally, bolted joints may creep and move when subject to varying forces. Some design considerations for assembly of fabricated flexures are discussed in chapter 8. In general, if there appear to be discrepancies between the model and flexure behavior, the above assumptions are likely sources.

References

De Bono, 1971, *The Use of Lateral Thinking*, Pelican Books

De Bono, 1974, *Po: Beyond Yes and No*, Pelican Books

Carlson H., 1978, *Spring Designer's Handbook*, Marcel Dekker, Inc., NY

Ertas A. and Jones J.C., 1996, *The Engineering Design Process*, John Wiley and Sons Inc., NY, chpt. 1.

Evans C.J., Hocken R.J. and Estler W.T., 1996, Self-calibration: Reversal, redundancy, error separation and 'absolute testing', *Annals of the CIRP*, **45**(2), 617-636

Geary P.J., B.S.I.R.A. Research Report M18, (1954),

-B.S.I.R.A., Research Report R249, (1960)

Howard L.P. and Fu J., 1997, Accurate force measurements for miniature mechanical systems; a review of progress, *Proc. SPIE*, **3225**, 2-11.

Jones F.D., 1932, *How to Sketch Mechanisms*, The Industrial Press, NY.

Jones R.V., 1987, *Instruments and Experiences*, J. Wiley and Sons, London.

Love A.E.H., 1927, *A Treatise on the Mathematical Theory of Elasticity*, 4th ed., Dover Publications Inc (1983), NY, chapter 1 provides an historical overview, the rest of the book is for the serious scholar.

Baron Rayleigh, J.W.S. Strutt, 1894, *Theory of Sound*, Dover Publications (1945 edition), NY, §149, page 243

Rivin E., 1994, *Stiffness in Design*, ASPE tutorial notes.

Slocum A.H., 1992, *Precision Machine Design*, Prentice Hall, NJ.

Sobel D., 1995, *Longitude: The Story of a Lone Genius Who Solved the Greatest Scientific Problem of His Time*, Walker and Company, NY, as the title suggests, this is light reading but nevertheless interesting from a flexure design perspective.

Smith S.T. and Chetwynd D.G., 1992, *Foundations of Ultraprecision Mechanism Design*, Gordon and Breach, London, UK.

Sydenham P.J., 1984, Elastic design of fine mechanism in instruments, J. Phys. E: Sci. Instrum., **17**, 922-930, see also, Mechanical design of instruments, parts A & B: Putting elasticity to use, Measurement and Control, **14**, 179 - 185 & 219 - 227.

Timoshenko S.P., 1953, *History of Strength of Materials*, Dover Publications Inc (1983), NY.

Wahl A. M., 1964, *Mechanical Springs*, McGraw-Hill Book Co., NY.

2 Essentials

2.0 Overview

Being concerned in this book with the achievement of precise motion from elastic distortions upon application of controlled forces, an appreciation of stress and strain relationships is necessary throughout. This chapter reviews some essential concepts of elasticity and the various techniques for the prediction of stresses, strains and subsequent distortions. Because reliability of flexure designs is a common concern, overviews of failure criteria for brittle and ductile materials as well as simple fatigue calculation methods are presented. In general, this chapter is a compilation of relevant physical laws and a selection of the commonly used mathematical techniques extracted, for the most part, from texts on elasticity and strength of materials. Finally, flexure mechanisms are used to achieve a prescribed displacement that can be characterized by one or more independent coordinates. Analysis of the number of independent coordinates and their loci forms the subject of kinematics. For any flexure mechanism, it is necessary to determine the number of independent coordinates, or freedoms. Considering a mechanism as a kinematic chain of rigid bodies subject to constraints imposed by the flexure elements, the number of freedoms can be readily assessed using the concept of mobility analysis presented at the end of this chapter.

2.1 Basic elasticity

Hooke (1635-1703) was the first to observe a linear relationship between the extension of a wire due to a force applied along its axis and the proportionate scaling with the length. Mathematically, from this simple relationship it is possible to define the two parameters stress, σ, and strain, ε, given by

$$\sigma = \frac{F}{A}$$
$$\varepsilon = \frac{l - l_o}{l_o} = \frac{\Delta l}{l_o} \tag{2.1}$$

where F is the applied force normal to a surface of area A and l is the length of the wire after the applied force and l_o is the original length, see figure 2.1.

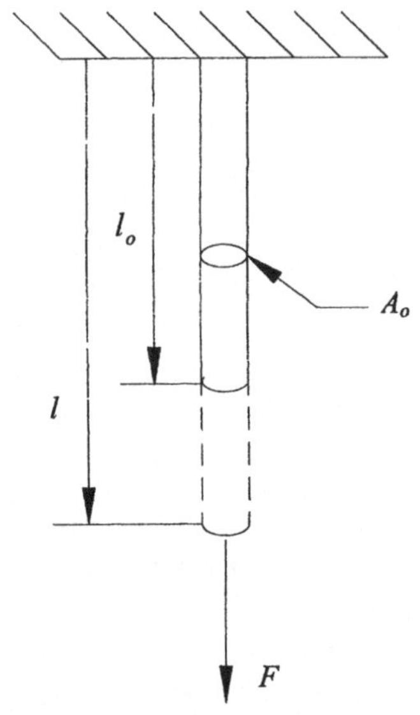

Figure 2.1: Experimental arrangement for the measurement of a thin wire due to an applied load

In practice, equations (2.1) apply only to thin wires in tension under relatively light loads. It is also necessary to have a more rigorous definition of the quantities being measured. In the first of the above equations, the true stress σ_{true} should be considered as the ratio of tensile force in the thin wire to that of the instantaneous cross section of the wire. Inherent to this equation is the assumption that the stress is uniformly distributed across the wire (a reasonable assumption in this instance). Under these circumstances the *true stress* is given by

$$\sigma_{true} = \frac{\text{Applied force}}{\text{Instantaneous area}} = \frac{F}{A_i} \tag{2.2}$$

Unfortunately, even when measuring something as simple as a wire, it is often a lot more convenient to measure the dimensions of the wire before testing and use the original area for the calculation of stress. Being a more pragmatic measurement, the computed ratio of force to area is known as the *engineering stress* and is given by

$$\sigma_{eng} = \frac{\text{Applied force}}{\text{Original area of wire}} = \frac{F}{A_o} \tag{2.3}$$

Similarly, when considering the *engineering* strain in the wire given by equation (1.1), it is assumed that the length does not change significantly. For a true assessment of strain it becomes necessary to integrate incremental changes in length divided by the instantaneous length giving

$$\begin{aligned} \varepsilon_{true} &= \int_{l_o}^{l_o+\Delta l} \frac{dl}{l} = \ln\frac{l_o + \Delta l}{l_o} = \ln\left(1 + \frac{\Delta l}{l_o}\right) \\ &= \ln\left(1 + \varepsilon_{eng}\right) \\ &= \frac{\Delta l}{l_o} - \frac{1}{2}\left(\frac{\Delta l}{l_o}\right)^2 + \frac{1}{3}\left(\frac{\Delta l}{l_o}\right)^3 \cdots \end{aligned} \tag{2.4}$$

Clearly, for small deformations, the distinction between *engineering* and *true* stress and strains is not significant. Fortunately, throughout this book we are concerned with the design of elastic mechanisms. For a flexure to be repeatable and reliable it is necessary to keep stresses small and therefore we may reasonably ignore this distinction between 'true' and 'engineering' stress and strain and thus drop the subscripts.

2.2 Behavior of materials

Equations (2.1) to (2.4) above impose no limits on the possible values of stress or strain. In reality, all materials have limits, the values of which define the strength of a given structure. To evaluate these limits, it is common to apply a tensile or compressive load of known value to a specimen of uniform cross section and known length and measure subsequent extension. From such measurements it is relatively straightforward to compute the engineering stress/strain curves. In general it is found that there are two quite distinct types of behavior. For almost all metals there is an initial linear-elastic (note that elastic does not necessarily imply linear) region in which the specimen will return to its original length if the load is released. This is followed by a rapid extension with increasing load as the stress is increased beyond a certain value. In this latter stage of deformation the material has undergone an irreversible (or plastic) flow. The ability to plastically deform without fracture is called *ductility* and is characteristic of many metallic solids. Other materials that might be used for flexure applications, such as many ceramics, exhibit brittle failure due to the growth of cracks in the plane of the maximum tensile stress. Because of these differences, the responses to simple uni-axial loads of metals and non-metals are discussed separately below.

2.2.1 Metals

Figure 2.2 shows a typical stress verses strain curve for a medium carbon steel specimen. At low loads, it can be seen that there is a relatively linear region and, upon unloading, the specimen will return very nearly to its original length. Above a certain value (point *A* in figure 2.2) there is a relatively sudden plastic transition, the upper yield stress, after which the apparent stress drops to a lower yield stress that continues at a relatively constant value for a finite range of strain. At strains above point *C* the stress then continues to steadily increase with a relatively large proportion of plastic distortion. This reaches a peak at *D*, which is known as the ultimate strength after which there is a decline in stress with strain followed by catastrophic failure. Experimentally, the variability of the strain at which complete separation occurs renders this value of little utility.

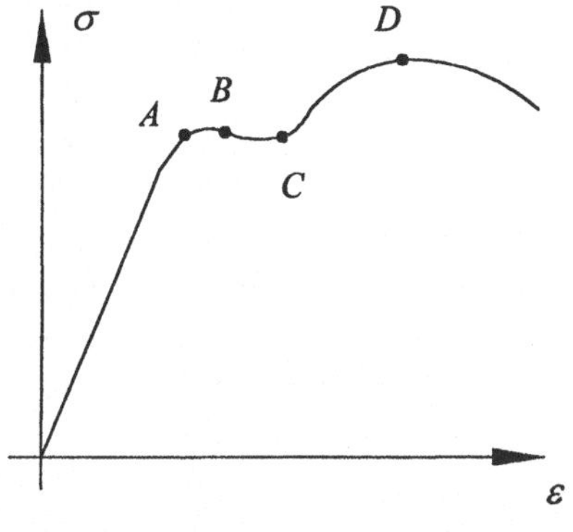

Figure 2.2: Typical stress-strain curve for a soft steel specimen

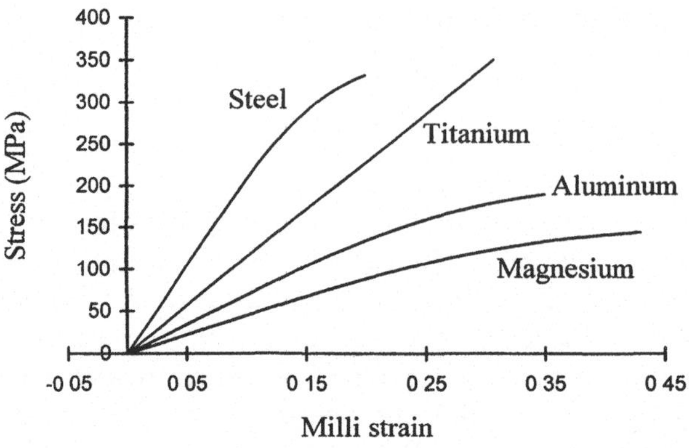

Figure 2.3: Typical stress strain curves for commercially available metal alloys

If the steel is loaded beyond the lower yield stress region and the load is then released, the specimen will tend to recover exhibiting a near elastic characteristic but with a permanent offset strain at zero applied stress. Upon subsequent testing it is often observed that the lower yield phenomena is absent resulting in a smooth transition from elastic to plastic behavior. Such curves are also characteristic of a large number of other metals and typical stress strain curves for a variety of these are shown in figure 2.3.

To design reasonable flexure mechanisms, it is important to clarify some of the parameters used to described the strength properties of metals. Possibly the most common measure is called the *yield stress*. Generally, this refers to the value of stress that must be applied to produce a certain percentage of permanent engineering strain. Most suppliers choose to use the stress at 0.2% permanent strain although 0.1% has been used in the past. Such a measure introduces a quandary to the flexure designer interested in high precision mechanisms that are usually required to perform with considerably better repeatability. Consequently, it makes more sense to seek a value that represents a considerably lower percentage permanent deformation. The *proportional* and *elastic limit* represent two alternative and more promising parameters that are often quoted. The proportional limit represents the maximum stress at which the stress-strain characteristic does not deviate from a straight line to within the capability of the instrumentation with which it is measured. Similarly, the elastic limit represents the maximum stress at which no permanent deformation can be detected upon removal of all applied stresses. In practice, for many metals, both values are nearly the same and the two are often considered synonymous. Unfortunately, both of these parameters are very dependent on the precision of the instrumentation with which they are measured. One particularly sensitive measure of the elastic limit is to apply a cyclic load and plot this continuously. Clearly, if there is any permanent deformation each cycle will exhibit a closed loop the area of which is a monitor of 'hysteresis' and thus permanent deformation in the stress-strain characteristic. Using such methods, for some materials the limiting shear stress for purely elastic behavior can be anything down to one hundredth that of the yield stress, Dieter, 1986.

As a consequence of the above arguments, parameters such as the true elastic limit are of little utility for most designs and alternative rule of thumb parameters must be sought. For steels, it is known that below the *endurance stress*, cyclic loads will not introduce significant fatigue damage. For a wide range of steels the endurance limiting stress, S_e, can be reasonably found from the ultimate tensile stress, S_{ut}, using the relationship, Shigley and Mische, 1989

$$S_e = \begin{cases} 0.504 S_{ut} & S_{ut} < 1400 \text{ MPa} \\ 700\text{MPa} & S_{ut} > 1400 \text{ MPa} \end{cases} \tag{2.5}$$

Unfortunately there does not appear to be a straightforward relationship between the ultimate tensile and yield strengths in simple tension, the latter parameter being more commonly measured. However, it is *always* safe to assume that the endurance stress is more than half of the yield stress and often greater than two thirds of this value.

For other metals, the onset of plastic deformations in microscopic regions within a stressed solid is again difficult to predict and has been subject to extensive investigations over the years. Unfortunately, no simple conclusions can be drawn from such studies. However, the stress at the first detectable onset of plastic deformation in metals, often called micro-slip, is often dependent on the grain size and can have values of anything from 25 to 75% of the macroscopic yield stress commonly supplied by manufacturers, Brentnall and Rostoker, 1965. Consequently, a reasonably conservative value of one quarter of the yield stress might be more appropriate for ultra-precision applications requiring a high degree of repeatability.

A measure of a materials capacity for the storage of strain energy is the *resilience*, U_R. This can be relatively simply derived from the work done per unit volume in an element on the point of yielding given by

$$U_R = \frac{\sigma_Y \varepsilon_Y}{2} = \frac{\sigma_Y^2}{2E} \tag{2.6}$$

Clearly, a good material would possess a high yield stress and low modulus of elasticity. Work-hardened and heat-treated beryllium copper is a good example of such a material. Resilience represents the ability of a material to return energy after elastic deformation. *Toughness* is the ability to absorb energy during plastic deformation. This could be calculated from the area under the stress-strain curve between the strain at yield and at failure. However, the variability of the failure strain renders this an unreliable measure and a more rigorous definition based on fracture mechanics is often adopted. For flexure applications it is necessary to avoid irreversible deformation and, as a consequence, behavior of materials in the plastic region is not considered further. However, it should be noted that the toughness is a measure of the degree to which an overstressed flexure might withstand catastrophic failure and might be important for safety critical applications.

2.2.2 Non-metals

Considering the experimental difficulties associated with the determination of the elastic properties of metals, brittle materials at first appear a lot more amenable to a simple elastic analysis. Figure 2.4 shows the results from a bending test on a simply supported beam machined from single crystal silicon, a characteristic typical of many 'brittle' materials. For convenience, the ordinate has been scaled to give an estimate of the maximum tensile stress on the surface of the beam. In general, for many 'brittle' materials, the stress-strain graph is extremely linear up until catastrophic failure.

Because of this near perfect linear, elastic behavior, such materials would appear ideal. Certainly, for applications to flexures requiring extremes of precision this is indeed so. Brittle materials tend to *either* exhibit superb linear elastic properties *or* they break. As a consequence they can often be considered the mechanical equivalent of the electric fuse. A spring constructed from a brittle material can be trusted to either work as designed or not at all. In contrast, a metal spring may undergo plastic deformation. When this occurs, the spring characteristic becomes non-linear and hysteretic. Note also that the magnitude of the hysteresis is dependent on the strain history. As a consequence, satisfactory analytic models for the prediction of force-displacement characteristics for springs undergoing plastic deformation have yet to be developed.

As usual there are catches, brittle materials are often difficult to both manufacture and assemble/fasten, will not tolerate accidental overloads and, like metals, the point at which catastrophic failure occurs is somewhat difficult to predict. For example, during testing of silicon beams, the maximum stress in bending varied from around 50 up to 400 MPa dependent upon the surface damage introduced during manufacture. However, the upper values of stress could only be obtained after a more than 80 μm of the as-ground surface had been removed by

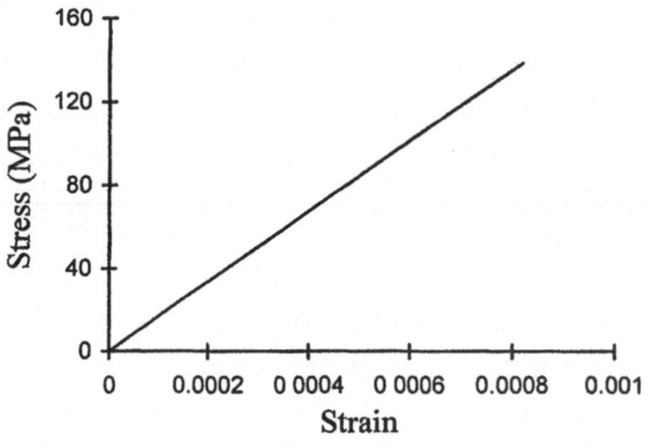

Figure 2.4: Stress versus strain curve for a single crystal silicon beam measured in a three point bending apparatus

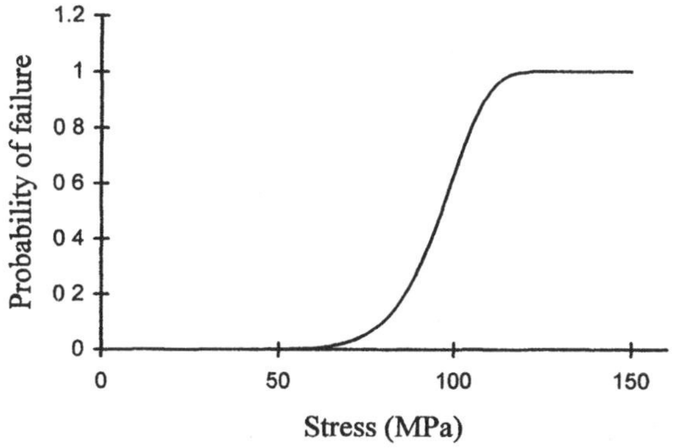

Figure 2.5: Cumulative probability of failure corresponding to a Weibull distribution with $m = 10$ and $\sigma_0 = 100$ MPa

chemical etching, Smith *et al.*, 1991. It is now generally accepted that surface flaws are responsible for the reduction in the strength of a broad range of brittle materials represented by predominantly covalent bonded solids and many ceramics. The effect of a particular flaw on the surface strength has again been studied in considerable detail and can now be considered relatively well understood, see for example Lawn, 1993. Unfortunately, the measurement or prediction of flaw sizes is not possible for a vast range of engineering processes. For applications where the breaking strength must be consistent, it is common to follow up the rough machining of ceramics, usually grinding, with finishing processes such as wet etching, lapping and polishing. The random nature of the physical process tends to result in a random flaw size enabling use of statistical approaches for strength prediction. For a large number of ceramic materials, it has been found that the probability of failure for a given stress with a material produced using a specific process will closely follow a Weibull distribution. Consequently, the probability of failure, *P*, follows a cumulative distribution of the form

$$P = 1 - e^{-(\sigma/\sigma_o)^m} \tag{2.7}$$

where σ is the stress acting to open the crack (referred to as mode I type stress in fracture mechanics), σ_0 is called the scaling stress and m is the Weibull modulus. Figure 2.5 shows a graph of the above equation for $m = 10$ (typical for many ceramics) and $\sigma_0 = 100$ MPa. The probability density given by the derivative of (2.7) is shown in figure 2.6. For example, if a failure rate of 1 percent is considered acceptable the design stress can be either measured directly from the graph or computed from the equation

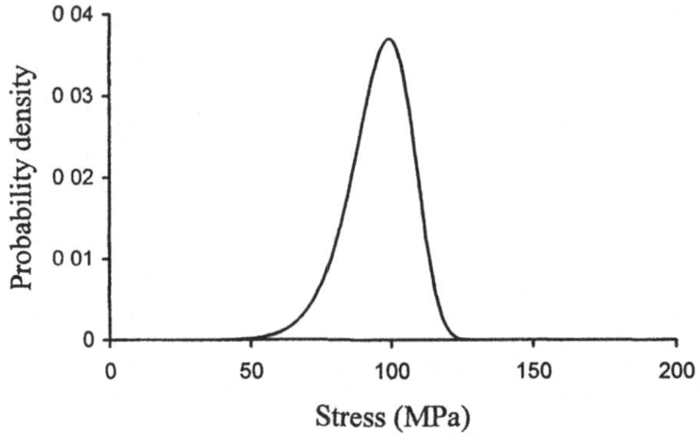

Figure 2.6: Probability density of the Weibull distribution for the parameters used in figure 2.5

$$\sigma = \sigma_0 \sqrt[m]{\ln(1-P)} \tag{2.8}$$

which for this example corresponds to a working tensile stress of 63 MPa, a value not unreasonable for single crystal silicon.

Associated with the almost binary performance of these materials is the fact that they can be survival tested after manufacture by applying the maximum stress that they are likely to encounter during use. This is known as proof testing and has the effect of eliminating specimens with unacceptably large flaws. For a probability of failure during a proof testing P_p, the new cumulative probability distribution of the remaining specimens, P', is given by

$$P' = \frac{P - P_p}{1 - P_p} \tag{2.9}$$

If a suitable proof stress is chosen, by eliminating the low strength tail of the distribution, such a procedure can result in a substantially reduced variability with the loss of a small percentage of total production.

2.3 Principal stresses and strains

Clearly, for solid bodies of arbitrary geometry and applied loads, to determine the state of stress at a specific point in a solid, it becomes necessary to resort to the more advanced mathematical theory of elasticity. It is not the intention of this book to provide a text for such a study and therefore we shall restrict ourselves to a brief outline of the relevant formulae and definitions of the terms contained herein.

The linear relationship between load and strain first observed by Hooke suggests a constant ratio of stress to strain for a given material called the *modulus of elasticity, E,* given by

$$E = \frac{\sigma}{\varepsilon} \tag{2.10}$$

This simple linear relationship is of enormous utility to engineers for the estimation of stresses in solid bodies of homogeneous, isotropic materials when subject to externally applied loads. It should be emphasized at this juncture that the elastic modulus is a material property and, unlike measured parameters such as stress and strain, will be prone to breakdown at extreme values, to be discussed in sections 2.3 and 2.4. To extend this linear relationship between stress and strain to three dimensional objects it is informative to discard the thin wire model in favor of a vanishingly small cube of material subject to simple, uniform stresses on each of its three orthogonal pairs of faces as shown in figure 2.7. In the first of these figures, the cube has been selected at an arbitrary angle. Under these circumstances the possible types of stresses *applied* at each face can be reduced to a direct stress plus two shear stresses acting in the plane of the element. The subscripts used to denote the shear stresses indicate the plane and direction respectively and a counter-clockwise applied shear stress is considered to be positive. The second of figures 2.7 represents the stresses on a cube that has been oriented in the direction of the principal stresses. In this orientation the shear stresses are zero. To emphasize the unique values of the principal stresses these are denoted with the subscripts 1, 2 and 3 instead of x, y and z used to denote three orthogonal stresses at an arbitrary orientation. Obviously, if the

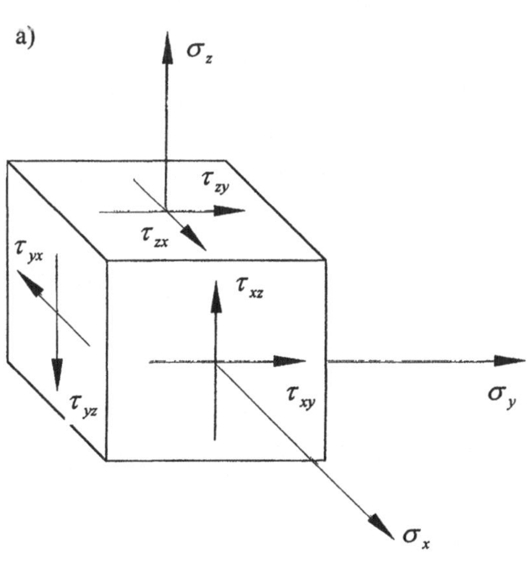

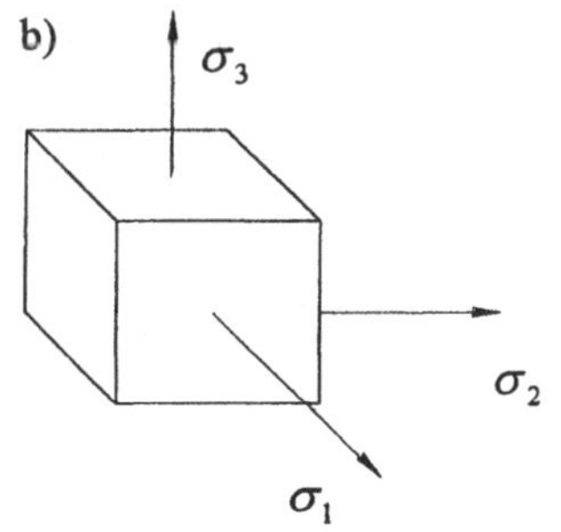

Figure 2.7: Stresses on the faces of an elemental cube of material subject to external loads, a) stresses on a cube of arbitrary orientation, b) stresses on a cube oriented so that the shear stresses are zero and, by definition, the direct stresses correspond to principal stresses. Note that equilibrium requires $\tau_{ij} = \tau_{ji}$

cube becomes vanishingly small, the stresses on each of the three opposing faces will be equal. Considering the effects of either a tensile or compressive stress on one face of the cube, it is obvious that there will be a corresponding strain in the same direction. However, close inspection will reveal strains of opposite sign on the other two faces. Experimentally it can be shown that these strains are proportional to the applied stress from which it is possible to deduce the relationship

$$-\upsilon = \frac{E\varepsilon_2}{\sigma_1} = \frac{E\varepsilon_3}{\sigma_1} \tag{2.11}$$

or

$$\varepsilon_2 = \varepsilon_3 = -\upsilon\varepsilon_1 = -\upsilon\frac{\sigma_1}{E} \tag{2.12}$$

where the dimensionless material property υ and is known as Poisson's ratio in honor of its originator S.D. Poisson (1781-1840).

2.3.1 Biaxial stress

Because the relationships between stress and strain are all linear, it is possible to apply the principle of superposition and state that the total strain is simply the sum of all contributions. Consider the case where there are only two applied stresses. The forces generating these stresses are in a single plane with zero stress perpendicular to this. Such a state of stresses is common to many engineering structures (particularly in flat plates and shells) and is known as plane stress where it is usually assumed that the zero stress is that normal to any free surface. In the presence of forces in directions 1 & 2 of figure 2.7(b) the three components of strain are simply given by the sum due to individual stresses

$$\begin{aligned} \varepsilon_1 &= \frac{\sigma_1}{E} - \upsilon\frac{\sigma_2}{E} \\ \varepsilon_2 &= \frac{\sigma_2}{E} - \upsilon\frac{\sigma_1}{E} \\ \varepsilon_3 &= -\upsilon\frac{\sigma_2}{E} - \upsilon\frac{\sigma_1}{E} \end{aligned} \tag{2.13}$$

It is informative to consider the special case where the strain in any of the directions is zero. Ignoring the trivial case in which the two stresses are both zero, the first strain will be zero if the stress in its direction is exactly the same as the product of Poisson's ratio and the stress in plane two and both stresses are of the same sign. One can easily imagine this by considering firstly, compressing one side of the cube in direction 2 say so that there is a corresponding expansion in direction 1. Applying a stress, σ_1 of magnitude $\upsilon\sigma_2$ will then return this face of the cube to its original dimension thus reducing the strain to zero. Similar arguments follow for the strain in direction 2 while for the strain in the unconstrained direction it is obvious that this can only be reduced to zero through the application of equal stresses having opposite sign. In words, this occurs when the strain due to compressive stress in one direction is canceled by an equal and opposite tensile stress in the other.

Equations (2.13) can be relatively simply rearranged to yield the components of stress in terms of the strains

$$\sigma_1 = \frac{E(\varepsilon_1 + \upsilon\varepsilon_2)}{1-\upsilon^2} \qquad \sigma_2 = \frac{E(\varepsilon_2 + \upsilon\varepsilon_1)}{1-\upsilon^2} \tag{2.14}$$

Equation (2.14) can be useful when strain gages are being used to measure the stresses *provided that they are oriented in the direction of the two orthogonal applied forces.* The reason for this proviso is because we are presently considering an element that is oriented in the direction of the *principal stresses* of which the more general case will be discussed in section 2.5.

2.3.2 Triaxial stresses

Based upon the reasoning for the derivation of the strain due to biaxial stresses it is a relatively obvious extension to write the strains due to a state of triaxial stresses

$$\varepsilon_1 = \frac{1}{E}(\sigma_1 - \upsilon(\sigma_2 + \sigma_3)) \tag{2.15}$$

with similar expressions for ε_2 & ε_3

Again, with some algebraic manipulation it can be shown that the explicit relationship for principal stresses in terms of principal strains is given by

$$\sigma_1 = \frac{E\varepsilon_1(1-\upsilon) + \upsilon E(\varepsilon_2 + \varepsilon_3)}{1-\upsilon-2\upsilon^2} \tag{2.16}$$

with similar expressions for σ_2 and σ_3

2.4 Non-principal stresses

2.4.1 Plane stress

So far we have only considered an elemental cube of material subject to forces normal to its surface. Assuming that the directions of these principal stresses are not known in advance it is likely that the cube would be subject to forces at an angle to the surfaces. Under these circumstances, it is possible to simplify the problem by resolving these forces into components normal and coplanar with each surface. The normal forces are given the prefix x, y or z while the planar components of force when normalized by the area of the plane are known as the shear stresses and are denoted by τ_{ij}, where i is the plane (i.e. the plane i = constant) and j indicates the direction. Figure 2.7 shows the complete system of stresses acting on an elemental cube. Although not of great utility in this book, to be consistent, the shear stresses are considered to be positive when acting counter-clockwise.

A complete analysis of three dimensional stresses and strains for arbitrary applied loads tends to become rather involved and is adequately covered in such texts as Timoshenko and Goodier, 1970, or Popov, 1976. For the purposes of this book, the relevant formulae and techniques for the visualization of stresses (Mohr's circle) will be introduced. The primary purpose of this section is to indicate that any combination of stresses can be reduced, by suitable rotation of the coordinates of the element, to an orientation at which the shear stresses are zero! Such a reduction then enables the prediction of stresses likely to produce failure as well as the likely orientation of either shear deformation or crack propagation depending on whether the material acts in ductile or brittle manner.

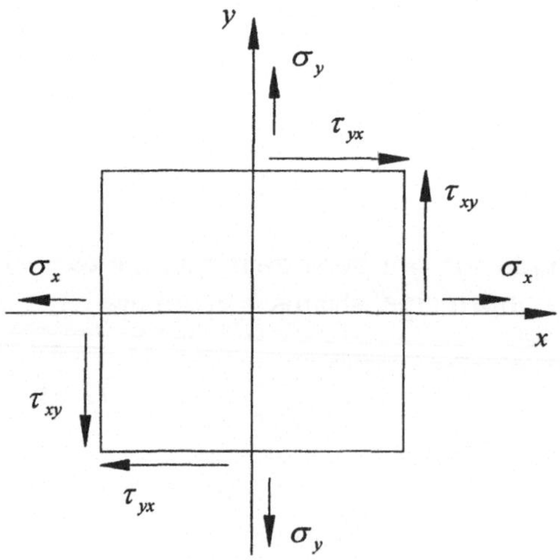

Figure 2.8: Stresses acting on the faces of an element for which stresses in the plane of the z axis are zero

Ignoring, for the time being, both the direct and shear stresses in the z plane, it is possible to reduce our analysis to a two dimensional problem of a thin elemental shell subject to stresses only in the x-y plane. Such a plane stress approximation is applicable to a broad range of engineering problems. A thin shell element subject to plane stress is depicted in figure 2.8. Clearly, if the element is not going to rotate, equilibrium

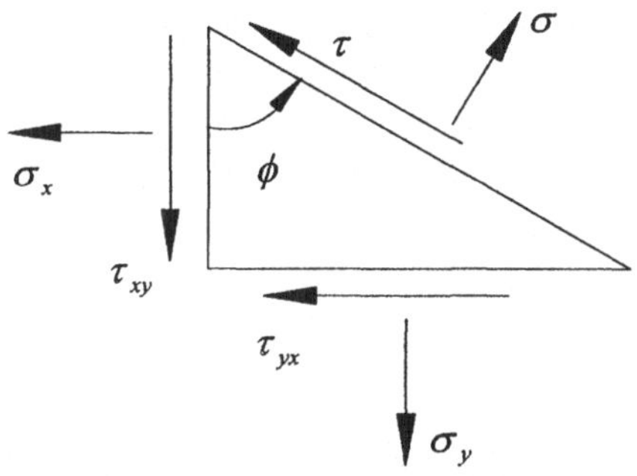

Figure 2.9: Stresses acting on the faces of a triangular element cut from that of figure 2.8 at an angle ϕ

requires that

$$\tau_{yx} = \tau_{xy} \tag{2.17}$$

Cutting the element of figure 2.8 at an arbitrary angle ϕ, see figure 2.9, and resolving forces results in the equation

$$\sigma ds = \sigma_y \sin(\phi)dx + \sigma_x \cos(\phi)dy + \tau_{xy}\left(\sin(\phi)dy + \cos(\phi)dx\right) \tag{2.18}$$

However, noting that

$$\begin{aligned} dx &= \sin(\phi)ds \\ dy &= \cos(\phi)ds \end{aligned} \tag{2.19}$$

equation (2.18) can be rearranged, with some trigonometric manipulation, to give the direct stress in the explicit form

$$\sigma = \frac{\sigma_x + \sigma_y}{2} + \frac{\sigma_x - \sigma_y}{2}\cos(2\phi) + \tau_{xy}\sin(2\phi) \tag{2.20}$$

Again, with some algebraic manipulation, the shear stress at an arbitrary angle ϕ can be shown to be

$$\tau = \frac{\sigma_x - \sigma_y}{2}\sin(2\phi) + \tau_{xy}\cos(2\phi) \tag{2.21}$$

The angle at which the direct stress is a maximum or minimum can be obtained from the equation

$$\frac{d\sigma}{d\phi} = 0 = -\left(\sigma_x - \sigma_y\right)\sin(2\phi) + 2\tau_{xy}\cos(2\phi) \tag{2.22}$$

from which we have

$$\tan(2\phi) = \frac{2\tau_{xy}}{\sigma_x - \sigma_y} \tag{2.23}$$

Because the right hand side of this equation represents the stresses on the element at in the direction of the fixed coordinate system, these represent known values. Consequently, solving for the tangent term on the left-hand side results in two values for ϕ which are shifted by $\pi/2$. Similarly, there are two angles at 90 degrees to each other at which shear stresses are a maximum given by

$$\tan(2\phi) = \frac{-(\sigma_x - \sigma_y)}{2\tau_{xy}} \tag{2.24}$$

Before discussing the implications of this equation in more detail, it is informative to investigate the values of this shear stress when the direct stresses are at a maximum and vice versa. Equation (2.24) (ϕ when the direct stresses are a maximum) can be rearranged to produce

$$\sin(2\phi) = \frac{2\tau_{xy}\cos(2\phi)}{\sigma_x - \sigma_y} \tag{2.25}$$

Substituting this into the equation for the shear stress (2.21) at this angle gives

$$\begin{aligned} \tau &= \frac{-(\sigma_x - \sigma_y)}{2}\frac{2\tau_{xy}}{(\sigma_x - \sigma_y)}\cos(2\varphi) + \tau_{xy}\cos(2\varphi) \\ &= 0! \end{aligned} \tag{2.26}$$

This is particularly important for analysis of plane stress problems in that it indicates that there will always be an angle at which we can orient our coordinate system, so that, only principal direct stresses are acting on the element. Under these conditions equations (2.10) to (2.16) can then be used to convert between stress and strain and these will be used in the next section for the prediction of failure of materials subject to three-dimensional stresses.

It can also be shown that the value of the stress normal to the plane of maximum shear stress is the same in each plane and is given by

$$\sigma = \frac{\sigma_1 + \sigma_2}{2} \tag{2.27}$$

Note that the subscripts *1* and *2* in equation (2.27) indicate that these are *principal* stresses. This leads us to a general principle for plane stresses, which can be extended to three-dimensional analyses, and this is

> *For an elemental volume of material subject to any combination of stresses it is possible to find an orientation in which the element is subject to triaxial principal stresses only*

In 1882 Otto Mohr (pronounced 'more') showed that the above rather complex trigonometric relationships could be graphically represented as a circle in which the *x*- axis represents direct stresses and the *y*-axis represent shear stresses. First of all squaring and then adding equations (2.20) and (2.21) gives

$$\left(\sigma - 1/2\left(\sigma_x + \sigma_y\right)\right)^2 + \tau^2 = \left(1/2\left(\sigma_x - \sigma_y\right)\right)^2 + \tau_{xy}^2 \tag{2.28}$$

which is an equation of a circle of the form

$$\left(x - x_0\right)^2 + y^2 = R^2 \tag{2.29}$$

where $x = \sigma$ and $y = \tau$ and

$$R = \frac{1}{2}\sqrt{\left(\sigma_x - \sigma_y\right)^2 + 4\tau_{xy}^2} \tag{2.30}$$

$$x_0 = \frac{\left(\sigma_x + \sigma_y\right)}{2} \tag{2.31}$$

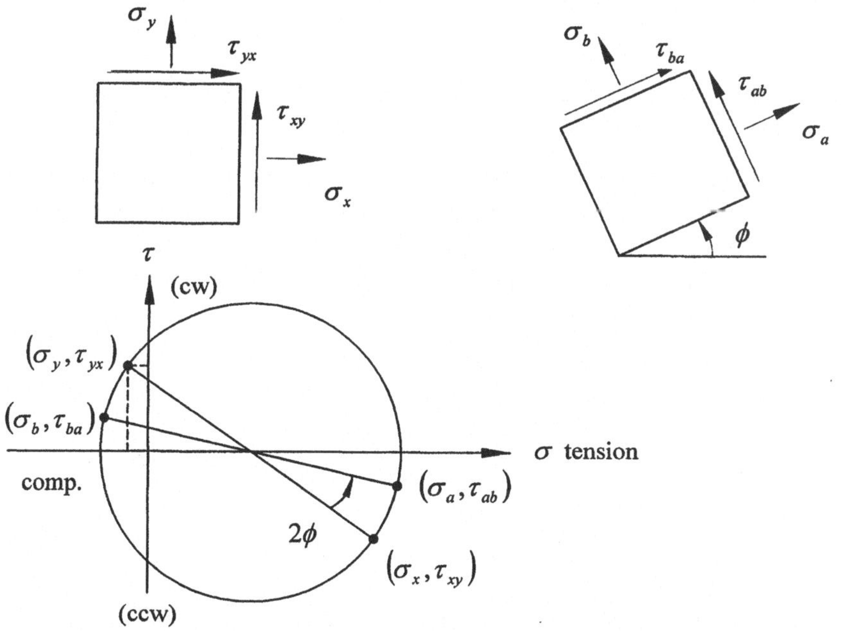

Figure 2.10: Geometric representation of stresses on an element at an arbitrary angle ϕ using Mohr's circle

A graphical representation of this circle is shown in figure 2.10. This is called Mohr's circle for stress. Clearly, this can be easily plotted for any plane stress problem if two orthogonal values of direct stress plus a shear stress are known at any orientation. Construction of this circle from such data is covered extensively in undergraduate courses and will not be reproduced here. However, it is worth indicating the general steps for the determination of the magnitude and orientation of the principal stresses using Mohr's circle. These are

Determine the direction and magnitude of the *applied* direct and shear stresses in the plane of interest

For the two faces (i.e. x and y planes) determine the nature of the stresses. These being either tensile or compressive in the case of the direct stresses and clockwise and counter-clockwise for the shear stresses.

Draw the axes for Mohr's circle. In this case, the horizontal coordinate represents values of the direct stresses, with tensile stresses to the right and compressive stresses to the left of the origin. In opposition to the signs of the shear stresses, these are represented by the vertical axis with clockwise (or negative) shear stresses being above the horizontal line and counter-clockwise shear stresses being below.

Plot the two points corresponding to the coordinates (σ_x, τ_{xy}) and (σ_y, τ_{yx}).

There will then be only one possible circle that bisects these two points, draw this.

To determine the stresses at an angle ϕ rotate around the perimeter of the circle an angle 2ϕ in the *same* direction as on the element. Note that, if the counter-clockwise shear is plotted above the origin, it would be necessary to rotate in the opposite direction on Mohr's circle to that on the element.

From equation (2.26) the principal stress occurs when the shear stress is zero. Consequently, from (2.28) the principal stresses for this plane stress problem are given by

$$\sigma_1, \sigma_2 = \frac{\sigma_x + \sigma_y}{2} \pm \frac{1}{2}\sqrt{(\sigma_x - \sigma_y)^2 + 4\tau_{xy}^2} \tag{2.32}$$

Again from (2.28) the maximum shear stress *in the plane under consideration* is given by

$$\tau_{max} = \pm\frac{1}{2}\sqrt{(\sigma_x - \sigma_y)^2 + 4\tau_{xy}^2} \tag{2.33}$$

2.4.2 Three dimensional stresses

In the above analyses, although not stated, it is assumed that stresses applied to the two-dimensional thin shell represent critical design values. However, it has already been stated that, for a general three-dimensional element, it is possible to determine an orientation such that the element is subject to three orthogonal principal stresses. Under these circumstances it can be shown, Ford, 1977, that three single Mohr's circles can be used to represent the stresses at an arbitrary orientation within the element. The Mohr's circle representation of stresses for the elemental cube of figure 2.7 is shown in figure 2.11. Although beyond the scope of this monograph to analyze such a system of stresses, it is informative to review the main conclusions from such an analysis. These are

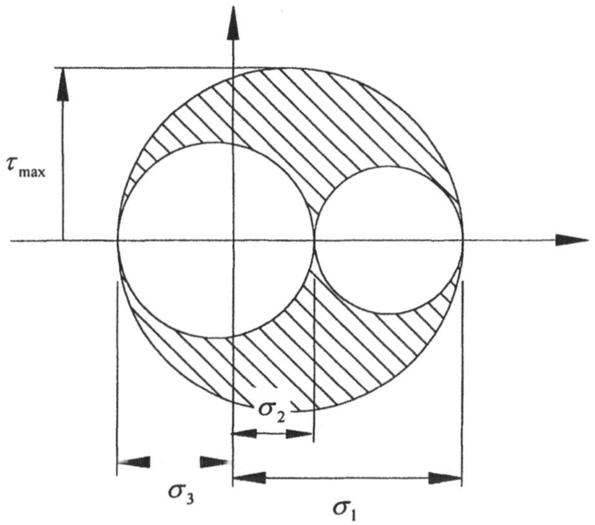

Figure 2.11: Mohr's circle representation of three-dimensional stresses

For an arbitrary orientation within the element, the stress will be within the shaded region of figure 2.11

To assess maximum and minimum stresses in *any* element (including plane stress) it is necessary to use values from the Mohr's circle of largest radius. For example, in a plane stress problem in which the principal stresses are equal $\sigma_1 = \sigma_2$, the Mohr's circle becomes a point. However, because $\sigma_3 = 0$ it is relatively easy to visualize that the maximum shear stress in the element is $\sigma_1 / 2$. It is obviously most important to keep this point in mind when assessing the stresses in a solid

For a plane stress problem, the zero stress normal to the plane of the two dimensional element is a principal stress

2.4.3 Shear stresses and shear strain

So far, because the major effort of the above analysis has been concerned with the determination of principal stresses in which the shear terms are zero, the effects of shear stress on the distortion of an element have been ignored. However, shear distortion can be significant in a number of applications, in particular, for bodies subject to torsion loading.

Following Timoshenko and Goodier, 1970, it is informative once again to consider a two dimensional element subject to both linear and angular distortions as shown in figure 2.12. It can be shown that the components of strain on an element at any point in our Cartesian coordinate system are simply given by the equations

$$\varepsilon_x = \frac{\partial u}{\partial x}$$
$$\varepsilon_y = \frac{\partial v}{\partial y} \qquad (2.34)$$
$$\varepsilon_z = \frac{\partial w}{\partial z}$$

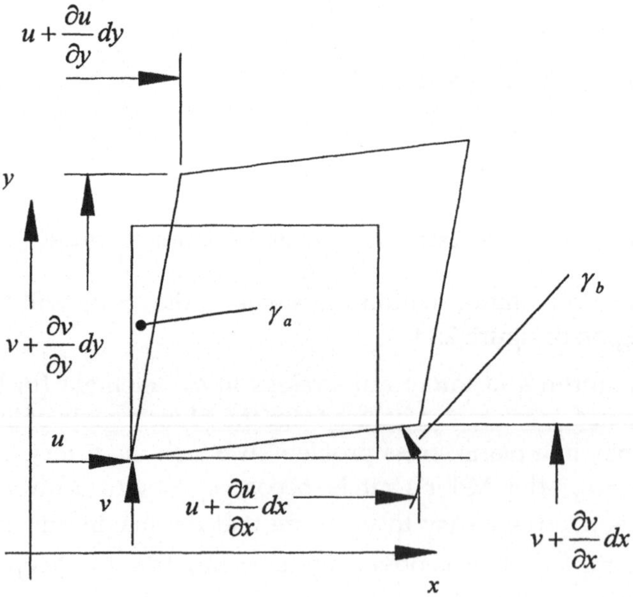

Figure 2.12: Distortions of an element subject to biaxial stress

u, v, w are the displacements of particles in the body caused by the strains. Through analogous reasoning, it can also be shown, Timoshenko, 1970, the shearing strains, γ (measured in radians) can be obtained from

$$\begin{aligned}\gamma_{xy} &= \frac{\partial u}{\partial y} + \frac{\partial v}{\partial x} \\ \gamma_{xz} &= \frac{\partial u}{\partial z} + \frac{\partial w}{\partial x} \\ \gamma_{yz} &= \frac{\partial v}{\partial z} + \frac{\partial w}{\partial y}\end{aligned} \tag{2.35}$$

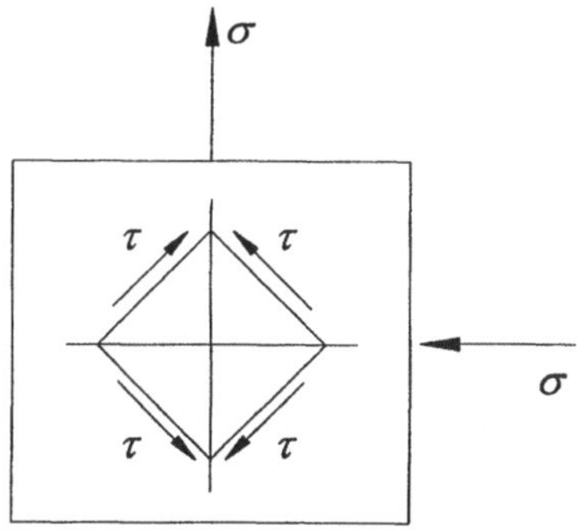

Figure 2.13: Stresses acting on the faces of a square element within a solid subject to equal and opposite direct stresses of similar magnitude

Considering Mohr's circle once again, it is obvious that the only condition in which it is possible to have pure shear on the faces of this element and zero normal stresses is if the average direct principal stress is zero. For our plane stress element this corresponds to the two principal stresses being of equal magnitude with one compressive and the other tensile. It is also relatively easy to see that these will be at 90 degrees to the element on which we are considering the shear stress. Consequently, an element under pure shear can be equally assessed by considering an element that is at an angle of 45 degrees within a large shell subject to simple tensile and compressive stresses of equal magnitude, σ, as shown in figure 2.13. Again following the treatment of Timoshenko and Goodier, 1970, the shear stress on the element is given by

$$\tau = \frac{1}{2}(\sigma_1 - \sigma_2) = \sigma \tag{2.36}$$

From simple geometric considerations it can be shown that if γ is the total angular distortion of the square element then it can be shown

$$\begin{aligned}\tan\left(\frac{\pi}{4} - \frac{\gamma}{2}\right) &= \frac{\tan(\pi/4) - \tan(\gamma/2)}{1 + \tan(\pi/4)\tan(\gamma/2)} = \frac{1+\varepsilon_1}{1+\varepsilon_2} \\ &\approx \frac{1-(\gamma/2)}{1+(\gamma/2)}\end{aligned} \tag{2.37}$$

Because we are considering biaxial principal stresses

$$\varepsilon_1 = -\frac{(1+\nu)\sigma}{E}$$
$$\varepsilon_2 = \frac{(1+\nu)\sigma}{E} \tag{2.38}$$

From equations (2.37) and (2.38) it is relatively simple to derive the relationships for shearing stresses and strains in the form

$$\gamma = \frac{2(1+\nu)}{E}\sigma = \frac{2(1+\nu)}{E}\tau \tag{2.39}$$

Often it is useful to define the constant relationship between shearing strain and shearing stress using the notation

$$G = \frac{E}{2(1+\nu)} \tag{2.40}$$

from which

$$\gamma = \frac{\tau}{G} \tag{2.41}$$

where G is called the *modulus of rigidity*

Assuming that the element was initially square, the shearing strain is the angle at which the elemental cube *differs* from its initial value of 90 degrees. Equation (2.41) can be extended to yield the general relationship

$$\gamma_{xy} = \frac{\tau_{xy}}{G}$$
$$\gamma_{yz} = \frac{\tau_{yz}}{G} \tag{2.42}$$
$$\gamma_{zx} = \frac{\tau_{zx}}{G}$$

The importance of the above equations will become apparent when we consider the torsion of prismatic beams in section 2.8

2.5 Yield criteria

As an introduction to the field of failure prediction, it is worth contemplating the remarkable graph of figure 2.14 (reproduced from Spain, 1987). This shows the change in the volume of a single crystal silicon specimen as it is subjected to hydrostatic pressures up to 100 GPa, as measured using a fascinating instrument known as a diamond anvil cell. These extreme pressures are achieved by compressing the specimen between the flats of two diamond anvils with a metallic seal to prevent lower stresses perpendicular to the applied stress (for a popular account of the development of this technology see Hazen, 1993). Contrast this with the simple bending test shown in figure 2.5 indicating

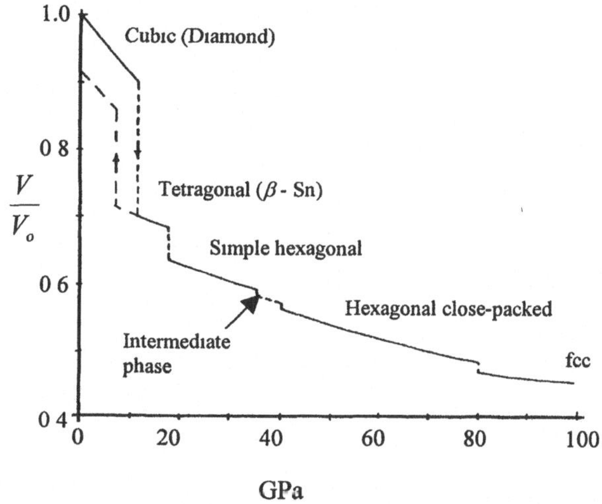

Figure 2.14: Relative volume of silicon as a function of pressure. Dashed line indicates metastable state on slow pressure to the metallic state. Dashed dot line represents volume on release of pressure to a metastable phase (reproduced from Spain, 1987).

catastrophic failure at an applied stress of 138 MPa! It is obvious from these two graphs that it is not simply the magnitude of stress that causes failure. Although it has only recently been possible to demonstrate materials responses to such high stresses, it has been known for many years that it is the *deviation* from hydrostatic stress that results in failure. For ductile materials there are two criteria for predicting the onset of yield, these are

1. For metals, failure occurs when the maximum shear stress exceeds that which is observed to cause yielding in a simple tensile test (Tresca's criteria). For brittle materials, failure is predicted when the tensile stress at the surface rises above a specified value.

2. Alternatively, failure may be predicted when the *distortion* strain energy in an elemental volume becomes comparable to that present in a simple tensile test at the onset of yield (von Mises criteria)

2.5.1 Ductile material failure criteria

Considering the case of metals, the maximum shear stress is simply obtained from the radius of the largest Mohr's circle for a given element. Fortunately, the value of the maximum shear stress is easily measured in the laboratory using a simple tensile test. Such a test imposes a uniform, principal stress over the cross section of the specimen with zero stress in the other two orthogonal planes. Consequently, the maximum Mohr's circle passes through both the origin and the yield stress. Clearly, under these conditions the maximum shear stress is

$$\tau_{\max} = \sigma_Y / 2 \tag{2.43}$$

For arbitrary stresses, the failure condition can be obtained by computing the largest difference between principal stress values. For example if $\sigma_1 > \sigma_2 > \sigma_3$ then failure will occur when

$$\frac{(\sigma_1 - \sigma_3)}{2} \geq \tau_{\max} \tag{2.44}$$

Alternatively, failure is predicted upon the condition

$$(\sigma_1 - \sigma_3) \geq \sigma_Y \tag{2.45}$$

In general, this condition can be written

$$\max.\left[|\sigma_1 - \sigma_2|, |\sigma_2 - \sigma_3|, |\sigma_3 - \sigma_1|\right] = \sigma_Y \tag{2.46}$$

where σ_y is the yield stress measured using a simple tension test

This theory of failure was proposed by Tresca in 1868 and often bears his name.

An alternative failure criteria based on distortion strain energy appears to have been first proposed by James Clerk Maxwell. Ignorant of this, the same hypothesis was independently presented over subsequent decades by a number of researchers including E. Beltrami, M. T. Huber, R. von Mises and H. Hencky, (for a discussion of the development of this theory see Timoshenko, 1983). For historic reasons, it is now more commonly associated with von Mises. Stated simply, it is predicted that failure will occur when the *distortion* strain energy of an element of material exceeds that at the onset of yield in a simple tensile test. Based on this, it is necessary to consider the distortion of an elemental cube of material. These distortions can be resolved into two separate components. One corresponds to the average strain in each axis that results in a change of *size* of the elemental cube. To get to the final distorted shape, it is necessary to add additional strain energy corresponding to deviations of stress about the average value (often called the deviatoric stresses) resulting in a *distortion* of the *shape* of the element. Based on this criterion, failure is predicted to occur when the differences in stress become equal to the yield stress in the relation

$$(\sigma_1 - \sigma_2)^2 + (\sigma_2 - \sigma_3)^2 + (\sigma_3 - \sigma_1)^2 = 2\sigma_Y^2 \qquad (2.47)$$

In many applications, beams or other components are subject to torsion and bending loads. This introduces a simple direct stress plus shear and, depending on whether the von Mises or Tresca criteria are used, failure is predicted when

$$\sigma_Y^2 \leq \sigma_x^2 + 3\tau_{xy}^2 \qquad \text{von Mises} \qquad (2.48)$$

$$\sigma_Y^2 \leq \sigma_x^2 + 4\tau_{xy}^2 \qquad \text{Tresca} \qquad (2.49)$$

Although both of the above results are identical for the prediction of yield in a simple direct tensile test, there is a difference in the criteria for failure due to additional effects of shear. For example, if a torsion test is applied to a rod of material, failure due to shear is predicted to occur at the point at which

$$\tau_{xy} = \frac{\sigma_Y}{\sqrt{3}} \qquad \text{von Mises} \qquad (2.50)$$

$$\tau_{xy} = \frac{\sigma_Y}{2} \qquad \text{Tresca} \qquad (2.51)$$

Experimentally it has been found that the von Mises criteria more accurately predicts the onset of yield while the Tresca criteria provides a more conservative estimate of the allowable design stress. When subject to cyclic loads, ductile materials can fail at stresses below the yield stress in an apparently brittle manner, whence

2.6 Fatigue

2.6.1 S/N curves

It is the intention of this short section to provide a minimal outline of fatigue analysis. Consequently, it will not be possible to discuss the more intricate methods that are required for full solutions to problems in which accurate predictions may be critical to the safe use of a particular design.

Although metals are known to fail due to shear under slowly increasing load conditions, if a cyclic stress is applied to a material, it is possible that a pre-existing crack in the material will steadily grow until it reaches a critical value after which the component will fail catastrophically. Such failure is characteristic of brittle materials and can be analyzed in terms of fracture mechanics. However, for our purposes, it is more informative to present a brief overview of more conventional engineering approach to the problem of fatigue prediction and, more importantly, avoidance. More complete introductory discussions of this topic can be found in Sherratt, 1982, Hertzberg, 1989 and Zahavi and Torbilo, 1996. Consider the behavior of a small specimen subject to a stress that cycles slowly between equal values of tension and compression. For many commonly used metals, it is found that the number of cycles to failure is dependent on the magnitude of the applied stress. If the stress is relatively high in comparison with the ultimate tensile stress, failure can occur after a few thousand cycles or less. Components with such short lives are analyzed in what is called the Low Cycle Fatigue region. For flexure designs, it is unlikely that a design will be subject to stresses so close to the yield stress. It is more common to design with limiting stresses for which the component will survive a large number of cycles to failure (called the High Cycle Fatigue region) or a stress that is below a value that can be endured indefinitely. Such a stress value for a particular material is known as its endurance limit.

Unfortunately, the fatigue life of a particular component will be dependent upon the nature of the applied load, the material of its construction and its geometry. However, in many applications, a component is subject to either simple direct stresses or it is possible to predict the major principal stresses. Consequently, it is reasonable to apply simple tensile and compressive stresses of magnitude S to a specimen and count the number of cycles, N, to failure. The upper case S is used to denote the engineering stress as opposed to true stresses, which may be significantly different when measuring the ultimate tensile stress S_{ut}. Plotting the engineering stress against cycles to failure on a log-log graph often results in the plot shown in figure 2.15. In this graph it can be seen that there is a linear portion between approximately 10^4 and 10^6 cycles after which failure is not observed. Clearly, in the linear region of this graph or high cycle fatigue, there is a relationship between number of cycles to failure, N, and stress, S, of the form

$$N = aS^{-b} \tag{2.52}$$

where

$$S = \frac{S_{\max} - S_{\min}}{2} \tag{2.53}$$

a and *b* are material constants.

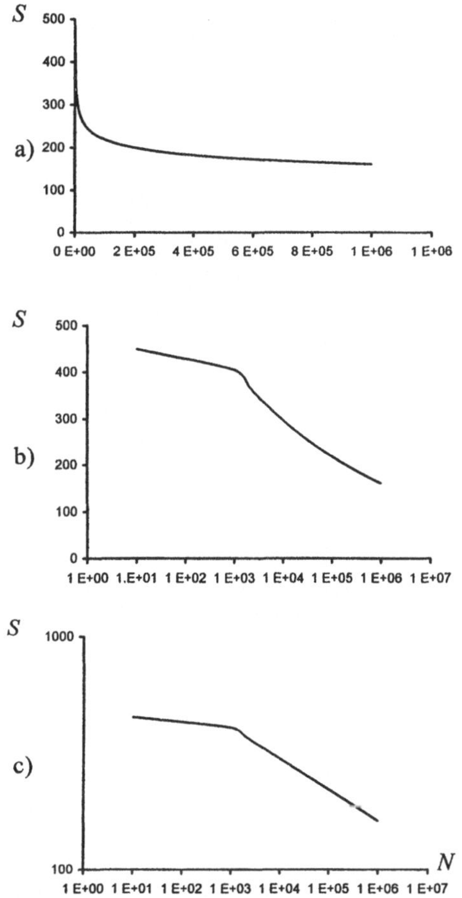

Figure 2.15: Plots of $S-N$ data on different scales, a) linear, b) Log-linear, b) log-log scales

Looking at the S - N curves for metals other than steel it will soon be apparent that this simple picture may provide faulty predictions. For example, although aluminum does not have an endurance limit it is usual to quote the stress at which it will survive 500 million cycles. Table 2.1 at the end of this chapter lists the commonly used endurance limits for a number of metals as provided in the Atlas of Fatigue Curves edited by Howard E. Boyer.

At this stage in the discussion it is worth spending a little time indicating some of the errors that may be introduced when following this analysis route. Great care has been taken to indicate that we are considering only the engineering stresses on a specimen. A paradox becomes apparent when it is realized that we are assessing a phenomenon in which a crack is gradually growing within the stressed zone. As a consequence, as the crack grows the area supporting the applied loads reduces, thereby increasing the true stress in this region. For small specimens, this will probably not have a significant effect but care should be employed when analyzing components in which the stress extends over more than a few tens of millimeters in any one direction. Techniques are available using strain control that, potentially, overcomes this particular problem. A discussion of strain-based analysis is outside of the scope of this book. Another related problem is that of accidental overload. If this overload stress is beyond the high cycle fatigue limit a significant reduction in life might ensue after which the analysis is no longer valid. Again, techniques to cope with such scenarios do exist. However, the complexity of their implementation means that it is often more economic to monitor for overloads and replace components after such an event has occurred. Another, more frequently encountered, and easily analyzed, fatigue life reduction mechanism is discussed in the following section.

2.6.2 Effects of notches

Upon inspection of components that have failed by fatigue, it is often apparent that this has occurred as a consequence of the growth of a crack from a region where there is a sudden change of geometry such as a notch or hole. For some pretty spectacular examples see Hertzberg, 1989. Clearly, this is due to the concentration of stress in the region near to the notch. Reasonably, the effect of the notch could be analyzed by calculating the maximum stress due to this sudden change of geometry and using it as the basis for subsequent calculations. For this purpose it is usual to define a *stress concentration factor,* K_t, given by

$$K_t = \frac{\text{Maximum stress in the notch}}{\text{A nominal stress away from the notch}} \tag{2.54}$$

For example, if a small hole is drilled into a wide flat plate that is subject to a uniform tensile stress there will be a local threefold increase in stress at two points on the periphery of the hole (i.e. $K_t = 3$). Clearly, it would seem reasonable to measure the stress somewhere remote from the hole and simply multiply it by the appropriate factor to predict the fatigue life. However, subsequent tests on notched specimens often reveal that such an approach is rather conservative and specimens will last longer than predictions based on the stress concentration factor, see figure 2.16. Consequently, for notched specimens it is usual to use a *strength reduction factor,* K_f, which is always less than or equal to the stress concentration factor and given by the ratio of allowable stresses for specimens with and without notches i.e.

$$K_f = \frac{\text{Allowable range of stresses without notch}}{\text{Allowable range of stresses with notch}} \tag{2.55}$$

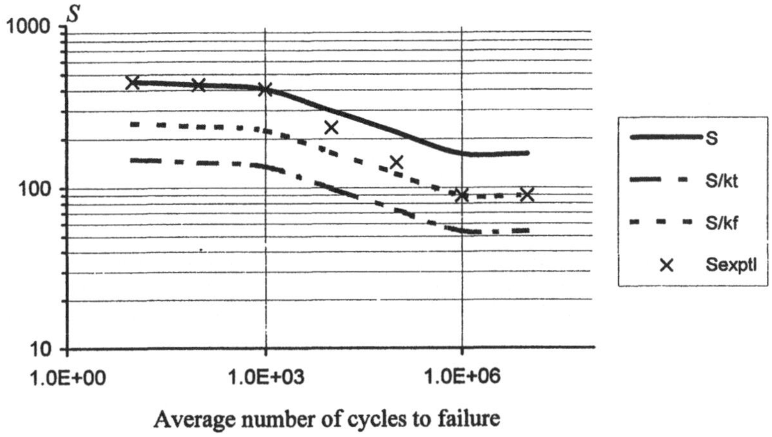

Figure 2.16: Plots of $S-N$ data for a specimen containing a local notch

Often the allowable stresses are simply given as the endurance limits for the notched and notch-free specimens. Unfortunately, there are two problems with such an approach. Firstly, the strength reduction factor is only valid if the curve for the notched specimen is parallel to that for the notch-free case as shown in the figure. In practice, the notches have little effect in the low cycle fatigue region. The influence tends to increase with number of cycles as shown in the figure. Consequently, the use of a constant value for the strength reduction factor is only appropriate for predicting the limiting endurance stress in the presence of notches. Secondly, the strength reduction factor is not a material property and must therefore be experimentally determined for each design, a costly process. Clearly, the strength reduction factor must be in some way related to the radius of the notch, r, the stress concentration factor and some property of the material. A further factor known as the *notch sensitivity, Q,* is given by the relationship, Pilkey, 1997

$$Q = \frac{1}{1 + p/r} = \frac{K_f - 1}{K_t - 1} \tag{2.56}$$

where p is a material constant.

Consequently, using graphs of the notch sensitivity of metals as a function of notch radius and stress concentration factor, it is possible to determine the appropriate value for the strength reduction factor. A comprehensive collection of stress concentration graphs for a broad range of geometry's and load conditions can be found in Pilkey, 1997, and no respectable library is without a copy.

2.6.3 Effects of mean stress

In many applications, pre-loads due to fastening, internal residual stresses, self weight or constant applied forces are unavoidable. Consequently, stresses often vary about a mean value, given by

$$S_m = \frac{S_{\max} + S_{\min}}{2} \tag{2.57}$$

Under these conditions it is usual to use the allowable stress amplitude, S_a, as predicted from (2.52) for a given life as a function of the mean load given by

$$S_a = S_0\left(1 - \left(\frac{S_m}{S_{ut}}\right)^n\right) \tag{2.58}$$

where S_0 represents the equivalent stress at zero mean offset and is commonly measured in standard fatigue tests, n is a constant with common values being

1 for general use and this is known as the Goodman criteria

2 for a more accurate representation of experimental data for a broad range of materials and is known as the Gerber parabola

1.5 is commonly used for steels

Sometimes, the ultimate tensile stress in the Goodman version of equation (2.58) is replaced by the yield stress and this is known as the Soderberg criteria which is also the most conservative estimate of the effect of mean stress.

It is assumed, and necessary, that stresses, even in notched regions, are below the yield stress. Provided that the requisite conditions are not violated, the above analysis will enable the computation of a reasonable expected fatigue life for a component if the stresses are accurately known in advance. For a more precise estimate of expected number of cycles to failure, a statistical approach should be used to account for the scatter in experimental data and the reader is again referred to Sherratt, 1982, or Zahavi and Torbilo, 1996, for a brief overview.

2.6.4 Damage assessment

Stress variations on a component are rarely going to be of uniform amplitude or mean value. Consequently, it is necessary to assess the damage induced from an arbitrary load cycle. To do so, it is first necessary to break the stress down into individual cycles having individual values of mean and amplitude. From each of these, it is then possible to determine the number of cycles to failure. Assuming that each cycle contributes in a linear way to the total damage, we may compute the damage of the ith cycle from the simple relationship

$$D_i = \frac{1}{N_i} \tag{2.59}$$

where N_i is the value for the number of cycles to failure for this particular cycle

It is then a simple matter to sum over all cycles to produce an estimate of the total damage, D_{tot} given by the Palmgren-Miner equation

$$D_{tot} = \sum_i \frac{1}{N_i} \tag{2.60}$$

A value at which the component has reached the end of its useful life is again in the hands of the designer although a total damage of 1 is reasonable and the more conservative value of 0.3 also commonly used.

As a final word before moving on to new ground, it is important that the counting of stress cycles is correctly evaluated. For example, there will often be irregular stresses that vary rapidly about more substantial variations. Using a routine that simply looks at adjacent peaks and valleys, larger fluctuations over the longer time periods would be ignored. Such an approach would be particularly erroneous because it would fail to include the stress variations that induce the most damage. A more robust algorithm called 'rainflow counting' should be employed to determine stress cycles from an arbitrary stress history. In this, cycles are identified using a peak detection routine that identifies the small stress variations within larger peaks and valleys. Successive cycles are eliminated from the stress history until the last cycle represents the maximum difference in stress over the complete history. Again, more detailed explanations can be found in the references.

2.7 Bending of symmetric beams

2.7.1 The bending equation

So far, the evaluation of stress and strain has been restricted to a point on the surface or within a component. A great many of the flexures presented in this book will require the prediction of maximum stresses and strains as well as deflections relative to a fixed frame. In particular, a broad range of flexures utilizing deflection of slender, symmetric beams of uniform cross section can, with relative ease, be either fabricated or machined from solid. Consequently, it is necessary to review those techniques used for the prediction of stresses and subsequent deflections of simple beam elements. Consider the slender beam of figure 2.17 subject to a pure bending moment, M_z, that induces a deflection having a curvature, $1/R$. Clearly, this will result in a simple direct stress, σ_x,

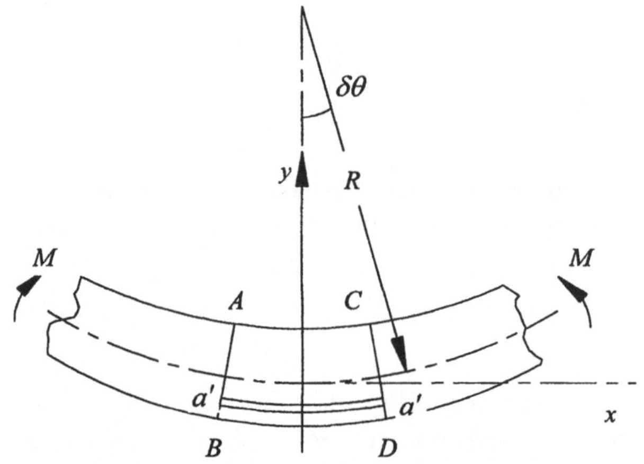

Figure 2.17: A symmetric beam subject to a pure bending moment

acting in a direction along the beam axis. For the beam to remain stationary, it is necessary that the total force due to the distributed stress is zero. Consequently, the total sum of compressive stresses must equal the tensile ones. Also it is apparent that there will be a compressive stress at the upper surface and a tensile stress on the lower one. Therefore, there must be a plane running along the axis of the beam in which the stress is zero, appropriately called the *neutral plane*. If we were to take a cross section through the beam, the neutral plane would intersect along a line in the z direction and this is known as the *neutral axis*. For a symmetric beam (all the beams considered in this book) planes parallel to the neutral plane represent contours of equal stress and, therefore, strain. Looking at one such plane, *a'a'* in figure 2.17, it is readily verified that the strain for a given curvature is given by

$$\varepsilon_x = \frac{(R+y)\delta\theta - R\delta\theta}{R\delta\theta} = \frac{y}{R} \qquad (2.65)$$

Because this is a principal strain, the relationship between stress and strain is immediately written

$$\sigma_x = E\frac{y}{R} \qquad (2.66)$$

Summing the forces across the complete cross section gives

$$\int_A \sigma_x dA = \frac{E}{R}\int_A y dA = 0 \qquad (2.67)$$

From this we conclude that the neutral axis coincides with the *centroid* of area of the cross section. It is also apparent that the total bending moment about the neutral axis can be obtained through the summation of all products of stress, elemental areas and the distance y from the neutral axis i.e.

$$M = \int_A \sigma_x y dA = \frac{E}{R} \int_A y^2 dA = \frac{EI_{zz}}{R} \tag{2.68}$$

Combining this with (2.66) produces the bending equation

$$\frac{\sigma}{y} = \frac{E}{R} = \frac{M}{I} \tag{2.69}$$

From this, we observe that the stress varies linearly with perpendicular distance from the neutral axis of bending and that the curvature of the beam is proportional to the applied bending moment and inversely proportional to the *flexural rigidity*, *EI*. *I* is always the second moment of area about the neutral axis.

2.7.2 Deflection of beams

Rearranging equation (2.69) in terms of beam curvature gives

$$\frac{1}{R} = \frac{M}{EI} \tag{2.70}$$

For the coordinate system shown in figure 2.17, the curvature is positive and can be calculated from the equation

$$\frac{1}{R} = \frac{\frac{d^2y}{dx^2}}{\left[1 + \left(\frac{dy}{dx}\right)^2\right]^{3/2}} \tag{2.71}$$

For small deflections, the denominator of equation (2.71) is approximately 1 leaving the relatively simple relationship

$$\frac{1}{R} = \frac{d^2y}{dx^2} \tag{2.72}$$

Substituting (2.70) into the above gives the governing differential equation for elastic deflection of symmetric beams

$$EI_{zz} \frac{d^2y}{dx^2} = M_z(x) \tag{2.73}$$

The notation explicitly infers the fact that we are considering the distribution of moments about the z-axis along the axis of the beam. However, in future equations it is assumed from the context of the discussion. Consequently, subscripts will be dropped in future analyses.

2.7.2.1 Sign convention for bending moments

In all of our subsequent analyses any beam will be assigned a right hand coordinate system. To determine the sign of the bending moment, consider an arbitrary position x along a cantilever beam subject to a concentrated moment at the free end. It is relatively simple to determine the direction of the *applied* bending moment at that position looking either to the left or the right. Imagine that the observer has cut the beam at this position and is holding it together. Our sign convention is

> *a bending moment is considered to be negative if it acts to induce a tensile stress in the upper face of the beam*

A simple example would be a cantilever beam with the left side fixed and with a constant clockwise bending moment of magnitude M_0 at the free end. Clearly this will be a negative bending moment from which (2.73) gives

$$EI\frac{d^2y}{dx^2} = -M_0 \tag{2.74}$$

Integrating this twice and substituting in the boundary conditions requiring that both slope and deflection are zero at the fixed end on the left (at $x = 0$), after a little rearranging, we obtain

$$\frac{EI}{M_0 L^2} y = -\frac{X^2}{2} \tag{2.75}$$

where $X = x/L$

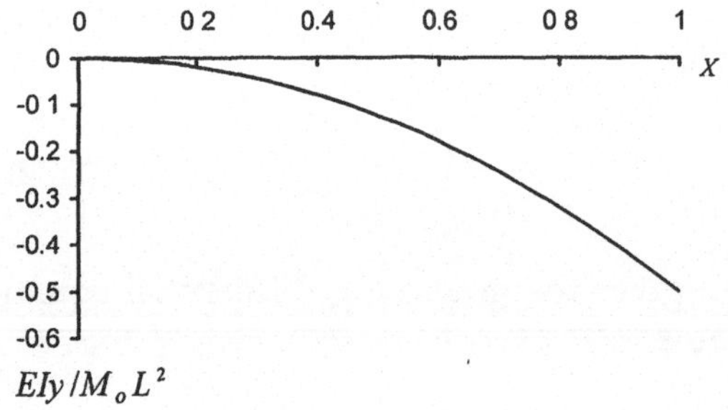

Figure 2.18: Cantilever beam with a bending moment applied at the free end

Equation (2.75) is plotted in figure 2.18 in terms of the dimensionless lengths and displacements.

In fact, it is not necessary to integrate (2.74). It has already been shown that the curvature of a beam subjected to pure bending is constant and the right hand side of (2.75) is simply the truncation of the expression for y as a function of x for a small portion of a large circle i.e.

$$EIy = -R(1 - \cos(x/R)) \approx -\frac{1}{M_0}\frac{M_0^2 x^2}{2} \tag{2.76}$$

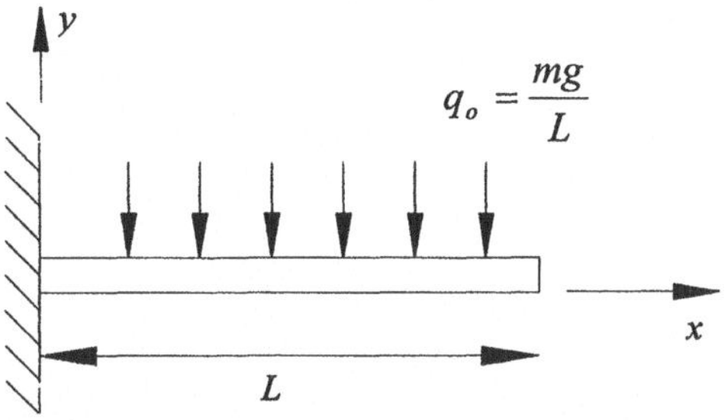

Figure 2.19: Cantilever beam subject to self-weight load

As a slightly less obvious example consider the cantilever beam of figure 2.19. In this example it has been assumed that self-weight cannot be ignored. The self-weight is given by a force per unit length, q_0, which is simply the product of gravity acceleration, density and area of cross section. For the purpose of computing the bending moment, the distributed load has been lumped to a point load at its center of action which is half way from the observer at x to the free end of the beam. With gravity, the force due to self-mass will act downwards resulting in a clockwise, applied bending moment to the right of x. Being of negative sign, the moment at position x is given by

$$M(x) = -q_0 \frac{(L-x)^2}{2} \tag{2.77}$$

Substituting this into (2.73) gives

$$EI\frac{d^2y}{dx^2} = -q_0 \frac{(L-x)^2}{2} \tag{2.78}$$

Integrating (2.78) twice with respect to x

$$\frac{EI}{q_0}y = -\frac{(L-x)^4}{24} + C_1x + C_2 \tag{2.79}$$

Substituting the boundary conditions of zero slope and deflection at the fixed ends enables the determination of the two constants after which the expression for deflection is given overleaf by

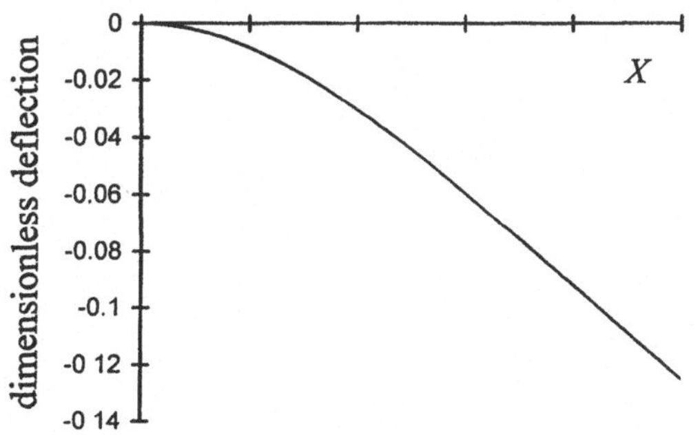

Figure 2.20: Deflection of a cantilever beam under a uniform vertical load

$$\frac{EIy}{q_0 L^4} = Y = \left[\frac{X^3}{6} - \frac{X^2}{4} - \frac{X^4}{24}\right] \tag{2.80}$$

In this instance it is less clear that the beam is sagging under its own weight, but this is readily verified by looking at the plot of this equation in figure 2.20.

2.7.3 Moment, shear force and rate of loading relationships

In the above example for a uniform cantilever beam subject to its own self-weight loading, it was assumed that the load was distributed evenly and, therefore, the load per unit length could be represented by the constant value. Considering the case of a constant distribution of load, q say, acting in the positive direction, it is relatively simple to show that the total force acting at a position x is

$$V = \int_x^L q\,dx = q(L - x) \tag{2.81}$$

Considering now the bending moment due to this shear force. Starting at the fixed end and moving in a distance x, the bending moment is simply the product of all incremental moments along the beam from x to the free end. Consequently, the moment for a constant load per unit length is

$$\begin{aligned} M &= \text{(load at centroid)(distance from point } x \text{ to centroid)} \\ &= q(L-x)\frac{(L-x)}{2} = q\frac{(L-x)^2}{2} \end{aligned} \tag{2.82}$$

Comparison between (2.82) and (2.81) implies the relationship

$$M = -\int_x^L V\,dx \tag{2.83}$$

Equations (2.83) is generally valid for all beams and can be written in the form

$$V = -\frac{dM}{dx} = -EI\frac{d^3y}{dx^3} \tag{2.84}$$

and

$$q = -\frac{dV}{dx} = EI\frac{d^4y}{dx^4} \tag{2.85}$$

In summary, the relationship between displacement, slope, bending moment and shear force is given by

$$\begin{aligned} &\text{Deflection} && y \\ &\text{Slope} && \frac{dy}{dx} \\ &\text{Bending moment} && EI\frac{d^2y}{dx^2} \\ &\text{Shear force} && V = -\frac{dM}{dx} = -EI\frac{d^3y}{dx^3} \\ &\text{Rate of loading} && q = -\frac{dV}{dx} = EI\frac{d^4y}{dx^4} \end{aligned} \tag{2.86}$$

This leads to a generalized method for the solution of beam deflections in terms of the load distribution and is outlined in the section below. It is important to note that the equations for deriving moments, slopes and displacements do not require any changes in sign when integrating the rate of loading equation (2.85). Normally, for the purpose of deriving the shape of a beam, it is necessary that the equation satisfies the boundary conditions at each end (or at each support/constraint). The simplest and often most usable conditions are those having zero value (i.e. zero displacement → pinned, zero shear → free, zero moment → clamped).

2.7.4 Singularity functions

Straightforward integration of the moment distribution along a beam will result in the deflection shape of a beam for any load conditions. However, in many instances, the beam will be supported at discrete points, or subject to bending moments from attachments occupying a short span of the total beam length. Under these conditions it can be assumed that the load or bending moment appears at a point along the beam. This can be included in the bending equation and integrated to determine the displacement. Consider for example the simply supported beam of figure 2.21 with a pure applied bending moment at mid span. Being a clockwise moment of magnitude, *M*, it is obvious that the beam will be bent in some sort of *'S'* shape. A problem arises when we try to express the moment along the beam. More importantly, it is even more difficult to envisage, let alone model, detailed mechanisms in these small regions. For example, any pure bending moment acting at a point will induce an infinite shear stress. Clearly, the

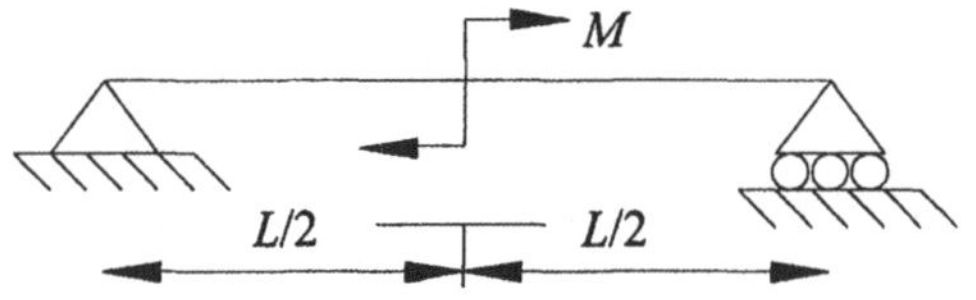

Figure 2.21: A simply supported beam with a centrally applied pure bending moment

brackets and connections around this point will locally distort, a discussion of which is outside the scope of this text. A powerful method for solving such beam bending problems was originally developed by an engineer named Macauley and is sometimes named after him. Today, this is more commonly referred to as the method of *singularity functions*. In this, a singularity function is defined so that; it is always zero if its contents are less than zero, is undefined if its contents are zero and is equal to its contents when they are greater than zero. A singularity function is discriminated by enclosing it in sideways vee brackets, i.e. ⟨contents of singularity function are here⟩. This makes it possible to describe a broad range of load conditions that are easily integrated in compliance with the rules outlined below.

Table 2 (p58) shows the various expressions for the rate of loading on a beam due to a variety of applied loading conditions. It will be worth spending a little time looking at the rather unusual properties of the singularity functions as they appear in this table. When the singularity function is raised to the power 0, it does not become unity nor does the integral of the integral of the singularity function raised to the power -1 become a logarithm. Because of these integral properties, the reader is advised to proceed with great care when using singularity functions. Another small but important point, the keen observer will have noticed that in the last row the expression for the singularity function at $x \geq a$ becomes a regular equation, a point indicated by the change in parentheses.

Returning now to our problem of the simply supported beam with a central bending moment, the rate of loading equation can be easily written as

$$q = R_1 < x >^{-1} + M < x - a >^{-2} + R_2 < x - L >^{-1} \tag{2.87}$$

Integrating (and noting that the $-\infty$ or even zero limit makes all singularity functions zero)

$$-V = R_1 < x >^{0} + M < x - a >^{-1} + R_2 < x - L >^{0} \tag{2.88}$$

Again, integrating

$$M(x) = R_1 < x >^{1} + M < x - a >^{0} + R_2 < x - L >^{1} \tag{2.89}$$

At the right hand support the condition for zero bending moment gives

$$M(L) = R_1(L) + M = 0 \tag{2.90}$$

From which

$$R_1 = -R_2 = -M / L \tag{2.91}$$

Equation (2.89) can therefore be written

$$M(x) = -\frac{M}{L} < x >^1 + M < x - a >^0 - R_2 < x - L >^1 \tag{2.92}$$

Taking moments about each end could have more easily solved for the moment equation (2.92). Substituting the above into (2.86) and setting $a = L/2$ gives

$$EI\frac{d^2y}{dx^2} = M(x) = M < x - L/2 >^0 - Mx/L \tag{2.93}$$

Integrating twice for the vertical deflection

$$EIy = M\frac{< x - L/2 >^2}{2} - Mx^3/6L + C_1x + C_2 \tag{2.94}$$

From the boundary conditions of zero deflection at either end the constants can be eliminated to yield

$$\frac{EI}{ML^2}y = \frac{< X - 1/2 >^2}{2} - \frac{X^3}{6} + \frac{X}{24} \tag{2.95}$$

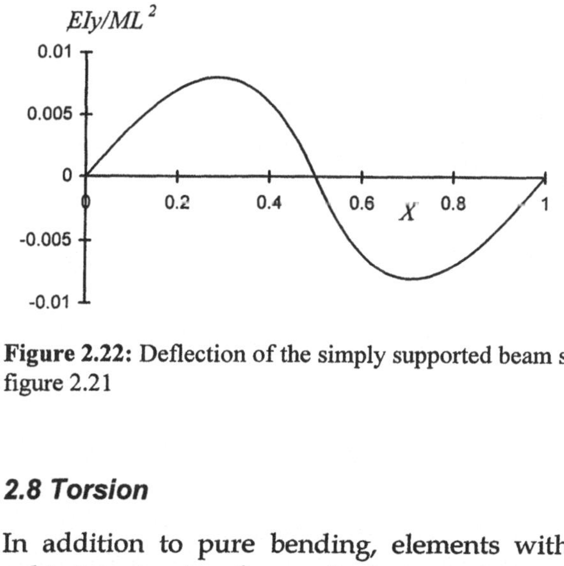

Figure 2.22: Deflection of the simply supported beam shown in figure 2.21

A dimensionless deflection along the beam is plotted in figure 2.22. Although a relatively simple model, it is applicable for the modeling of many flexure mechanisms and variants on this are discussed in more detail in chapter 5.

2.8 Torsion

In addition to pure bending, elements within flexure mechanisms are often subject to torsion forces. In fact, we have already seen that any generalized applied force can be reduced to simple direct plus shear stresses. Additionally, if a body is subject to an arbitrary but small force, this can be resolved into forces and torsional moments. Using the principle of superposition, it is then possible to analyze the distortions due to each load in isolation and sum these to derive the complete distortion due to the original load. Consequently, we shall look only at applied shear stresses due to pure torsional loads and leave it to further chapters to demonstrate the use of superposition to determine full solutions for practical load cases. For the purposes of this book it will be adequate to restrict attention to the analysis of shear stresses and subsequent distortion due to pure torsional moments applied about the axis of prismatic beams of circular or rectangular cross-section.

2.8.1 Torsion of a prismatic beam of circular cross section

In the case of a pure torsion applied along the axis of a beam of circular cross section, the shear stress is zero along the axis and increases linearly to a maximum at the outside edge resulting in an angular distortion of a square element on the surface of angle

$$\gamma = \frac{\tau_{\max}}{G} \tag{2.96}$$

The angle of twist of the cross section, θ, is simply the integral of the angular distortion along the beam so that

$$\theta = \int_0^L \gamma dx = \frac{\tau_{\max} L}{G} \tag{2.97}$$

Exploiting the linear relationship between shear stress and radial distance from the axis r and the fact that the angle of twist of the cross section is a constant, it is therefore possible to derive the shear stress on the cross section at any radial position r by the equation

$$\frac{\tau}{r} = \frac{G\theta}{L} \tag{2.98}$$

Based on the above it is a simple matter to derive the torsional load to produce this deflection given by

$$T = \frac{\pi R^4}{2}\frac{G\theta}{L} = J\frac{G\theta}{L} \tag{2.99}$$

where R is the outer radius of the rod, J is the polar second moment of area and T is used to represent the torsional moment about the x-axis to distinguish it from the bending moment, M_z in the previous analyses.

Combining equations (2.98) and (2.99) produces the general torsion equation for beams of circular cross-section

$$\frac{T}{J} = \frac{G\theta}{L} = \frac{\tau}{r} \tag{2.100}$$

For a beam of tubular section with inside and outside radii, R_i and R_o respectively, this equation can still be used with the modified polar second moment of area given by

$$J = \frac{\pi\left(R_o^4 - R_i^4\right)}{2} \tag{2.101}$$

2.8.2 Torsion of a prismatic beam of rectangular cross section

For a rectangular bar the shear distribution over the cross-section is not as straightforward. Solutions for the maximum shear stress and applied torque are derived using more advanced elasticity theory, see for example Timoshenko and Goodier, 1970[1].

For a beam of thickness, *t*, and width, *d*, the maximum shear stress in a beam of length *L* subject to an angular twist θ is given by

$$\begin{aligned}\tau_{\max} &= \frac{G\theta t}{L} - \frac{8G\theta t}{\pi^2 L}\sum_{n=1,3,5,}^{\infty}\frac{1}{n^2\cosh(n\pi d/2t)} \\ &= k\frac{G\theta t}{L}\end{aligned} \tag{2.102}$$

where *k* is a constant dependent upon the ratio *d/t*.

For a thin rectangular strip, the ratio *d/t* is large and the above reduces to

$$\tau_{\max} = \frac{G\theta t}{L} \tag{2.103}$$

Similarly, it can be shown that the torque *T* to produce an angular deflection θ is given by

$$\begin{aligned}T &= \frac{t^3 d}{3}\frac{G\theta}{L}\left(1 - \frac{192}{\pi^5}\frac{t}{d}\sum_{n=1,m3,5,.}^{\infty}\frac{1}{n^5}\tanh\left(\frac{n\pi d}{2t}\right)\right) \\ &= k_1\frac{G\theta}{L}\frac{t^3 d}{3}\end{aligned} \tag{2.104}$$

Combining equations (2.103) and (2.104) to eliminate θ, the maximum shear stress as a function of the applied torque is given by

---

[1] Note that in the Timoshenko and Goodier reference θ is in units of radians per meter (i.e. θ/L in our text) and $t = 2a$ and $d = 2b$

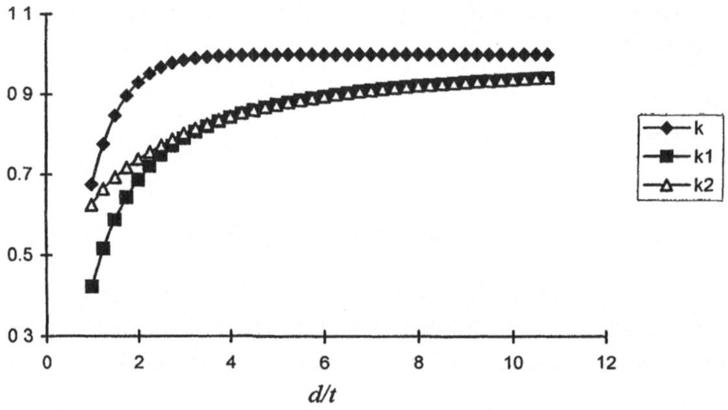

Figure 2.23: Graph of the torsion constants for a rectangular bar

$$\tau_{\max} = \frac{k}{k_1} \frac{3T}{t^2 d} = \frac{3T}{k_2 t^2 d} \tag{2.105}$$

For convenience, all three of these constants are plotted in figure 2.23. All of the above series converge rapidly so that, for computation of reasonably accurate values, it is only necessary to use three or four terms in any of the series expansions.

2.9 Mobility

Flexure mechanisms considered throughout this book will be modeled as a chain of rigid links connected by flexible elements that are relatively compliant to some forces or moments and very stiff in other directions. Consequently, flexure elements can be modeled as connections between the rigid bodies that impose a specific number of constraints. In mobility analysis, each of these connections is called a *joint*. For example, a leaf type flexure connecting adjacent links might produce large displacements of the links due to applied bending moments whereas all other forces may have little effect. In this case, the leaf type flexures would be modeled as 'ideal' hinges providing a single rotation freedom while constraining relative motion between the links in all other 5 directions (i.e. 2 rotations and 3 linear translations). For simple mechanisms, the possible motions of the links are obvious. For example, a simple pendulum consisting of a rigid link attached to a rigid structure by a leaf spring would be free only to rotate about its 'hinge' axis. Hence the pendulum is said to have a mobility of 1. As the number of links and joints increases, the subsequent mobility of the mechanism is less easily assessed. Using generalized methods of kinematic analysis relatively simple equations for mobility analysis are developed below.

Consider first the analysis of planar mechanisms. In this case it is apparent that a rigid link in a mechanism has only three degrees of freedoms corresponding to rotation in the plane and translation in x and y directions (for a detailed discussion of this subject see the book of Phillips, 1984). Consequently, the mobility can be determined by first of all separating each link in which case the freedom of the complete ensemble is simply 3 times the number of links, n, minus the one that is necessarily fixed. Joining each successive link together through a joint, i, having between 1 and 2 freedoms reduces the mobility by the number of constraints, c, imposed. This leads to an expression for the planar mobility, M_3, given by

$$\begin{aligned} M_3 &= 3(n-1) - \sum_{i=1}^{j} c_i = 3(n-1) - \sum_{i=1}^{j} (3 - f_i) \\ &= 3(n-j-1) + \sum_{i=1}^{j} f_i \end{aligned} \tag{2.106}$$

where n is the number of links in the mechanism, j is the number of joints and f is the number of freedoms provided by the ith joint. In figures the value of f is indicated as a superscript and should not be confused with the normal mathematical convention for raising variable to a power. For example, a simple parallel hinge consisting of four links with four single degree of freedom joints (hinges) can be readily shown to possess a single mobility. To consider three-dimensional motion, it is necessary to consider each link possessing 6 unconstrained freedoms leading to mobility M_6 given by

$$M_6 = 6(n-j-1) + \sum_{i=1}^{j} f_i \tag{2.107}$$

Subsequent analysis of our simple linear spring mechanism reveals a mobility of -2 implying that such a mechanism is *over constrained.* This is indeed the case and this was obviated in the planar analysis by the intrinsic assumption that all of the hinges had axes parallel and normal to the plane of motion. Due to finite manufacturing tolerances, this will not be possible and there will be some out of plane torsional moments about two or more of the hinges.

A single degree of freedom mechanism is obtained if three further freedoms are added. This could be achieved by replacing three of the one degree of freedom hinges with those having two. In many practical situations over constraint is unlikely to produce stresses that are unacceptably large and, consequently, it may be assumed that planar analysis is adequate for planar mechanisms. This being the case, consider a mechanism in which each joint possesses a single degree of freedom. This could be a slideway (not necessarily linear), a simple single degree of freedom joint or a mechanism composed from a subset of links and joints. Clearly, under these conditions, the sum of freedoms is equal to the number of joints and equation (2.106) reduces to Grubler's equation

$$M = 3(n-1) - 2j \tag{2.108}$$

Equations (2.106) to (2.108) are used extensively in chapter 5.

References

Howard E. Boyer (ed.), 1986, *Atlas of Fatigue Curves*, Metals Park, Ohio: American Society for Metals

Brentnall W.D. and Rostoker W., 1965, Some observations on microyielding, *Acta Metallurgica*, **13**, 187-198

Dieter G. E., 1987, Mechanical Metallurgy, McGraw-Hill, NY, chapter 4.

Ford, H., 1977, *Advanced mechanics of materials*, in four parts ; part IV with the collaboration of J. M. Alexander, 2d ed., Halsted Press, NY

Hazen R. M., 1993, *The New Alchemists; Breaking Through the Barriers of High Pressure*, Times Books, NY

Hertzberg R. W., 1989, *Deformation and Fracture Mechanics of Engineering Materials*, John Wiley and Sons, NY, chapter 1

Lawn B. R., 1991, *Fracture of Brittle Solids*, Cambridge University Press

Pilkey W.D., 1997, *Peterson's Stress Concentration Factors*, 2nd ed., Wiley and Sons Inc., NY.

Phillips J., 1984, *Freedom In Machinery*, Cambridge University Press.

Popov E. P., 1976, *Mechanics of Materials*, Prentice-Hall, Inc., NJ

Sherratt F., 1982, Fatigue life estimation: A review of traditional methods, *J. Soc. Envir. Engs.*, December, 23-30

Shigley J. E. and Mischke C. R., 1989, Mechanical Engineering Design, McGraw-Hill, NY

Smith S.T., Chetwynd D.G. and Jackson D., 1991, Manufacture of large scale devices in single crystal silicon by high speed grinding, *SPIE*, **1573**, 53-61

Spain I.L., 1987, Semiconductors at high pressure: New physics with the diamond anvil cell, *Contemp. Phys.*, **28**, 523-546

Timoshenko S.P. and Goodier J. N., 1970, *Theory of Elasticity*, McGraw-Hill, NY (note that in both Timoshenko and this book the total shear strain is used. Many other texts use tensor, or index, notation that leads to symmetric matrix representations that are often mathematically concise and computationally more efficient. However, in tensor notation $2\varepsilon_{ij} = \gamma_{ij} \quad i \neq j$, see chapter 1 section 7 of this reference)

Timoshenko S.P., 1983, *History of Strength of Materials*, Dover Publications, Inc., NY. The discussion of distortion energy theory appears on pages 368-372.

Zahavi E. and Torbilo V., 1996, *Fatigue Design; Life Expectancy of Machine Parts*, CRC Press, Boca Raton

Table 2.1: Strength characteristics of some metals

Material	Condition	Tensile strength (MPa)	Yield stress (MPa)	Endurance limit (MPa)
Steel alloys (endurance limit based on 10 million cycles)				
1015	Cold drawn - 0%	455	275	240
1015	Cold drawn - 60%	710	605	350
1040	Cold drawn - 0%	670	405	345
1040	Cold drawn - 50%	965	855	410
4340	Annealed	745	475	340
4340	Quenched and tempered (204 C)	1950	1640	480
4340	Quenched and tempered (427 C)	1530	1380	470
4340	Quenched and tempered (538 C)	1260	1170	670
HY140	Quenched and tempered (538 C)	1030	980	480
D6AC	Quenched and tempered (260 C)	2000	1720	690
9Ni-4Co-0.25C	Quenched and tempered (315 C)	1930	1760	620
300M		2000	1670	800
Aluminum alloys (endurance limit based on 500 million cycles)				
1100-0		90	34	34
2014-T6		483	414	124
2024-T3		483	345	138
6061-T6		310	276	97
7075-T6		512	503	159
Titanium alloys (endurance limit based on 10 million cycles)				
Ti-6Al-4V		1035	885	515
Ti-6Al-2Sn-4Zr-2Mo		895	825	485
Ti-5Al-2Sn-2Zr-4Mo-4Cr		1185	1130	675
Copper alloys (endurance limit based on 100 million cycles)				
70Cu-30Zn brass	Hard	524	435	145
90Cu-10Zn		420	370	160
Magnesium alloys (endurance limit based on 100 million cycles)				
HK31A-T6		215	110	60-80
AZ91A		235	160	70-90

Table 2.2: Macauley functions describing the rate of loading along a beam as a function of the load type

Load type	Singularity function
Pure bending moment M acting at a position *a*	$M\langle x-a\rangle^{-2}=0 \quad x\neq a$ $\int_{-\infty}^{x}\langle x-a\rangle^{-2}dx=\langle x-a\rangle^{-1}$ $\langle x-a\rangle^{-2}=\pm\infty \quad x=a$
Concentrated force R acting at point *a* and commonly used to model a reaction force	$R\langle x-a\rangle^{-1}=0 \quad x\neq a$ $\int_{-\infty}^{x}\langle x-a\rangle^{-1}dx=\langle x-a\rangle^{0}$ $\langle x-a\rangle^{-1}=\pm\infty \quad x=a$
Unit step at *a* corresponding to a uniform load to the right of this point	$\langle x-a\rangle^{0}=\begin{cases}0 & x<a\\ 1 & \mathrm{x}\geq \mathrm{a}\end{cases}$ $\int_{-\infty}^{x}\langle x-a\rangle^{0}dx=\langle x-a\rangle^{1}$
Ramp starting at *a*	$\langle x-a\rangle^{1}=\begin{cases}0 & x<a\\ x-a & \mathrm{x}\geq \mathrm{a}\end{cases}$ $\int_{-\infty}^{x}\langle x-a\rangle^{1}dx=\frac{\langle x-a\rangle^{2}}{2}$
Parabolic load starting at *a*	$\langle x-a\rangle^{2}=\begin{cases}0 & x<a\\ (x-a)^{2} & \mathrm{x}\geq \mathrm{a}\end{cases}$ $\int_{-\infty}^{x}\langle x-a\rangle^{2}dx=\frac{\langle x-a\rangle^{3}}{3}$

3 Rigid body dynamics

3.0 Overview

The objective of this chapter is to introduce sufficient mathematical tools to enable a generalized approach to dynamic systems with sufficient breadth to include solutions for most of the situations encountered in the analysis of flexure systems. Consequently, it is already assumed that the reader is familiar with the basic concepts of dynamic systems and it is hoped that this chapter will provide a comprehensive and concise approach to dynamics that will build on techniques already developed during a typical undergraduate science based degree.

It almost goes without saying that the two most important characteristics of a flexure system are motion repeatability and dynamic response. Synonymous with dynamic performance is the appearance of the first mode natural frequency when the flexure provided with an impulse which, in the absence of external energy sources, will eventually decay. By monitoring the periodic oscillation and the rate at which the system eventually comes to rest, it is often possible to deduce the dynamic performance of the flexure.

To a large extent the prediction of free oscillation of flexures undergoing relatively small displacements with negligible energy dissipation will form the focus of the proceeding pages. This simplifies the analysis in two important ways. First, the small oscillations will, in general, not influence the overall geometry of the flexure system. Subsequently the equations governing motion reduce to simultaneous, second order, linear differential form. Second, because most systems do not dissipate significant energy, the natural frequencies of the damped and undamped system are likely to be comparable. Consequently, for many purposes in which the fundamental frequency is of prime importance, a reasonable estimate can be determined from an analysis of an undamped system. Such an approximation avoids the complexity associated with forces due to energy dissipation, which, in all but the simplest forms, considerably increases modeling complexity.

To derive the equations governing motion of flexure systems, use will be made of the equivalent, energy-based approaches known as Hamilton's principle and Lagrange's equations. Though abstract in their derivation, it will be found that these readily produce equations of motion for both continuous and lumped models of arbitrary complexity. For systems involving more than two degrees of freedom Lagrange's equation are of great utility. For small amplitude vibrations, the equations of motion invariably reduce to simultaneous, linear, differential form that can be represented in matrix form. Solutions to these equations have formed a field of study for hundreds of years and can be readily computed using eigen-analysis. To this end, matrix techniques for analysis of multi-degree of freedom systems will be presented.

In many instances, the equations of motion will become too complex to be able to produce simple analytic expressions for the natural frequency. Under these circumstances it may be adequate to estimate the frequency based on an assumed *deflection shape. This is known as Rayleigh's method and has been shown to provide a reasonable estimate for the natural frequency provided that the estimated shape satisfies, at least most, and desirably all, of the boundary conditions. Rayleigh's method has two important features. Firstly, the solutions often resolve to a simple formula from which numerical values for the first mode natural frequency can be computed using a calculator or spreadsheet. Secondly, although the solutions can only claim to be approximate resulting in a slightly high value for the natural frequency, for any reasonable guess for the mode shape, the errors are usually within a few percent. In most designs, errors of manufacture and, in particular, assembly, lead to significantly larger variability in frequency. Consequently, the 'approximate' equations are almost always sufficiently close to the 'real' value for engineering purposes.*

At the end of this chapter, a case study considers a simple, two-freedom, planar flexure. Using the tools of this chapter, it is shown how geometric and applied damping parameters may be selected to produce a required frequency response.

3.1 Generalized coordinates

In the following section it will be shown that all reversible mechanical systems can be described in terms of a simple principle known, rather obscurely, as the 'principle of least action'. Before moving on to this, it is worthwhile reviewing the concepts of generalized coordinates. First, consider a system of N infinitesimally small particles, the position of each can be determined in rectangular, or Cartesian, coordinates from the values

$$x_i, y_i, z_i \qquad i = 1, 2, \ldots N \tag{3.1}$$

Alternatively, we might wish to express the positions of our particles through different coordinates of the form

$$q_1, q_2, \ldots q_{3N} \tag{3.2}$$

For example, for a single particle, it might be more desirable to use spherical coordinates. The relationship between the two systems is given by

$$x = r\sin\theta\cos\phi$$
$$y = r\sin\theta\sin\phi \tag{3.3}$$
$$z = r\cos\theta$$

In general

$$x_1 = f_1(q_1, q_2, \ldots q_{3N})$$
$$\cdot \tag{3.4}$$
$$\cdot$$
$$z_N = f_{3N}(q_1, q_2, \ldots q_{3N})$$

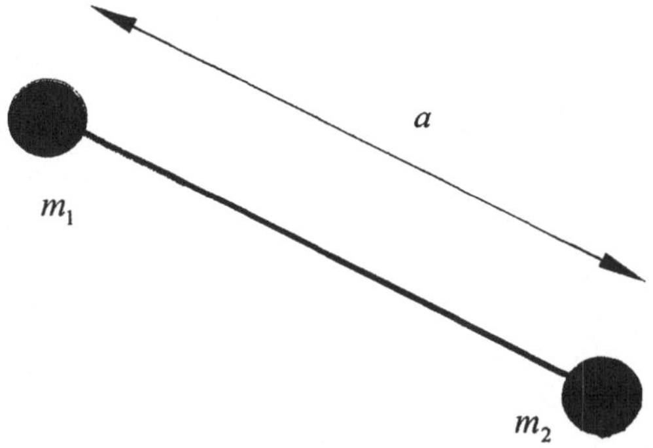

Figure 3.1: Two rigid point masses connected by a rigid massless link of length a

For a system of free particles $3N$ is the number of degrees of freedom. Clearly, to control the motion of a mechanical system, there must be some constraints. Consider an idealized molecule made up from two rigidly connected atoms a fixed distance a apart, figure 3.1.

Because of the rigid coupling between the two masses, there is an additional constraint such that

$$(x_1 - x_2)^2 + (y_1 - y_2)^2 + (z_1 - z_2)^2 = a^2 \tag{3.5}$$

Consequently, if five coordinates are known, the sixth can be readily determined. Those having previously studied kinematic design would consider this a six-degree of freedom rigid body in free space. It must be remembered, however, that point masses do not have rotational inertial properties and therefore rotation of the atoms about the axis of the link can be ignored. In reality it would be necessary to know the rotation of any finite mass about this axis to fully characterize this system dynamically (because these are point masses, rotation about this axis involves no momentum). To save further labor over such semantic terminology, upon the addition of a further atom to our molecule, an additional coordinate would be necessary to determine the positions of all particles and this would then become a rigid body with six degrees of freedom. For engineering purposes, a rigid body will be assumed to consist of a large number of rigidly connected atoms, therefore requiring six coordinates to describe its position in space.

Returning to our simplified rigid bodies, it can be stated that, in general, if there are m constraints, the number of independent coordinates and, therefore, degrees of freedom n can be determined from the equation

$$n = 3N - m \tag{3.6}$$

Consequently, if the q's are to represent independent generalized coordinates, equation (3.4) can be written in the form

$$\begin{aligned} x_1 &= f_1(q_1, q_2, \ldots q_n) \\ &\cdot \\ &\cdot \\ z_N &= f_{3N}(q_1, q_2, \ldots q_n) \end{aligned} \tag{3.7}$$

In subsequent discussions, only holonomic systems will be considered. Under such conditions, n represents both the number of degrees of freedom *and* the number of coordinates necessary to specify the potential and kinetic energy of the system. Exceptions to this rule usually involve rolling systems (e.g. a sphere rolling on a smooth surface has three rotational degrees of freedom but will require five coordinates to specify its position, Pars, 1961).

Let us now consider a single particle whose coordinates are given by

$$\begin{aligned} x &= f_x(q_1, q_2, q_3) \\ y &= f_y(q_1, q_2, q_3) \\ z &= f_z(q_1, q_2, q_3) \end{aligned} \tag{3.8}$$

To apply the principle of virtual work, consider an incremental displacement of the coordinates to $q_1 + \delta q_1$, $q_2 + \delta q_2$ and $q_3 + \delta q_3$. This can be transformed to an increment in rectangular coordinates using the equation

$$\begin{aligned} \delta x &= \frac{\partial x}{\partial q_1}\delta q_1 + \frac{\partial x}{\partial q_2}\delta q_2 + \frac{\partial x}{\partial q_3}\delta q_3 \\ &= \sum_{k=1}^{n} \frac{\partial x}{\partial q_k}\delta q_k \end{aligned} \qquad n = 3 \tag{3.9}$$

similarly for y and z.

In general for an arbitrary number of particles

$$\delta x_i = \sum_{k=1}^{n} \frac{\partial x_i}{\partial q_k}\delta q_k \tag{3.10}$$

δ is used to indicate a *virtual* displacement. Although this *must* satisfy the *constraints* of the system, it does *not necessarily* conform to the equations governing actual motion. An actual incremental motion would be designated dq_i. Apart from this distinction, δ is considered to act as a normal differential operator.

Consider the necessary force, $\mathbf{F}$, acting on a single particle to impart a virtual displacement $\delta\mathbf{R}$. The work done, W, for such a displacement is

$$\delta W = \mathbf{F}.\delta\mathbf{R} = F_x\delta x + F_y\delta y + F_z\delta z \tag{3.11}$$

From (3.9), this can be expressed in terms of the generalized coordinates in the form

$$\begin{aligned}\delta W &= \sum_{k=1}^{n}\left(F_x\frac{\partial x}{\partial q_k} + F_y\frac{\partial y}{\partial q_k} + F_z\frac{\partial z}{\partial q_k}\right)\delta q_k \\ &= \sum_{k=1}^{n} Q_k\delta q_k\end{aligned} \tag{3.12}$$

Consequently, it can be seen that the generalized force can be obtained from forces is given by

$$Q_k = F_x\frac{\partial x}{\partial q_k} + F_y\frac{\partial y}{\partial q_k} + F_z\frac{\partial z}{\partial q_k} \tag{3.13}$$

Where each of the forces corresponds to that which is applied in the direction of its subscript and results in a virtual displacement measure in the *same* direction as indicated in equation (3.12). As a further extension, let us assume the work done can be expressed in terms of some differentiable commodity, U, that is a function only of the displacements i.e.

$$\delta W = \delta U \tag{3.14}$$

where

$$U = u(q_1, q_2, \ldots q_n) \tag{3.15}$$

Differentiating this

$$\delta U = \sum_{k=1}^{n}\frac{\partial U}{\partial q_k}\delta q_k \tag{3.16}$$

If this function is work being done by the system, it will be negative and is denoted V given by

$$V = -U \tag{3.17}$$

V is called the potential energy of the system and, from equations (3.12) and (3.16), this component of the generalized force is given by

$$Q_k = \frac{\partial U}{\partial q_k} = -\frac{\partial V}{\partial q_k} \tag{3.18}$$

This is of fundamental significance in mechanics. A general response to the question, 'what is force?' could be

'Force is a consequence of, and proportional to, a potential gradient and always acts in the direction of steepest gradient'

This can be extended to different topics such as; heat transfer where temperature is the potential, fluid mechanics where pressure is the potential and even diffusion where the potential is concentration (although in this case the 'force' is due to 'random walks'!).

This generalized approach to modeling is often referred to as Fields Theory. For our purposes, it is only important to recognize that

> *Q_k is the generalized force acting to produce a change in the coordinate q_k as is consistent with the definition in equation (3.12)*

Before moving on to consider mechanics from Hamilton's perspective it is worth recapitulating some mathematical tools.

3.2 Properties of variational operators

3.2.1 Commutation

It can be shown that, Lanczos, 1970

$$\frac{d(\delta y)}{dx} = \delta \frac{dy}{dx} \tag{3.19}$$

also

$$\delta \int f(x)dx = \int \delta f(x)dx \tag{3.20}$$

3.2.2 Minima of a function $L = f(q, \dot{q}, t)$

Consider the general function

$$L = f(q, \dot{q}, t) \tag{3.21}$$

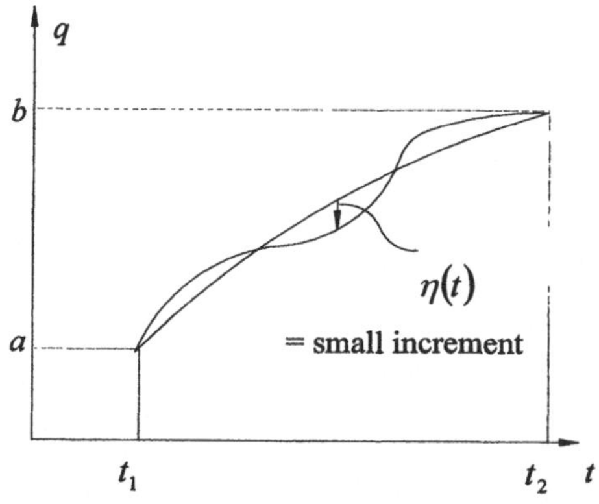

Figure 3.2: Deviation of path corresponding to a small increment in the Lagrangian δL as system transforms from states A to B for a particle

Assume that the function remains a fixed relationship of the three variables but undergoes incrementally small changes as it transforms between states A and B caused by correspondingly small increments in the two variables q and $\dot{q}$. Consequently, the deviation from the actual path can be expressed by the relationship

$$\delta L = f(q + \eta(t), \dot{q} + \dot{\eta}(t), t) - f(q, \dot{q}, t) \tag{3.22}$$

Assuming that the system is a point mass (of unit value, although other values can be included with little added complexity, Longair 1984) moving over time between positions a and b corresponding to times t_1 and t_2. This is illustrated in figure 3.2 in which the function L is identified as the Lagrangian and is the difference between kinetic and potential energies of the system ($L = T - V$). It is now possible to state Hamilton's principle

For the true, or actual, motion of a system, the time integral of the Lagrangian of the system as it transforms from fixed state A to another B will always be a minimum

Mathematically this can be expressed as

$$\delta \int L dt = \int \delta L dt = 0 \tag{3.23}$$

To fully appreciate this remarkable theory the reader is recommended to read Feynman's enthusiastic account of his personal discovery of the simplicity and universality of this principle, Feynman *et al.*, 1964.

Proceeding, it is possible to expand the expression of equation (3.22) using Taylor's (named after Brook Taylor, 1685-1731) series and, ignoring second order and higher terms in η (which we assume is small anyway), (3.23) is given by

$$\int \delta L dt = \int \left(\frac{\partial f}{\partial q} \eta + \frac{\partial f}{\partial \dot{q}} \dot{\eta} \right) dt \tag{3.24}$$

To eliminate the fluxion of the function η we can integrate the right hand term in parentheses by parts to give

$$\begin{aligned} \int_{t_1}^{t_2} \frac{\partial f}{\partial \dot{q}} \dot{\eta} dt &= \left[\frac{\partial f}{\partial \dot{q}} \eta \right]_{t_1}^{t_2} - \int_{t_1}^{t_2} \frac{d}{dt} \left(\frac{\partial f}{\partial \dot{q}} \right) \eta dt \\ &= - \int_{t_1}^{t_2} \frac{d}{dt} \left(\frac{\partial f}{\partial \dot{q}} \right) \eta dt \end{aligned} \tag{3.25}$$

It is clear from figure 3.2 that the deviation from the path η is always 0 at the two end points, thus justifying the disappearance of the definite integral above. Substituting (3.25) into (3.24)

$$\begin{aligned} \delta \int_{t_1}^{t_2} L dt &= \int_{t_1}^{t_2} \left(\frac{\partial f}{\partial q} - \frac{d}{dt} \left(\frac{\partial f}{\partial \dot{q}} \right) \right) \eta dt \\ &= \int_{t_1}^{t_2} \left(\frac{\partial L}{\partial q} - \frac{d}{dt} \left(\frac{\partial L}{\partial \dot{q}} \right) \right) \eta dt = 0 \end{aligned} \tag{3.26}$$

Because η is arbitrary, and, for conservative systems, the forces are always either inertial or potential, Lagrange's equation for a system with no externally applied forces is simply given by

$$\left(\frac{\partial L}{\partial q} - \frac{d}{dt} \left(\frac{\partial L}{\partial \dot{q}} \right) \right) = 0 \tag{3.27}$$

3.3 Hamilton's principle

Starting off with the original formulation of Hamilton's principle, equation (3.23) can be expressed by the equation, Goldstein, 1980,

$$\delta \int L dt = \int \delta L dt = \int (\delta T - \delta V) dt = 0 \tag{3.28}$$

It is most important to realize that this analysis is derived without any consideration of virtual work. The principle of least action can be considered as a fundamental property of dynamic systems that will always follow the path that minimizes the exchange between kinetic and potential energy.

Considering a system of masses constrained to motion in a single linear coordinate then the work done in changing the kinetic energy of a system is

$$\delta W_T = -\sum_i m_i \ddot{y}\delta y \tag{3.29}$$

Or in the limit

$$\delta W_T = -\int m_l \ddot{y}\delta y dx \tag{3.30}$$

where m_l is the mass per unit length along the direct x.

Substituting (3.29) into (3.27) and multiplying through by -1 it can be seen that

$$\int m_l \ddot{y}\delta y dx + \delta V = 0 \tag{3.31}$$

Alternatively, an expression for the kinetic energy may be written in the form

$$T = \frac{1}{2}\sum_i m_i \dot{x}_i^2 \tag{3.32}$$

For a small increment in x to $x + \delta x$ equation (3.32) becomes

$$\begin{aligned} T + \delta T &= \frac{1}{2}\sum_i m_i (\dot{x} + \delta\dot{x})^2 = \frac{1}{2}\sum_i m_i \left(\dot{x}^2 + 2\dot{x}\delta\dot{x} + \delta\dot{x}^2\right) \\ &\approx T + \sum_i m_i \dot{x}\delta\dot{x} = T + \sum_i m_i \dot{x}\frac{d\delta x}{dt} \end{aligned} \tag{3.33}$$

Substituting δT from equation (3.33) into the equation (3.28)

$$\begin{aligned} \int(\delta T - \delta V)dt &= \int\left(\sum_i m_i \dot{x}\delta\dot{x} - \delta V\right)dt \\ &= \int\left(\sum_i m_i \dot{x}\frac{d\delta x}{dt} - \delta V\right)dt \end{aligned} \tag{3.34}$$

Integrating the summation by parts yields

$$\begin{aligned} \int\left(\sum_i m_i \dot{x}\frac{d\delta x}{dt} - \delta V\right)dt &= \left[\sum_i m_i \dot{x}\delta x\right]_{t_1}^{t_2} + \int\left(-\sum_i m_i \ddot{x}\delta x - \delta V\right)dt \\ &= \int\left(-\sum_i m_i \ddot{x}\delta x - \delta V\right)dt = 0 \end{aligned} \tag{3.35}$$

Equation (3.35) is D'Alembert's principle, Lanczos, 1970.

Example 1: A simple spring mass system

As a first example, a simple spring mass system shall be considered. For such a system, it would be far quicker to apply Newton's laws directly. However, this will serve to illustrate some of the basic steps in the application of Hamilton's principle.

As usual, the spring is considered to have insignificant mass while the mass is rigid. Additionally, it will be assumed that the motion of the mass is constrained to motion in a single linear direction x for which the potential and kinetic energies are given by

$$\begin{aligned} T &= \frac{1}{2} m\dot{x}^2 \\ V &= \frac{1}{2} kx^2 - mgx \end{aligned} \tag{3.36}$$

Finite variations of the energy terms can be readily obtained

$$\begin{aligned} T + \delta T &= \frac{1}{2} m(\dot{x} + \delta\dot{x})^2 \\ &\approx T + m\dot{x}\delta\dot{x} \end{aligned} \tag{3.37}$$

Similarly,

$$\delta V \approx kx\delta x - mg\delta x \tag{3.38}$$

From Hamilton's principle the motion of the system is governed by

$$\int (\delta T - \delta V) dt = \int (m\dot{x}\delta\dot{x} - kx\delta x + mg\delta x) dt = 0 \tag{3.39}$$

Again, to eliminate the fluxion of x, the first term in (3.39) can be integrated by parts so that

$$\begin{aligned} \int (m\dot{x}\delta\dot{x} - kx\delta x + mg\delta x) dt &= [m\dot{x}\delta x]_{t_1}^{t_2} - \int m\ddot{x}\delta x dt - \int (kx\delta x - mg\delta x) dt \\ &= \int (-m\ddot{x} - kx + mg)\delta x dt = 0 \end{aligned} \tag{3.40}$$

By definition, the boundary term of the integral vanishes at the two limits. Because δx is arbitrary, it is necessary that the expression in parentheses in (3.40) vanishes over all times so that the equation that must be satisfied is

$$-m\ddot{x} - kx + mg = 0 \tag{3.41}$$

This result will be familiar to anyone who has taken an elementary course in mechanics.

Example 2: Lateral vibration of a bar

This is a more complex situation, but nevertheless can be similarly treated. The beam will be considered to have uniform cross-section and be subject to small displacements.

The potential energy in an element of length dx will be equal to the work done in distorting it to a radius R by bending the element through an angle θ. Consequently, the potential can be obtained from

$$\begin{aligned} V &= \sum \frac{1}{2} BMd\theta = \sum \frac{EI}{2R}\frac{ds}{R} \\ &\approx \int_0^l \frac{EI}{2}\left(\frac{\partial^2 y}{\partial x^2}\right)^2 dx \end{aligned} \tag{3.42}$$

where BM is the bending moment and ds is the length of the element, which is considered to be equal to x in view of the straightness of the beam, and the radius has been approximated for small slope.

For an element of length dx the kinetic energy is comprised from two separable components, these being due to linear and angular velocities. For planar motion

$$\begin{aligned} T &= \frac{m}{2l}\int \dot{y}^2 dx + \frac{mj^2}{2l}\int \left(\frac{d}{dt}\frac{\partial y}{\partial x}\right)^2 dx \\ &= \frac{m}{2l}\int \dot{y}^2 dx + \frac{mj^2}{2l}\int \dot{\theta}^2 dx \end{aligned} \tag{3.43}$$

where j is the radius of gyration of the element, which is a constant for our prismatic beam (i.e. a beam of uniform cross section), and m is the total mass of the beam. For present purposes, it will be assumed that the contribution due to angular rotation can be neglected, thus reducing the kinetic energy term to the first of the integrals in (3.43).

Again, the variation of potential can be readily deduced from the equation

$$\begin{aligned} V + \delta V &= \frac{EI}{2}\int \left(\frac{\partial^2 (y + \delta y)}{\partial x^2}\right)^2 dx \\ &\approx V + EI\int \left(\frac{\partial^2 y}{\partial x^2}\frac{\partial^2 \delta y}{\partial x^2}\right) dx \end{aligned} \tag{3.44}$$

Integrating the above equation by parts *twice*, the variation of potential is given by

$$\delta V = \left[EI\frac{\partial^2 y}{\partial x^2}\frac{\partial \delta y}{\partial x}\right]_0^l - \left[EI\frac{\partial^3 y}{\partial x^3}\delta y\right]_0^l + EI\int\left(\frac{\partial^4 y}{\partial x^4}\delta y\right)dx \tag{3.45}$$

After integration by parts, δT is

$$\frac{m}{l}\int_0^l \ddot{y}\delta y dx \tag{3.46}$$

Consequently, the variational equation of motion is

$$\left[EI\frac{\partial^2 y}{\partial x^2}\frac{\partial \delta y}{\partial x}\right]_0^l - \left[EI\frac{\partial^3 y}{\partial x^3}\delta y\right]_0^l + \int\left(-\frac{m}{l}\ddot{y} - EI\frac{\partial^4 y}{\partial x^4}\right)\delta y dx = 0 \tag{3.47}$$

Because the boundary terms must separately vanish (3.47) can be split into the equation which must be satisfied at all points along the beam

$$\frac{m}{l}\ddot{y} + EI\frac{\partial^4 y}{\partial x^4} = 0 \tag{3.48}$$

with the boundary conditions

$$\begin{aligned}\left[EI\frac{\partial^2 y}{\partial x^2}\frac{\partial \delta y}{\partial x}\right]_0^l &= 0\\ \left[EI\frac{\partial^3 y}{\partial x^3}\delta y\right]_0^l &= 0\end{aligned} \tag{3.49}$$

For example, a completely free beam will have an arbitrary slope and deflection at the free ends. Therefore, the boundary conditions given by equation (3.49) are

$$\begin{aligned}\left[EI\frac{\partial^2 y}{\partial x^2}\right]_0^l &= 0\\ \left[EI\frac{\partial^3 y}{\partial x^3}\right]_0^l &= 0\end{aligned} \tag{3.50}$$

The two terms in (3.50) correspond to zero bending moment and shear force at *each* end of the beam. This provides 4 constraints from which a complete solution for the motion of the beam can be obtained. From (3.49), it can be seen that the boundary conditions depend on combinations of specified displacements and three successive derivatives. Separately, these dictate the independent conditions corresponding to displacement, slope, bending moment and shear force. These correspond to pinned or clamped (zero displacement), clamped (zero slope), pinned (zero moment) and free (zero moment and/or shear) constraints respectively. In general, conditions for which solutions can be readily obtained can be drawn from any combination of these at either end of the beam. Solutions of the free vibration of beams for these boundary conditions are given towards the end of this chapter.

3.4 Lagrange's equation

D'Alembert's principle to can be extended to be expressed only in terms of generalized coordinates. Using methods of calculus, it is possible to equate the inertial term with an equivalent formulation involving only the kinetic energy. Dropping the vector notation from our previous analysis, we can write out d'Alembert's principle for N particles of mass m_i can be written in the form

$$\sum_{i=1}^{N}\left(m_i\left(\ddot{x}_i\delta x_i+\ddot{y}_i\delta y_i+\ddot{z}_i\delta z_i\right)\right)=\sum_{i=1}^{N}\left(F_{x_i}\delta x_i+F_{y_i}\delta y_i+F_{z_i}\delta z_i\right) \tag{3.51}$$

For a system with n degrees of freedom, there will be the functional relationships

$$\begin{aligned} x_i &= f_i(q_1,q_2,\ldots q_n) \\ y_i &= g_i(q_1,q_2,\ldots q_n) \\ z_i &= h_i(q_1,q_2,\ldots q_n) \end{aligned} \tag{3.52}$$

In general, a small increment in one of the generalized coordinates δq_s can be related to the Cartesian coordinates from the generalized relationships

$$\delta x_i=\frac{\partial f_i}{\partial q_1}\delta q_1+\ldots+\frac{\partial f_i}{\partial q_s}\delta q_s+\ldots+\frac{\partial f_i}{\partial q_n}\delta q_n \tag{3.53}$$

and so on for y and z.

Similarly

$$\dot{x}_i=\frac{dx_i}{dt}=\frac{\partial f_i}{\partial q_1}\dot{q}_1+\ldots+\frac{\partial f_i}{\partial q_s}\dot{q}_s+\ldots+\frac{\partial f_i}{\partial q_n}\dot{q}_n \tag{3.54}$$

and so on for y and z.

Differentiating (3.54) with respect to q_s immediately gives

$$
\begin{aligned}
\frac{\partial \dot{x}_i}{\partial q_s} &= \frac{\partial^2 f_i}{\partial q_s \partial q_1}\dot{q}_1 + \ldots + \frac{\partial\, f_i}{\partial q_s \partial q_n}\dot{q}_n \\
&= \frac{d}{dt}\left(\frac{\partial x_i}{\partial q_s}\right)
\end{aligned} \tag{3.55}
$$

Because the generalized coordinates are independent (3.54) can be differentiated with respect to $\dot{q}_s$ to produce

$$
\frac{\partial \dot{x}_i}{\partial \dot{q}_s} = \frac{\partial x_i}{\partial q_s} \tag{3.56}
$$

If only the sth coordinate is varied, the virtual work done by the system due to the inertia of the masses will be given by

$$
\begin{aligned}
&\sum_{i=1}^{N}\left(m_i\left(\ddot{x}_i \delta x_i + \ddot{y}_i \delta y_i + \ddot{z}_i \delta z_i\right)\right) \\
&= \sum_{i=1}^{N}\left(m_i\left(\ddot{x}_i \frac{\partial f_i}{\partial q_s} + \ddot{y}_i \frac{\partial g_i}{\partial q_s} + \ddot{z}_i \frac{\partial h_i}{\partial q_s}\right)\right)\delta q_s \\
&= -Q_s \delta q_s
\end{aligned} \tag{3.57}
$$

Canceling δq_s

$$
\begin{aligned}
Q_s &= \sum_{i=1}^{N}\left(-m_i\left(\ddot{x}_i \frac{\partial f_i}{\partial q_s} + \ddot{y}_i \frac{\partial g_i}{\partial q_s} + \ddot{z}_i \frac{\partial h_i}{\partial q_s}\right)\right) \\
&= S_s
\end{aligned} \tag{3.58}
$$

which is intrinsic to the concept of virtual displacements.

Because equation (3.58) contains both sets of coordinates, it is of little use. To eliminate the Cartesian coordinates system, the concept of kinetic energy, T, is introduced. In Cartesian coordinates, for a point mass, this is defined by

$$
T = \sum_{1=1}^{N} m_i\left(\dot{x}_i^2 + \dot{y}_i^2 + \dot{z}_i^2\right) \tag{3.59}
$$

Differentiating T with respect to $\dot{q}_s$ gives

$$
\frac{\partial T}{\partial \dot{q}_s} = \sum_{i=1}^{N}\left(-m_i\left(\dot{x}_i \frac{\partial f_i}{\partial q_s} + \dot{y}_i \frac{\partial g_i}{\partial q_s} + \dot{z}_i \frac{\partial h_i}{\partial q_s}\right)\right) \tag{3.60}
$$

Differentiating again with respect to time yields

$$\frac{d}{dt}\left(\frac{\partial T}{\partial \dot{q}_s}\right) = \sum_{i=1}^{N}\left(m_i\left(\ddot{x}_i\frac{\partial f_i}{\partial q_s}+\ddot{y}_i\frac{\partial g_i}{\partial q_s}+\ddot{z}_i\frac{\partial h_i}{\partial q_s}\right)\right)$$
$$\ldots+\sum_{i=1}^{N}\left(m_i\left(\dot{x}_i\frac{d}{dt}\left(\frac{\partial f_i}{\partial q_s}\right)+\dot{y}_i\frac{d}{dt}\left(\frac{\partial g_i}{\partial q_s}\right)+\dot{z}_i\frac{d}{dt}\left(\frac{\partial h_i}{\partial q_s}\right)\right)\right) \quad (3.61)$$
$$= -S_s+\sum_{i=1}^{N}\left(m_i\left(\dot{x}_i\frac{d}{dt}\left(\frac{\partial f_i}{\partial q_s}\right)+\dot{y}_i\frac{d}{dt}\left(\frac{\partial g_i}{\partial q_s}\right)+\dot{z}_i\frac{d}{dt}\left(\frac{\partial h_i}{\partial q_s}\right)\right)\right)$$

Equation (3.55) can be used to transform the last summation in (3.61) to give

$$\frac{d}{dt}\left(\frac{\partial T}{\partial \dot{q}_s}\right) = -S_s+\sum_{i=1}^{N}\left(m_i\left(\dot{x}_i\left(\frac{\partial \dot{f}_i}{\partial q_s}\right)+\dot{y}_i\left(\frac{\partial \dot{g}_i}{\partial q_s}\right)+\dot{z}_i\left(\frac{\partial \dot{h}_i}{\partial q_s}\right)\right)\right)$$
$$= -S_s+\frac{\partial T}{\partial q_s} \quad (3.62)$$

Equation (3.62) can be simply rearranged to yield

$$-S_s = \frac{d}{dt}\left(\frac{\partial T}{\partial \dot{q}_s}\right)-\frac{\partial T}{\partial q_s} \quad (3.63)$$

This has produced the desired effect of reducing the equation for the generalized inertial forces in terms only of the total kinetic energy of the system and the generalized coordinates. It is only a trivial step to introduce this back into d'Alembert's relationship in general coordinates, to produce

$$\sum_{s=1}^{n}\left(Q_s-\frac{d}{dt}\left(\frac{\partial T}{\partial \dot{q}_s}\right)-\frac{\partial T}{\partial q_s}\right)=0 \qquad s=1,2\ldots n \quad (3.64)$$

From (3.18) forces due to potential V, are given by

$$Q_m = -\frac{\partial V}{\partial q_m}$$

Consequently, the externally applied forces can be split into those due to potential and externally applied forces yielding

$$\sum_{s=1}^{n}\left(\left(\frac{d}{dt}\left(\frac{\partial T}{\partial \dot{q}_s}\right)-\frac{\partial T}{\partial q_s}+\frac{\partial V}{\partial q_s}\right)=Q_s\right) \quad (3.65)$$

This is called *Lagrange's* equation after its originator Joseph-Louis Lagrange (1736-1813).

Because the forces of potential are functions of position only, it is possible to define the Lagrangian $L = T - V$ and write the above in the abbreviated form

$$\sum_{s=1}^{n}\left(\left(\frac{d}{dt}\left(\frac{\partial L}{\partial \dot{q}_s}\right)-\frac{\partial L}{\partial q_s}\right)=Q_s\right) \tag{3.66}$$

For systems undergoing free vibration, there are no externally applied forces leading to the apparently simple equation

$$\sum_{s=1}^{n}\left(\left(\frac{d}{dt}\left(\frac{\partial L}{\partial \dot{q}_s}\right)-\frac{\partial L}{\partial q_s}\right)=0\right) \tag{3.67}$$

This is, of course, identical to equation (3.27) based on Hamilton's principle

3.4.1 Rayleigh's dissipation function

In the above equations it has been assumed that work done during motion of components of the system is due to changes in conserved quantities of energy. However, in *all* engineering systems there will be additional work due to mechanisms that dissipate energy. Generally, these involve frictional forces due to relative sliding of solids or shearing of fluids. Unfortunately, in all but the simplest cases this results in phenomena that cannot be simply modeled. By far the simplest dissipation mechanism is that due to forces in proportion to the relative velocity between two bodies. Under these circumstances, Rayleigh demonstrated that there is a dissipation function, D, of the form

$$D=\frac{1}{2}\sum_{i=1}^{n}\sum_{j=1}^{n}b_{ij}\dot{q}_1\dot{q}_2 \tag{3.68}$$

where the b's represent the damping force proportional to relative velocities between coordinates i and j. Under these conditions the equation governing motion of the system becomes

$$\sum_{s=1}^{n}\left(\left(\frac{d}{dt}\left(\frac{\partial T}{\partial \dot{q}_s}\right)-\frac{\partial T}{\partial q_s}+\frac{\partial D}{\partial \dot{q}_s}+\frac{\partial V}{\partial q_s}\right)=Q_s\right) \tag{3.69}$$

It is important to note that D is always positive as are each of its coefficients. This damping mechanism is characteristic of energy dissipation due to laminar shear of a massless, visçous fluid and is consequently referred to as *viscous* damping. It should also be noticed that the dissipation function has the same form as the kinetic energy and this is certainly the most frequently assumed form 'in books at any rate' to quote Rayleigh.

3.4.2 General use of Lagrange's equation

Determination of the equations of motion of an arbitrary system requires the following four steps

1. Decide on the most appropriate or meaningful coordinate system
2. Determine the potential and kinetic energy and the dissipation function for an arbitrary displacement
3. Determine the generalized forces applied to the system
4. Substitute the above into Lagrange's equation

The advantages of this approach are due to the relative simplicity that the *scalar* energy functions can be determined. There is also little room for error which often happens when attempting problems using vector and/or free body diagram approaches. The disadvantage of this approach is the difficulty of selecting the 'best' coordinate system for the problem.

Once the equations of motion have been derived they can be integrated either numerically or analytically to determine the subsequent velocities and displacements for each coordinate.

3.5 Linear systems theory

3.5.1 The simple, single degree of freedom, linear, spring-mass-damper system

Any arbitrary linear system can be considered as the superposition of simple single degree of freedom systems (this statement will be justified shortly). Consequently, an understanding of the general characteristics of a single degree of freedom spring-mass-damper system provides an intuitive feel for the more complex behavior and analytic techniques for handling systems having multiple degrees of freedom. To derive some more complete understanding of general linear systems, it is informative to analyze the simple idealized spring-mass-damper model of figure 3.3.

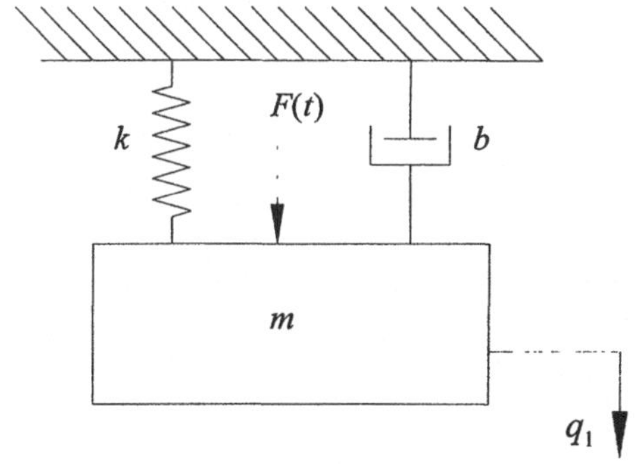

Figure 3.3: A single degree of freedom spring/mass/damper linear system

The equation governing motion q_1 of the mass m due to an applied force $F(t)$ is

$$m\ddot{q}_1 + b\dot{q}_1 + kq_1 = F(t) \tag{3.70}$$

Assuming that the applied force $F(t)$ repeats after some time T (which might be a very long while) we might consider it to be periodic and we can therefore express it as a Fourier series of the form

$$F(t) = a_o + \sum_{i=1}^{\infty}\left(a_i \cos(\omega_i t) + b_i \sin(\omega_i t)\right) \tag{3.71}$$

a_o is a constant over time and therefore only contributes a static load or deflection. As long as the flexure to which displacements are imposed is operating within its elastic limits, this will simply add to dynamic deflection and any subsequent stresses. Because we are only considering dynamic effects, this is not relevant to this study (this is not to say that it isn't important!) and will be ignored in all subsequent analysis. So, in general, the input is assumed to be of the form

$$\begin{aligned} F(t) &= \sum_{i=1}\left(a_i \cos(\omega_i t) + b_i \sin(\omega_i t)\right) \\ &= \sum_{i=1}\left(F_i \cos(\omega_i t + \varphi_i)\right) \end{aligned} \tag{3.72}$$

where the magnitude of the force F_i and phase φ_i at frequency ω_i can be derived from

$$\begin{aligned} F_i &= \sqrt{a_i^2 + b_i^2} \\ \varphi_i &= \tan^{-1}\frac{b_i}{a_i} \end{aligned} \tag{3.73}$$

and

$$\omega_i = \frac{i\pi}{T} \tag{3.74}$$

Without any justification, let it be assumed that, after all of the transient effects have decayed, the system will settle to a *steady state* described by a solution to equations (3.70) and (3.71) of the form

$$x_1(t) = \sum_{i=1}\left(H_i F_i \cos(\omega_i t + \varphi_i^*)\right) \tag{3.75}$$

where (for the time being) the H_i's and φ_i^*'s are also real constants

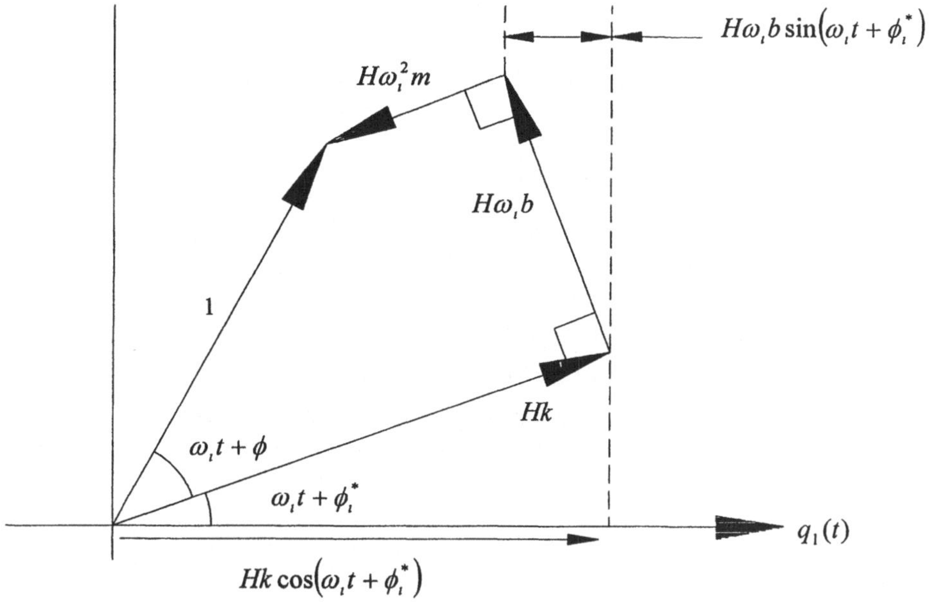

Figure 3.4: Vector representation of the individual terms in equation (3.77)

Substituting equation (3.75) into (3.70) results in the rather cumbersome expression

$$\sum_{i=1}\begin{bmatrix}-\omega_i^2 H_i F_i m\cos(\omega_i t+\varphi_i^*)-\omega_i bH_i F_i \sin(\omega_i t+\varphi_i^*)+\\ H_i F_i k\cos(\omega_i t+\varphi_i^*)=F_i\cos(\omega_i t+\varphi_i)\end{bmatrix} \tag{3.76}$$

Notice that the F_i's cancel so that we are left with the equation

$$\sum_{i=1}\begin{bmatrix}H_i\{-\omega_i^2 m\cos(\omega_i t+\varphi_i^*)-\omega_i b\sin(\omega_i t+\varphi_i^*)+\\ k\cos(\omega_i t+\varphi_i^*)\}=\cos(\omega_i t+\varphi_i)\end{bmatrix} \tag{3.77}$$

One method for a solution of the above equation is to draw each term in equation (3.77) as a vector diagram as shown in figure 3.4. From this, and using Pythagoras' theorem, the equation for the gain and relative phase difference between the input and output can be determined from the equations

$$|H_i|=\frac{1}{\left(\left(k-m\omega_i^2\right)^2+(b\omega_i)^2\right)^{1/2}} \tag{3.78}$$

$$\phi_\iota(\omega_\iota) = \varphi^*_{\iota)} - \varphi_\iota = -\tan^{-1}\left(\frac{b\omega_\iota}{k - m\omega_\iota^2}\right) \tag{3.79}$$

These two equations respectively represent the gain and phase response of this system to an input force.

3.5.2 Some equivalent definitions of a linear system

The above analysis relies on four very important assumptions governing the nature of this system. These are;

1. The system gain H_ι and relative phase lag ϕ_ι are **constant** for any given ω_ι. H_ι and ϕ_ι depend **only** upon the constant parameters *m, b, k* and ω_ι and are *independent* of the *amplitude* of the input.
2. The output is **linearly** proportional to the input at a given frequency.
3. The frequency at the output is **exactly** the same as that at the input and depends only upon inputs at that frequency.
4. Alternatively, the above might be stated that the **output** at a **given frequency** only depends upon inputs at that **same frequency**.

All of the above conditions individually define a linear dynamic system. At a first glance it is not apparent that, for example, statement 4 is the same as statement 3. It is worth spending a little time investigating why this is so. Let us imagine a simple system with a sinusoidal wave at the input with an unknown signal at the output. Clearly, if we were to amplify or attenuate this signal it will emerge as a sinusoid, although of a different amplitude, of the *same* frequency and will thus satisfy all of the above conditions. Let us now think of amplifying the input but changing the gain a little with amplitude. If the input signal is not changing sufficiently rapidly that we need to consider dynamic effects, this

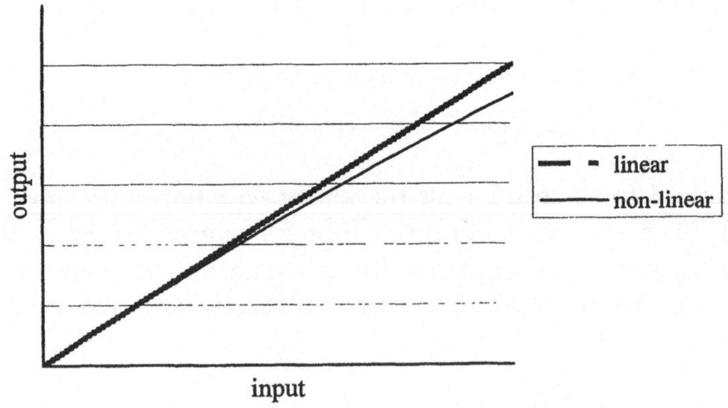

Figure 3.5: Input/output relationship of linear and weakly non-linear systems

might result in an input to output relationship shown in figure 3.5. Clearly, as the input increases in amplitude the output starts to 'flatten' out. Under these circumstances it is clear that the output will no be longer sinusoidal. There is obviously another shape present. If this were of the same frequency, it could not alter the shape of the signal but only change the phase. Therefore, the effect of deviating even slightly from a linear gain is to change the frequency *content* of the output which is simply a more involved way of restating condition 4.

Consequently, it is possible to use a black box approach to single degree of freedom linear systems and represent any system by the diagram of figure 3.6.

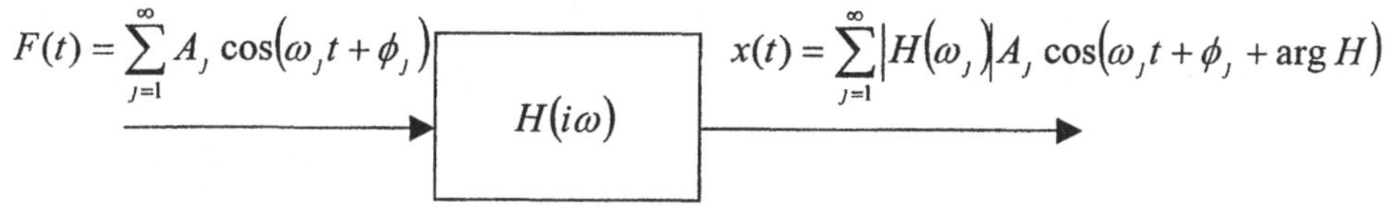

Figure 3.6: Input-output relationship for a single degree-of-freedom linear system

It is also important to note that the relative phase shift between input and output is independent of the phase of the input (condition 1). Consequently, to derive system responses, it makes no difference what value is chosen for the input phase. For mathematical convenience, the input phase will be assumed to be zero.

3.5.3 Frequency response functions

Based on the above arguments it is possible to model the input force to a linear system by the equation

$$F(t) = F_0 \cos(\omega t)$$

It turns out to be a lot more convenient to express this in the complex notation

$$F(t) = \mathrm{Re}\{F_0 e^{i\omega t}\} \tag{3.80}$$

where Re implies the extraction of the real part of the expression within the parentheses.

Correspondingly it is necessary to seek a *steady state* solution of the form

$$x(t) = \mathrm{Re}\{H(i\omega) F_0 e^{i\omega t}\} \tag{3.81}$$

The $i\omega$ in parentheses indicates that H is a *complex* function of the input frequency ω. Because it is a complex constant it must therefore be comprised of two constants (the real and imaginary values). Inserting equation (3.81) into the governing equation (3.70) and canceling the complex exponents yields

$$\mathrm{Re}\{-\omega^2 mH(i\omega)F_0(t) + i\omega bH(i\omega)F_0(t) + kH(i\omega)F_0(t) = F_0(t)\}$$

From this, the complex form of the transfer function, or frequency response, is given by

$$H(i\omega) = \frac{1}{k - m\omega^2 + i\omega b}$$

This can be rearranged into the more standard form

$$H(i\omega) = \frac{1/k}{\left(1 - \dfrac{\omega^2}{\omega_n^2}\right) + 2\xi i \dfrac{\omega}{\omega_n}} \tag{3.82}$$

Multiplying the top and bottom of equation (3.82) by the complex conjugate of its denominator gives

$$k.H(i\omega) = \frac{\left(1 - \dfrac{\omega^2}{\omega_n^2}\right) - 2\xi i \dfrac{\omega}{\omega_n}}{\left(1 - \dfrac{\omega^2}{\omega_n^2}\right)^2 + 4\xi^2 \dfrac{\omega^2}{\omega_n^2}} = \frac{C - iD}{C^2 + D^2} \tag{3.83}$$

Substituting equation (3.83) into (3.81) yields the steady state solution

$$\begin{aligned} x(t) &= \frac{F_0/k}{C^2 + D^2}\mathrm{Re}\{(C - iD)e^{i\omega t}\} \\ &= \frac{F_0/k}{C^2 + D^2}[C\cos\omega t + D\sin\omega t] \\ &= \frac{F_0/k}{\sqrt{C^2 + D^2}}[C\cos(\omega t + \tan^{-1}(D/C))] \\ &= \frac{F_0}{k}|H(i\omega)|\cos(\omega t + \phi(\omega)) \end{aligned} \tag{3.84}$$

where

$$|H(i\omega)| = \frac{1}{\left(\left(1 - \dfrac{\omega^2}{\omega_n^2}\right)^2 + 4\xi^2 \dfrac{\omega^2}{\omega_n^2}\right)^{1/2}} \tag{3.85}$$

and

$$\varphi = -\tan^{-1} \frac{2\xi \dfrac{\omega}{\omega_n}}{1 - \dfrac{\omega^2}{\omega_n^2}} \tag{3.86}$$

ξ is known as the critical damping ratio which should be familiar to anyone having attended an elementary course in vibrations. Substituting this back using the equations

$$\begin{aligned} \xi &= \frac{b}{b_c} = \frac{b}{2m\omega_n} = \frac{b}{2\sqrt{mk}} \\ \omega_n^2 &= \frac{k}{m} \end{aligned} \tag{3.87}$$

It can be seen that equations (3.85) and (3.86) are identical with equations (3.78) and (3.79) as expected. Essentially, ξ is a measure of the closeness of the damping value of the system to that for which the impulse response ceases to contain oscillatory components. The critical value of the system damping at which this occurs is $b = b_c$, at which point $\xi = 1$. In the absence of damping (i.e. $b = \xi = 0$), the response to an impulse will result in a harmonic oscillation continuing indefinitely. It should be emphasized at this juncture that critical damping does not imply the optimal value. In practice, the 'best' value is going to depend on the design application and can vary considerably. For example, vibration isolators or tuning forks may require a very low value ($\xi \approx 0.005$ - 0.00001 or less) while averaging indicators could be designed for values ξ of 5 - 100 or more. Control systems are usually designed to track or respond to an externally applied input and for such designs it is common to choose a damping ratio in the region of $\xi = 0.4$ to 0.8. Often, the compromise value of $\xi = 1/\sqrt{2}$ is used as a design goal for optimal response. Not only is this within the range of optimal response values, but it also happens to be at 45 degrees on a root locus diagram, see section 3.10. It is rare to require damping of $\xi = 1$ for two reasons. Firstly, it is not a value that that provides the fastest response for a control system and, secondly, it is a singular value. Therefore, errors due to tolerances in production will move this either into or out-of this singular region.

Even in the absence of designed damping, all systems intrinsically possess energy dissipation. These can arise due to a large variety of phenomena. These include; scattering of elastic waves in solids at grain boundaries or interfaces, internal energy dissipation as part of the materials properties manifest as a hysteresis in the stress strain characteristic, see Lazan, 1968 and section 8.2.4.1, contact with fluids or gases, etc. For simple mechanisms where no efforts have been made to alter the energy dissipation, the damping ratio will often be in the region $\xi = 0.1$ to 0.001.

The magnitude and phase between the applied force and output displacement is simply the magnitude and phase of is determined directly from equation (3.82). Because of its simplicity, this general method for deriving system responses is adopted throughout.

3.6 Measuring the critical damping ratio

Before leaving equation (3.85), it is worth looking at some of the characteristics of this second order response. In particular with many systems we are interested in the damping ratio, ξ. There are two ways of deriving this from the *transient* response, these being the *logarithmic decrement* method or, alternatively, fitting the decay characteristic to the transient solution of the second order equation using a non-linear curve fit. From this it is possible to determine the two parameters

$$\begin{gathered} \xi\omega_n \\ \omega_n\sqrt{1-\xi^2} \end{gathered} \tag{3.88}$$

The former of equations (3.88) is the inverse time-constant of the transient decay envelope and the latter is the damped natural frequency (in units of radians per second).

For steady state assessment, it is necessary to measure the response to an input containing a continuous bandwidth of frequencies encompassing the resonance. The resonance peak occurs at a frequency, ω_{res}, given by

$$\omega_{res} = \omega_n\sqrt{1-2\xi^2} \tag{3.89}$$

For low values of damping it can be seen that the resonant, damped and undamped natural frequencies approach the same value. Therefore the ratio of the input to output displacements at the resonant frequency can be approximated by canceling the left hand side of the denominator of (3.85) and assuming the maximum amplitude occurs at a frequency coincident to the undamped natural frequency to give

$$\frac{X_{max}}{F_0/k} = \frac{X_{max}}{X_{static}} = \frac{1}{2\xi} \approx Q \tag{3.90}$$

In principle, the above equation can be used to determine the damping ratio simply by measuring the maximum amplitude at the natural frequency. Unfortunately, this requires knowledge of both the applied force and the stiffness of the system and is very sensitive to errors of the applied frequency. In many cases either the parameters are not known or such a measurement is not feasible. Alternatively, calculation of the damping ratio based on the width of the resonance peak is often used. One method is to look at the width of the square of the ratio (3.85) at half of the maximum height. Clearly, the square of the output at any given frequency represents the power and therefore height of the resonant peak at half of the maximum value is known as the half power point which is given by

$$\frac{Q^2}{2} = \frac{1}{\left(1-\frac{\omega^2}{\omega_n^2}\right)^2 + 4\xi^2\frac{\omega^2}{\omega_n^2}} = \frac{1}{\left(1-\alpha^2\right)^2 + \frac{1}{Q^2}\alpha^2} \tag{3.91}$$

Expanding the denominator and rearranging yields a quadratic in α^2 given by

$$\begin{aligned}\alpha^2 &= \frac{2-\frac{1}{Q^2} \pm \sqrt{\left(2-\frac{1}{Q^2}\right)^2 - 4\left(1-\frac{2}{Q^2}\right)}}{2} \\ &\approx 1 \pm \frac{1}{Q}\end{aligned} \tag{3.92}$$

Clearly there are two solutions corresponding to the frequencies either side of the resonant peak at the half power points ω_1 and ω_2 say. Noting that these frequencies will also be near to the undamped natural frequency, we can subtract the two solutions to give

$$\begin{aligned}\alpha_1^2 - \alpha_2^2 &\approx \frac{\omega_1^2-\omega_2^2}{\omega_n^2} \approx \frac{\omega_1-\omega_2}{\omega_n}\frac{\omega_1+\omega_2}{\omega_n} \approx \frac{2}{Q} \\ \therefore \frac{\omega_1-\omega_2}{\omega_n} &\approx \frac{1}{Q} \approx 2\xi\end{aligned} \tag{3.93}$$

Consequently, it can be seen that, if a system has a relatively low damping ratio, it is possible to determine an approximate value for this based on the frequency difference at half the maximum square of the response function divided by the frequency at the peak.

A final, and particularly powerful, technique for assessing the damping ratio in the vicinity of resonant peaks can be derived from plots of the real and imaginary components of the frequency response. For a single spring/mass/damper system this is given by

$$H(i\omega) = \frac{X_{\max}}{F_o / k} = \frac{1}{\left(1 - \frac{\omega^2}{\omega_n^2}\right) + 2i\xi\frac{\omega}{\omega_n}}$$

$$= \frac{\left(1 - \frac{\omega^2}{\omega_n^2}\right) - 2i\xi\frac{\omega}{\omega_n}}{\left(1 - \frac{\omega^2}{\omega_n^2}\right)^2 + 4\xi^2\frac{\omega^2}{\omega_n^2}} \tag{3.94}$$

$$= \frac{C(\omega) - iD(\omega)}{C^2(\omega) + D^2(\omega)} = \frac{C(\omega) - iD(\omega)}{E(\omega)}$$

Clearly, C/E and $-D/E$ represent the real and imaginary parts of the response function.

Such a response function can be readily determined using modern signal analyzers. For a simple single degree of freedom spring mass damper system, the real and imaginary parts of the frequency response (3.94) are shown in figure 3.7. From these two graphs it is clear that the damping in the vicinity of the resonant peak can be determined from the characteristic shapes of these curves. The simplest piece of information to extract is the undamped natural frequency. This occurs at the zero crossing of the real component of the response function. To derive more information about this system, it is necessary to assess the values of the data at the peaks of each of these plots. For the real and imaginary parts of equation (3.94) the positions of zero slope are given by

$$\frac{d\,\mathrm{Re}\{H(i\omega)\}}{d\omega} = \frac{E\frac{dC}{d\omega} - C\frac{dE}{d\omega}}{E^2} = 0$$

$$\frac{d\,\mathrm{Im}\{H(i\omega)\}}{d\omega} = \frac{E\frac{dD}{d\omega} - D\frac{dE}{d\omega}}{E^2} = 0 \tag{3.95}$$

From the first of (3.95), the numerator is given by

$$E\frac{dC}{d\omega} - C\frac{dE}{d\omega} = \frac{2\omega}{\omega_n^2} - 8\xi^2\frac{\omega}{\omega_n^2} + \frac{2\omega^5}{\omega_n^6} - \frac{4\omega^3}{\omega_n^4} \tag{3.96}$$

Equating this to zero and solving this for ω, the two frequencies at the peak and valley of the real component of the frequency response are

$$\omega_{1,2}^2 = \omega_n^2(1 \mp 2\xi) \tag{3.97}$$

From this equation it is relatively simple to show that

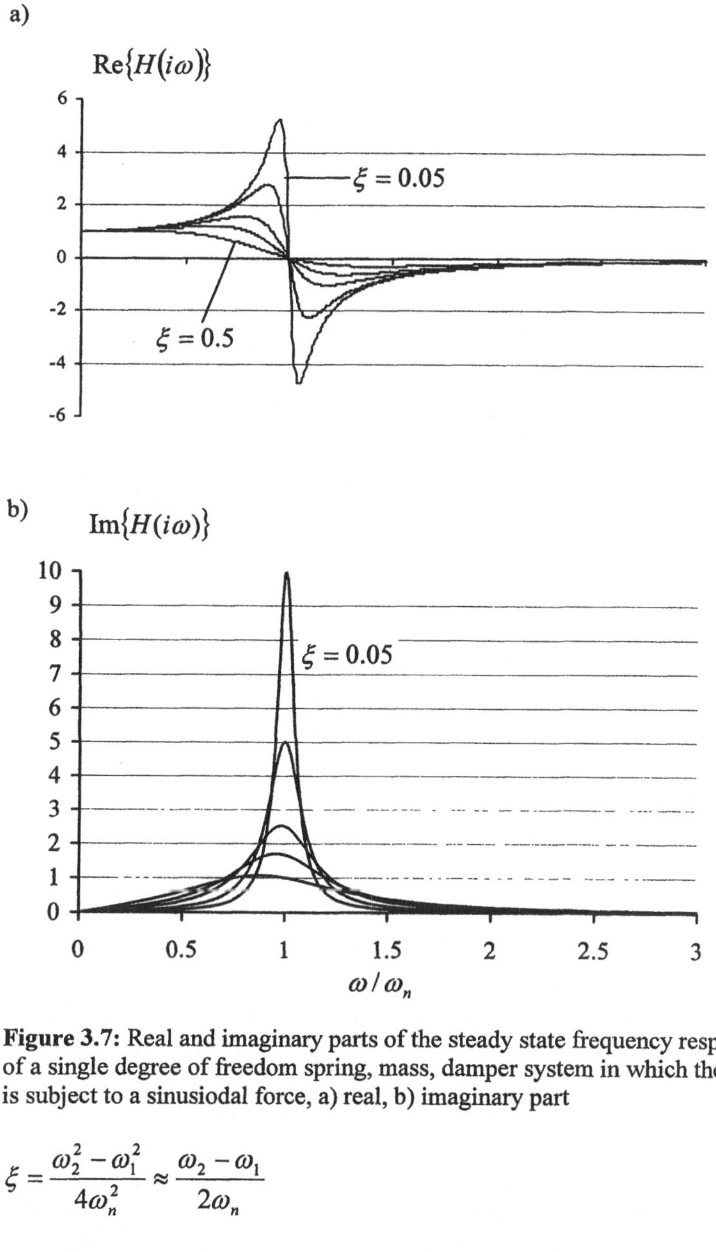

Figure 3.7: Real and imaginary parts of the steady state frequency response of a single degree of freedom spring, mass, damper system in which the mass is subject to a sinusiodal force, a) real, b) imaginary part

$$\xi = \frac{\omega_2^2 - \omega_1^2}{4\omega_n^2} \approx \frac{\omega_2 - \omega_1}{2\omega_n} \tag{3.98}$$

Because the frequencies are exactly the same as those of equation (3.93), this also provides the damping ratio in the same way as the half power points. In this case, the value for the damping ratio is independent of the actual values of the peaks, a most important result. Additionally, assuming that the frequency response is precisely determined, in contrast to the half power point approximate method, this equation provides an exact value for the damping ratio.

Another method can be derived by substituting these frequencies back into (3.94) after which the values of the real part of the frequency response are given by

$$\begin{aligned}\mathrm{Re}\{X(\omega_1)\} &= \frac{F_0}{4k\xi(1-\xi)} \\ \mathrm{Re}\{X(\omega_2)\} &= \frac{-F_0}{4k\xi(1+\xi)}\end{aligned} \tag{3.99}$$

From which, the peak to valley distance of the real part is

$$\mathrm{Re}\{X(\omega_1)\} - \mathrm{Re}\{X(\omega_2)\} = \frac{F_0}{2k\xi(1-\xi^2)} \tag{3.100}$$

At the undamped natural frequency (i.e. $\omega = \omega_n$), the amplitude of the imaginary component of the frequency response is given by

$$\mathrm{Im}\{X(\omega_n)\} = \frac{F_0}{2k\xi} \tag{3.101}$$

Dividing (3.100) by (3.101) and rearranging, it is possible to compute the critical damping ratio from

$$\xi = \sqrt{1 - \frac{\mathrm{Im}\{X(\omega_n)\}}{\mathrm{Re}\{X(\omega_1)\} - \mathrm{Re}\{X(\omega_2)\}}} \tag{3.102}$$

Note that this equation requires only the relative magnitudes of values of the frequency response measurement. Consequently, it is only required that the sensors used for measurement of the inputs and outputs have a constant gain (or sensitivity) over the narrow band of frequencies about the resonant peak. They do not need to be calibrated! Unfortunately, as ξ tends towards around 0.5, it becomes rather difficult to detect the peak values. However, reasonable values are obtained for values of less than 0.4 which is typical of many mechanical systems. For very low values of damping, it is sometimes difficult to discriminate the values at the peaks leading, once again, to inaccurate results.

Before leaving this topic, it is worth mentioning that the peak of the imaginary plot occurs at the undamped natural frequency, ω_d, and not the resonant frequency, ω_{res}, given by

$$\begin{aligned}\omega_d &= \omega_n\sqrt{1-\xi^2} \\ \omega_{res} &= \omega_n\sqrt{1-2\xi^2}\end{aligned} \tag{3.103}$$

At the peak, the *amplitude* of the imaginary term in (3.94) is given by

$$\frac{X_{\max}}{F_o / k} = \frac{-2\sqrt{1-\xi^2}}{4\xi - 3\xi^3} \tag{3.104}$$

For small ξ this, of course, tends towards equation (3.101)

3.7 General linear systems revisited

It is now possible to express mathematically the general input/output relationship for a linear dynamic system. Essentially, this represents any system that can be described by a linear differential equation of the form

$$\sum_i \sum_j \sum_n c_{ijn} \frac{d^n x_{ij}}{dt^n} = \sum_i \sum_j \sum_m d_{ijm} \frac{d^m y_{ij}}{dt^m} \tag{3.105}$$

A more compact form of the above can be achieved using the convention known as tensor notation. However, in most cases, the mechanical systems to be analyzed result in simultaneous equations that can be expressed as matrices having symmetric properties. Computational solutions to these have been extensively developed and are easily implemented. For this reason, matrix methods will be used in the following.

This general equation is for a system with i x j degrees of freedom. This can be simplified to a single degree of freedom by eliminating one subscript and making the other equal to one so that for a single degree of freedom system the relationship between a single output and the input can be written in the form

$$\sum_n c_n \frac{d^n x}{dt^n} = \sum_m d_m \frac{d^m y}{dt^m} \tag{3.106}$$

where c and d are real constants. This is the *only* equation that satisfies the conditions for linearity given in section 3.5.

As a second example, consider the frequency response of a single degree of freedom vibration isolation table. Assuming that we are interested in the displacement response of the table to an input displacement of the foundation $y(t)$. In the absence of any applied forces to the mass, the equation governing its motion is given by either

$$m\ddot{x} + b(\dot{x} - \dot{y}) + k(x - y) = 0 \tag{3.107}$$

or

$$m\ddot{x} + b\dot{x} + kx = b\dot{y} + ky \tag{3.108}$$

This is clearly of the form (3.106) and, therefore, must be a linear differential equation. Again to derive the frequency response in terms of the relationship between x and y, an input may be assumed to be of the form

$$y(t) = \mathrm{Re}\{Y_o e^{i\omega t}\} \tag{3.109}$$

with the subsequent response being given by

$$x(t) = \mathrm{Re}\{H(i\omega)Y_o e^{i\omega t}\} \tag{3.110}$$

Substituting (3.109) and (3.110) into the equation of motion (3.108) and rearranging we have

$$Y_o e^{i\omega t} H(i\omega)\left[-m\omega^2 + i\omega b + k\right] = Y_o e^{i\omega t}\left[i\omega b + k\right] \tag{3.111}$$

from which

$$\begin{aligned} H(i\omega) = \frac{X_{\max}}{Y_o} &= \frac{k + i\omega b}{k - m\omega^2 + i\omega b} \\ &= \frac{1 + i2\xi \dfrac{\omega}{\omega_n}}{1 - \dfrac{\omega^2}{\omega_n^2} + i2\xi \dfrac{\omega}{\omega_n}} \end{aligned} \tag{3.112}$$

Skipping the previous steps, the magnitude and phase of the transmissibility equation can be deduced directly from (3.112)

$$\begin{aligned} |H(i\omega)| &= \frac{X_{\max}}{Y_o} = \sqrt{\frac{1 + 4\xi^2 \dfrac{\omega^2}{\omega_n^2}}{\left(1 - \dfrac{\omega^2}{\omega_n^2}\right)^2 + 4\xi^2 \dfrac{\omega^2}{\omega_n^2}}} \\ \phi(\omega) &= \angle H(i\omega) = \tan^{-1}\left(2\xi \frac{\omega}{\omega_n}\right) - \tan^{-1}\left(\frac{2\xi \dfrac{\omega}{\omega_n}}{1 - \dfrac{\omega^2}{\omega_n^2}}\right) \end{aligned} \tag{3.113}$$

As a recap, the magnitude and phase of the transmissibility have been derived from the general complex number relationship

$$z_1 = a + ib = r_1 \angle \theta_1$$
$$z_2 = c + id = r_2 \angle \theta_2$$
$$\frac{z_1}{z_2} = \left|\frac{r_1}{r_2}\right| \angle(\theta_1 - \theta_2)$$

The transmissibility can be transformed to simple complex number form by multiplying top and bottom by the complex conjugate of the denominator. Subsequently, the expression for the magnitude and phase angle is

$$H(i\omega) = \frac{1 - \dfrac{\omega^2}{\omega_n^2} + 4\xi^2\dfrac{\omega^2}{\omega_n^2} + i2\xi\dfrac{\omega^3}{\omega_n^3}}{\left(1 - \dfrac{\omega^2}{\omega_n^2}\right)^2 + 4\xi^2\dfrac{\omega^2}{\omega_n^2}}$$
$$\phi(\omega) = -\tan^{-1}\left(\frac{2\xi\dfrac{\omega^3}{\omega_n^3}}{1 - \dfrac{\omega^2}{\omega_n^2} + 4\xi^2\dfrac{\omega^2}{\omega_n^2}}\right) \qquad (3.114)$$

It is also noted that the relationship (3.114) is identical to that for the ratio of forces transmitted to the foundation, F_{tr}, by forces applied to the mass, F_o, i.e.

$$\frac{X_{max}}{Y_0} = \frac{F_{tr}}{F_0} \qquad (3.115)$$

Plots of the transmissibility are shown in figure 3.8. It can be seen that attenuation of the motion between the foundation and mass (in vibration isolation the mass usually represents the platform that supports the instrument or machine) only begins at frequencies ω greater than $\sqrt{2}\omega_n$. The phase shift initially tends towards a lag of π. However, as the excitation frequency increases, so too will the forces due to viscous damping until eventually these become dominant. Consequently, at very high frequencies input and output tend towards a $\pi/2$ phase lag, the more rapidly for higher values of damping. Vibration isolation is always a compromise. On the one hand it is desired that the vibrations be attenuated as rapidly as possible, indicating as low a value of damper and natural frequency as possible. However, the resonant peak may cause some problems and therefore impose a limit on the damping ratio. Generally, vibration isolators with large mass, low support stiffness and reasonably low damping are desirable. Unfortunately, this is further complicated if there are resonant frequencies near to that of the isolator. This will be discussed shortly.

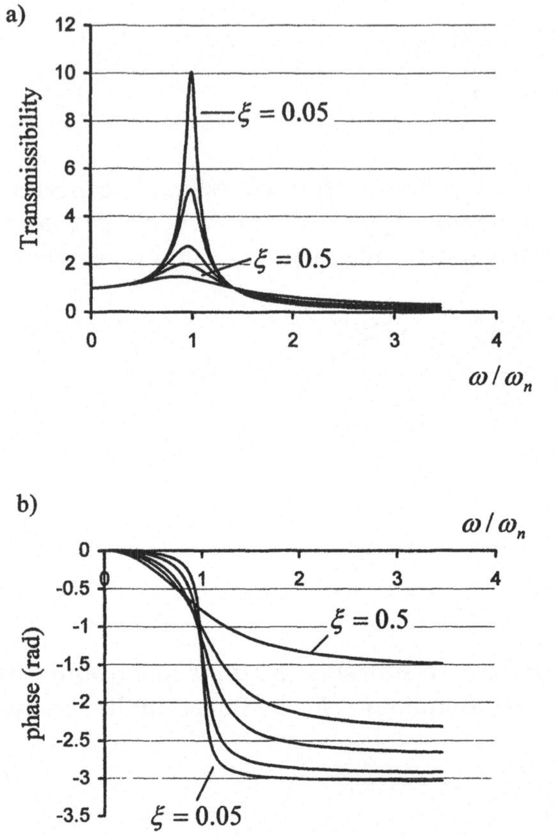

Figure 3.8: Transmissibility for a single degree of freedom spring/mass/damper system, a) magnitude response, b) phase response (ξ = 0.5, 0.3, 0.2, 0.1 and 0.05)

The above analysis is general to any single degree of freedom linear response function and will result in an output response *x(t)* to an input of magnitude $F_0(\omega)$ at any given frequency, ω, that can be expressed in the form

$$x(t) = F_o(\omega)|H(i\omega)|\cos(\omega t + \phi) \tag{3.115}$$

A curious question with regard to the representation of a phase lag arises at this point. For illustrative purposes let us assume that, at a particular input frequency, the gain is unity and there is a negative phase angle of $\pi/3$ radians. Plotting the input and output signals, figure 3.9 reveals that the output on the time axis moves to the right and intuitively gives the impression of a phase lead. However, if we place a point P on the input graph at time t_1 we can see that the

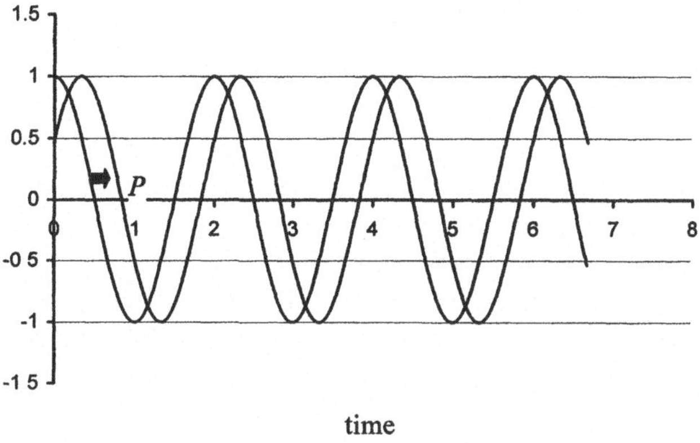

Figure 3.9: Representation of a cosine function with and without a phase lag ($\omega = 3.142$ and the phase lag = $\pi/3$)

output does not reach this point until a time $t' = \phi/\omega$ later. Consequently, a *positive* phase angle corresponds to a phase *lead* and a *negative* phase change a *lag*.

The velocity of the output for the same input is simply obtained by differentiating equation (3.115) with respect to time which is the only non constant variable on the right hand side, so

$$\begin{aligned}\dot{x}(t) &= -F_o\omega|H(i\omega)|\sin(\omega t + \phi(\omega)) \\ &= -F_o\omega|H(i\omega)|\cos(\omega t + \phi(\omega) - \pi/2) \\ &= F_o\omega|H(i\omega)|\cos(\omega t + \phi(\omega) + \pi/2)\end{aligned} \tag{3.116}$$

Clearly from this equation it can be seen that the velocity of the output leads the displacement by a factor of $\pi/2$. This transformation can be readily achieved by defining the velocity frequency response $H_v(i\omega)$ by the relationship

$$H_v(i\omega) = \frac{\dot{X}_{max}}{F_o} = i\omega H(i\omega) \tag{3.117}$$

Correspondingly, it can be shown that the acceleration $\ddot{x}$ for an input y is related by the acceleration response function

$$H_a(i\omega) = \frac{\ddot{X}_{max}}{F_o} = -\omega^2 H(i\omega) \tag{3.118}$$

It can readily be shown that the acceleration leads the velocity by $\pi/2$ and displacement by a factor π.

3.8 Multi-degree of freedom linear systems

It has already been stated that, at a single frequency, the output of a linear system is proportional to the input and that the total output is the sum of isolated responses to each input. This can be extended to an arbitrary number of inputs and outputs. Using the Fourier series, each of the inputs can be decomposed into a simple sum of separate frequencies. Based on the previous analysis, we may assume that for each input there will be a finite linear response at *all* of the outputs. Consequently, there will be a linear response function linking each input to all of the outputs. This is represented diagrammatically in figure 3.10 in which each output is independently and linearly related to all of the inputs. The subscripts i and j relate to the outputs and inputs respectively. Generally, for the mechanical systems being considered in this book, the total number of outputs j will be equal to the number of degrees of freedom of the system. Consequently, the response function $H_{ij}(i\omega)$ relates the response $x_j(t)$ to a sinusoidal input $y_i(t)$, or $F_i(t)$ in many cases. The outputs from the system can be simply obtained from the equation

$$\sum_{j=1}^{n}\left\{x_j(t)=\sum_{k=1}^{\infty}\sum_{i=1}^{m}Y_{jk}(\omega_k)\left|H_{ij}(\omega_k)\right|\cos\left(\omega_k t+\phi_k+\phi_{ij}(\omega_k)\right)\right\} \tag{3.119}$$

where m is the number of inputs, n is the number of degrees of freedom of the system and Y_{ik} are the inputs amplitudes (these can be forces or displacements) for the coordinate i and frequency ω_k. In practice, the number of inputs will nearly always be equal to the number of outputs. This corresponds to there being an input at each degree of freedom of the system. Exceptions are rare.

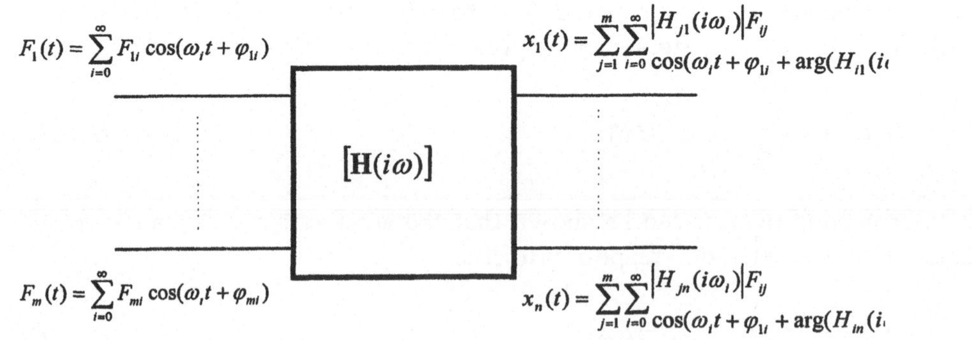

Figure 3.10: A 'black box' representation of a multi input/output linear system

At this stage, the reader is probably beginning to lose track of which components of the frequency response relate to which combination of inputs and outputs. Consequently, the meaning of individual terms tends to become obscured by the density of the equations. In general, the concept is less complex than the above mathematical representation appears. It is at this point that matrix notation is more concise.

Normally, it is the gain between an input force to a coordinate j and the subsequent output at coordinate i that is of interest. Fortunately, this can be obtained directly from equation (3.119) expressed in matrix notation

$$\{\mathbf{X}(\omega_k)\} = \begin{bmatrix} H_{11}(\omega_k) & - & H_{1j}(\omega_k) & - & H_{1m}(\omega_k) \\ | & & & & | \\ H_{i1}(\omega_k) & & & & H_{jm}(\omega_k) \\ | & & & & | \\ H_{n1}(\omega_k) & - & H_{nj}(\omega_k) & - & H_{nm}(\omega_k) \end{bmatrix} \{\mathbf{Y}(\omega_k)\} \tag{3.120}$$

(in this text curly parentheses, or braces, indicate a column vector)

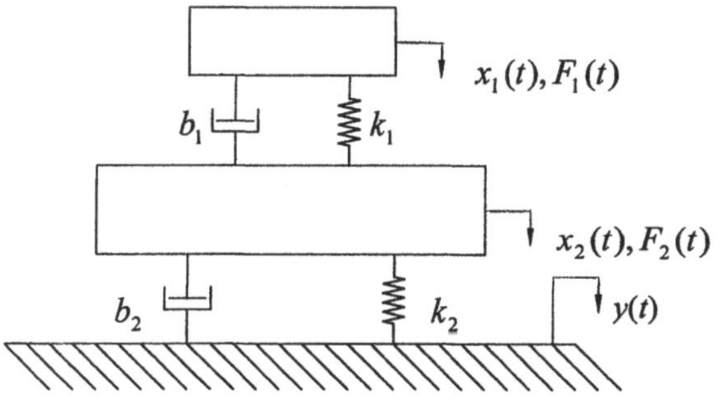

Figure 3.11: A vibration isolation table with a further spring/mass/damper system mounted on the table

Each response function $H_{ij}(\omega_k)$ of equations (3.119) and (3.120) gives the frequency response for the output at j for an input at i with all other inputs being equal to zero. To illustrate how various transfer functions may be deduced, consider the vibration isolation table of the previous example, only in this case with a further platform mounted on top, figure 3.11. In this particular example there are three possible inputs, these being the motion of the ground, and the two forces applied to each mass. Actually, this is the same as a three mass system with one of the masses being very heavy in comparison to the others (it could, for example, be the Earth). The equations governing motion of this system are

$$\begin{aligned} &m_1\ddot{x}_1 + b_1(\dot{x}_1 - \dot{x}_2) + k_1(x_1 - x_2) = F_1(t) \\ &m_2\ddot{x}_2 + b_1(\dot{x}_2 - \dot{x}_1) + b_2(\dot{x}_2 - \dot{y}) + k_1(x_2 - x_1) + k_2(x_2 - y) = F_2(t) \end{aligned} \tag{3.121}$$

The third equation is similar in form to the first of the above equations only we can imagine that the mass term is so large that it dominates the whole equation. For the first of our response functions it will be assumed that the input forces are equal to zero and our foundation is subject to a sinusoidal displacement (after all, subsequent motions of the masses of the table are not going to alter those of the Earth!) given by

$$y(t) = \mathrm{Re}\left\{Y_0 e^{i\omega t}\right\} \tag{3.122}$$

The responses are therefore given by

$$\begin{aligned} x_1(t) &= \mathrm{Re}\left\{H_{1y}(i\omega)Y_0 e^{i\omega t}\right\} \\ x_2(t) &= \mathrm{Re}\left\{H_{2y}(i\omega)Y_0 e^{i\omega t}\right\} \end{aligned} \tag{3.123}$$

From equations (3.121)-(3.123), we can derive the two simultaneous equations

$$\begin{aligned} -\omega^2 m_1 H_{1y}(i\omega) + i\omega b_1\left(H_{1y}(i\omega) - H_{2y}(i\omega)\right) + k_1\left(H_{1y}(i\omega) - H_{2y}(i\omega)\right) = 0 \\ -\omega^2 m_2 H_{2y}(i\omega) + i\omega b_1\left(H_{2y}(i\omega) - H_{1y}(i\omega)\right) + i\omega b_2\left(H_{2y}(i\omega) - Y_0\right). \\ k_1\left(H_{2y}(i\omega) - H_{1y}(i\omega)\right) + k_2\left(H_{2y}(i\omega) - Y_0\right) = 0 \end{aligned} \tag{3.124}$$

This provides two equations and two unknowns. Dropping the $i\omega$ terms in parentheses for the frequency response functions H, the first of equations (3.124) gives

$$H_{1y} = \frac{H_{2y}(k_1 + i\omega b_1)}{\left(-m_1\omega^2 + i\omega b_1 + k_1\right)} \tag{3.125}$$

Substituting (3.125) into the second of equations (3.124) and rearranging yields

$$H_{2y}\left[\begin{aligned} -\omega^2 m_2 + i\omega b_1\left(1 - \frac{(k_1 + i\omega b_1)}{\left(-m_1\omega^2 + i\omega b_1 + k_1\right)}\right) + i\omega b_2 \\ + k_1\left(1 - \frac{(k_1 + i\omega b_1)}{\left(-m_1\omega^2 + i\omega b_1 + k_1\right)}\right) \end{aligned}\right] = Y_o\left(i\omega b_2 + k_2\right) \tag{3.126}$$

This can be rearranged further to provide the response functions

$$\begin{aligned} \frac{H_{2y}}{Y_0} &= \frac{i\omega\left(b_1k_2 + b_2k_1 - m_1b_2\omega^2\right) + k_1k_2 - \omega^2\left(m_1k_2 + b_1b_2\right)}{\left(\begin{aligned} m_1m_2\omega^4 - i\omega\left(m_2b_1\omega^2 + m_1b_2\omega^2 + m_1b_1\omega^2 - b_2k_1 - b_1k_2\right). \\ -\omega^2\left(m_1k_2 + m_2k_1 + m_1k_1 + b_1b_2\right) + k_1k_2 \end{aligned}\right)} \\ &= \frac{A(\omega) + iB(\omega)}{C(\omega) + iD(\omega)} \end{aligned} \tag{3.127}$$

and

$$\frac{H_{2y}}{Y_0}=\frac{k_1k_2-\omega^2b_1b_2+i\omega(b_1k_2+b_2k_1)}{C(\omega)+iD(\omega)}$$
$$=\frac{E(\omega)+iF(\omega)}{C(\omega)+iD(\omega)} \tag{3.128}$$

It is interesting to note that the numerators of equations (3.127) and (3.128) not only contain lower powers of ω than in the denominator but both the real and imaginary terms generally contain smaller coefficients and will subsequently change less quickly. Consequently, it is the *common* denominator of both equations that is of interest. Before analyzing the complete response functions, it is informative to look at the limiting case when the system has zero damping. Under this condition and dividing through by the products of the stiffness of the two springs, equations (3.127) and (3.128) simplify to

$$\frac{H_{2y}}{Y_0}=\frac{1-\omega^2\frac{m_1}{k_1}}{\left(\frac{m_1m_2}{k_1k_2}\omega^4-\omega^2\left(m_1\left(\frac{1}{k_1}+\frac{1}{k_2}\right)+\frac{m_2}{k_2}\right)+1\right)} \tag{3.129}$$

and

$$\frac{H_{1y}}{Y_0}=\frac{1}{\left(\frac{m_1m_2}{k_1k_2}\omega^4-\omega^2\left(m_1\left(\frac{1}{k_1}+\frac{1}{k_2}\right)+\frac{m_2}{k_2}\right)+1\right)} \tag{3.130}$$

The most obvious thing about these new equations is that the complex components of both numerator and denominator have disappeared. Additionally, the denominator is a quadratic in ω^2. Consequently, two frequencies exist at which the denominator vanishes resulting in infinite displacements at masses 1 and 2. If we denote these frequencies λ_1 and λ_2, then we may also rewrite the above equations in the form

$$\frac{H_{2y}}{Y_0}=\frac{2\left(1-\omega^2\frac{m_1}{k_1}\right)}{\frac{m_1m_2}{k_1k_2}\left((\omega^2-\lambda_1^2)(\omega^2-\lambda_2^2)\right)} \tag{3.131}$$

and

$$\frac{H_{1y}}{Y_0} = \frac{2}{\frac{m_1 m_2}{k_1 k_2}\left(\left(\omega^2 - \lambda_1^2\right)\left(\omega^2 - \lambda_2^2\right)\right)} \tag{3.132}$$

These two equations are an indication that the transfer function of a system of arbitrary complexity can be relatively easily represented if we have a knowledge of the characteristic roots of the response function. This will be looked at in more detail in the following section. However, for now, to avoid misleading the reader, it is possible to make the statement that each term in parentheses represents the product of a complex conjugate pair and this accounts for the factor of two in the numerator. Finally, the values of the characteristic roots of the denominator (i.e. λ_1 and λ_2) are frequently encountered in matrix analysis and are called the *eigenvalues*.

Another fascinating consequence of equations (3.131) and (3.132) is that the numerator of the former equation is equal to zero at a frequency, ω given by

$$\omega = \sqrt{\frac{k_1}{m_1}} \tag{3.133}$$

This corresponds to the mass of the intermediate table being stationary relative to a fixed (or inertial) frame of reference while the second mass is oscillating at its individual natural frequency as if the second mass where not present. At first this seems to be a paradox because it is reasonable to ask how the mass knew to resonate if the second mass isn't moving. The answer is that the intermediate mass would move during the transient stage and will only eventually settle to a stationary value after these effects have decayed (there is always finite damping). Ultimately, in the *steady state* (which is the only condition being studied), when the foundation oscillates at this frequency, the upper mass will be applying an equal and opposite force on the upper side of the middle mass to that force due to the distortion of the lower spring. In the absence of damping, the two eigenvalues of this system are given by

$$\lambda_1^2, \lambda_2^2 = \frac{\frac{m_2}{k_2} + m_1\left(\frac{1}{k_1} + \frac{1}{k_2}\right) \pm \sqrt{\left(\frac{m_2}{k_2} + m_1\left(\frac{1}{k_1} + \frac{1}{k_2}\right)\right)^2 - 4\frac{m_1 m_2}{k_1 k_2}}}{2\frac{m_1 m_2}{k_1 k_2}} \tag{3.134}$$

It is also worth noting that the product of these two eigenvalues is

$$\lambda_1, \lambda_2 = \sqrt{2}$$

Turning our attention now to the frequency responses of the two masses having a force applied at each mass. For the first response we will assume $F_2(t)$ and y are zero and therefore we have a single input of the form

$$F_1(t) = \mathrm{Re}\{F_1 e^{i\omega t}\} \tag{3.135}$$

with the response

$$x_1(t) = \mathrm{Re}\{H_{11} F_1 e^{i\omega t}\} \tag{3.136}$$

Substituting equation's (3.135) & (3.136) into (3.121) and rearranging gives the responses

$$\begin{aligned}\frac{H_{2y}}{Y_0} &= \frac{i\omega b_1 + k_1}{\left(\begin{aligned}m_1 m_2\omega^4 - i\omega\left(m_2 b_1\omega^2 + m_1 b_2\omega^2 + m_1 b_1\omega^2 - b_2 k_1 - b_1 k_2\right).\\ -\omega^2\left(m_1 k_2 + m_2 k_1 + m_1 k_1 + b_1 b_2\right) + k_1 k_2\end{aligned}\right)}\\ &= \frac{i\omega b_1 + k_1.}{C(\omega) + iD(\omega)}\end{aligned} \tag{3.137}$$

and

$$\begin{aligned}\frac{H_{2y}}{Y_0} &= \frac{-m_2\omega^2 + i\omega(b_1 + b_2) + (k_1 + k_2)}{\left(\begin{aligned}m_1 m_2\omega^4 - i\omega\left(m_2 b_1\omega^2 + m_1 b_2\omega^2 + m_1 b_1\omega^2 - b_2 k_1 - b_1 k_2\right).\\ -\omega^2\left(m_1 k_2 + m_2 k_1 + m_1 k_1 + b_1 b_2\right) + k_1 k_2\end{aligned}\right)}\\ &= \frac{-m_2\omega^2 + i\omega(b_1 + b_2) + (k_1 + k_2)}{C(\omega) + iD(\omega)}\end{aligned} \tag{3.138}$$

Note also that, in the absence of damping, the displacement of the upper mass can be zero if the input force is at a single frequency given by

$$\omega^2 = \frac{k_1 + k_2}{m_2} \tag{3.139}$$

Similarly, it is possible to derive the transfer function response to a force applied at the middle mass

$$\frac{H_{22}}{F_2} = \frac{-m_1\omega^2 + i\omega b_1 + k_1}{\left(\begin{aligned}m_1 m_2\omega^4 - i\omega\left(m_2 b_1\omega^2 + m_1 b_2\omega^2 + m_1 b_1\omega^2 - b_2 k_1 - b_1 k_2\right).\\ -\omega^2\left(m_1 k_2 + m_2 k_1 + m_1 k_1 + b_1 b_2\right) + k_1 k_2\end{aligned}\right)} \tag{3.140}$$

and

$$\frac{H_{12}}{F_2} = \frac{i\omega b_1 + k_1}{\begin{pmatrix} m_1 m_2 \omega^4 - i\omega\left(m_2 b_1 \omega^2 + m_1 b_2 \omega^2 + m_1 b_1 \omega^2 - b_2 k_1 - b_1 k_2\right). \\ -\omega^2\left(m_1 k_2 + m_2 k_1 + m_1 k_1 + b_1 b_2\right) + k_1 k_2 \end{pmatrix}} \tag{3.141}$$

We can now write the general response matrix in the form

$$\begin{Bmatrix} X_1(i\omega) \\ X_2(i\omega) \\ Y(\omega) \end{Bmatrix} = \begin{bmatrix} H_{11}(i\omega) & H_{12}(i\omega) & H_{1y}(i\omega) \\ H_{21}(i\omega) & H_{22}(i\omega) & H_{2y}(i\omega) \\ 0 & 0 & 1 \end{bmatrix} \begin{Bmatrix} F_1(i\omega) \\ F_2(i\omega) \\ Y_o(i\omega) \end{Bmatrix} \tag{3.142}$$

Out of interest, equations (3.140) and (3.141) are plotted in figure 3.12. The eigenvalues, or the undamped natural frequencies, given by equation (3.134) are computed to be 6.2 and 16.2 (rad s^{-1}) respectively.

After having performed these rather lengthy and cumbersome algebraic manipulations, an obvious question at this point is how to generalize this analysis for systems having an arbitrary number of degrees of freedom and inputs. Unfortunately, the answer is not as simple as one would hope. Obviously, having obtained the general frequency response matrix (3.120), for our particular example, given (3.142), the only remaining task is to determine the frequencies and phases of the subsequent inputs. There are no simple, generalized methods for the determination of the frequency response functions. However, there are a number of short cuts outlined in the section below to enable plots of the frequency response to be computed for linear systems of arbitrary degrees of freedom. Also outlined are computational methods for calculation of characteristic roots of the denominator (also called eigenvalues or, because the frequency response shoots up to infinity for undamped systems, these are sometimes called poles) and associated mode shapes (also called eigenvectors).

3.8.1 Note on the graphical representation of frequency response functions

Three representations of the frequency response are shown in figure 3.12. Others could certainly be added. The objective of these graphical representations is to present three pieces of data

1. The magnitude of the gain between the input and output
2. The phase lag between input and output
3. The frequency at which the above occur

Figure 3.12 a)

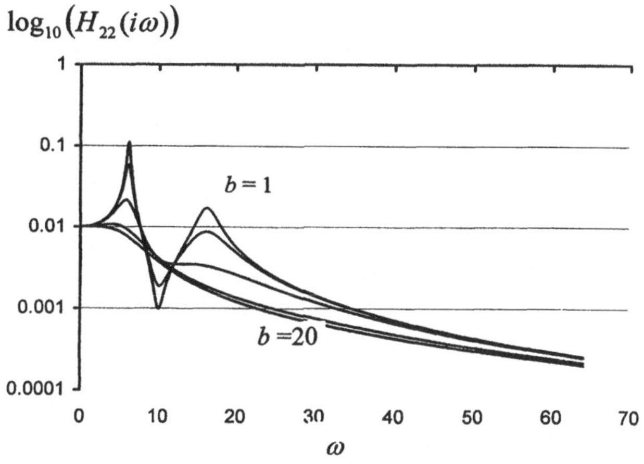

b)

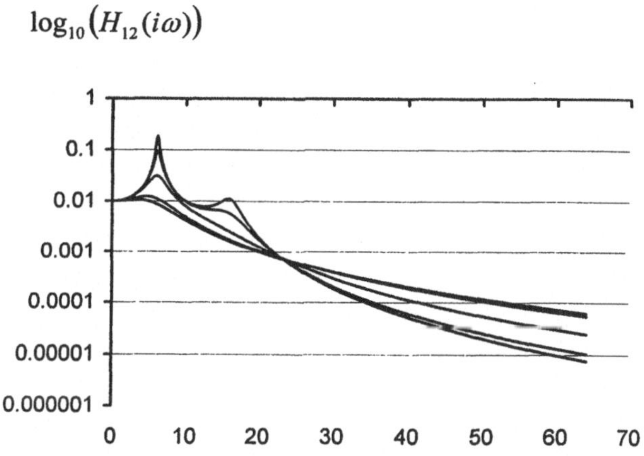

In the first plot of figure 3.12, the gain is plotted directly as a function of the frequency. However, the phase information is lost. Generally this should be included as a second plot often placed directly below to enable both gain and phase to be determined using a ruler or, at least, easily assessed 'by eye'. Often in electrical applications, linear systems are designed in a modular form. By insuring that the input impedance is high and output impedance is low, connecting successive systems results in a straightforward passing of the output from one system into the next. Consequently, the total gain of a system is simply the product of gains for all sub-systems through which the signal passes while phase shifts will simply add. Using a logarithmic scale for frequency and gain enables simple assessment of the effects of adding sub-systems together. As

e)

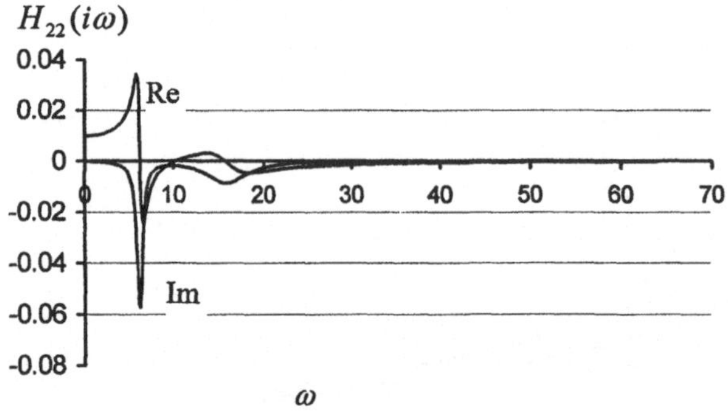

Figure 3.12: The frequency responses of the two masses of figure 11 to an input force at mass 2: a) The response of mass 2, b) the response of mass 1, c) & d) are the corresponding polar plots, e) real and imaginary parts of c) (parameters used for these plots are $m_1 = m_2 = 1$, $k_1 = k_2 = 100$, $b_1 = b_2 = 20, 16, 6, 2, 1$ respectively

plotted, the effect of series connection of sub-systems can be simply assessed by adding individual plots. If the gain is plotted in units of decibels this is called a *Bode* diagram. Unfortunately, for many mechanical systems the effects of adding further mechanical components will not correspond to a simple addition of system responses. Under these circumstances this method is not valid. The polar or *Nyquist* plots of figure 3.12 (c & d) indicate the gain and phase as radial vectors. However, the frequency information is lost. Because there are three pieces of information, these can only be represented in one diagram if it is three-dimensional. Three-dimensional polar diagrams are sometimes used in which the axes are the real and imaginary parts with frequency forming the third axis. In such a three-dimensional plot, the frequency response will be a *string*. The main advantage of the 2-D polar plot is that the phase shift can be readily evaluated irrespective of its value. For example, it can be seen from the figures that the phase of mass 1 tends to a 360-degree lag at high frequency while that of mass 2 tends towards 180 degrees. The real and imaginary plots of figure 3.12 e) are really only plots of the string in the 3-D Nyquist plots when viewed perpendicular to the frequency axis from the 'top' and 'side'. The advantages of this plot are that the damping at the individual resonance's can be readily assessed using equations (3.98) or (3.102).

3.9 General response function short cuts

Figure 3.12(cont'd): c)

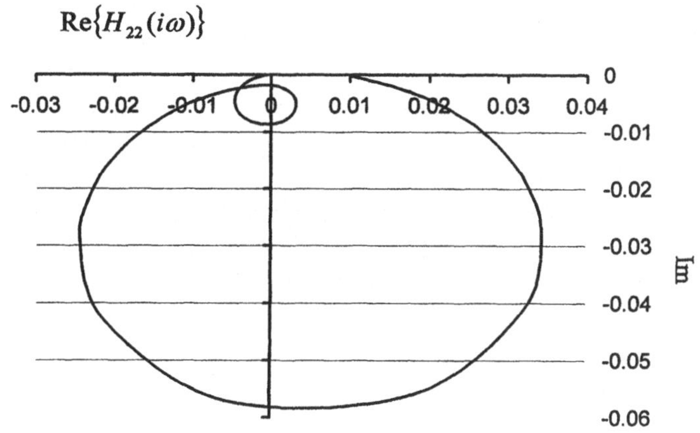

d)

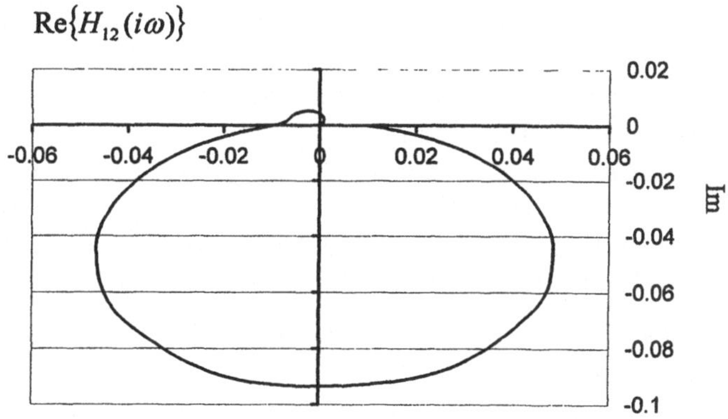

The first task is the derivation of a more direct method to obtain the numerator of the transfer function. It is important to realize that, after substitution of our assumed solution into our simultaneous differential equations, th a simple simultaneous equation emerges of the form

$$[A]\{H_i\} = \{1\} \qquad i = 1....n \qquad (3.143)$$

It is therefore possible to solve for each transfer function using Cramer's rule given by the ratio of determinants

$$H_i = \frac{|A_i|}{|A|} \tag{3.144}$$

Unfortunately it is common to use the same convention for the determinant of a matrix as for the magnitude of a complex number. Fortunately, these two situations rarely cross paths and, based on the context of the discussion, it is hoped that ambiguities will not arise. The upper determinant of equation (3.144) corresponds to the minor of A with the ith column replaced with the input vector containing a 1 in the position of the origin forcing function and zero's elsewhere. It is worth reviewing at this stage the method of determining the determinant of a square matrix. Taking a 3 x 3 square matrix of the form

$$[A] = \begin{bmatrix} a_{11} & a_{12} & a_{13} \\ a_{21} & a_{22} & a_{23} \\ a_{31} & a_{32} & a_{33} \end{bmatrix} \tag{3.145}$$

The determinant of this matrix is obtained by summing the values of successive sub-matrix determinants from the following four steps

1. Select a suitable row or column from the matrix (preferably one with lots of zero's in it)
2. Select successive elements from the selected row or column and strike out the row and column that it occupies
3. Find the determinant of the remaining matrix and multiply this by the currently selected element
4. Finally multiply this by $(-1)^{j+i}$. This is equivalent to multiplying successive elements by alternating positive and negative number.

For example, selecting the first row for step 1 above, the determinant of the above matrix is given by

$$|A| = a_{11}\begin{vmatrix} a_{22} & a_{23} \\ a_{32} & a_{33} \end{vmatrix} - a_{12}\begin{vmatrix} a_{21} & a_{23} \\ a_{31} & a_{33} \end{vmatrix} + a_{13}\begin{vmatrix} a_{21} & a_{22} \\ a_{31} & a_{32} \end{vmatrix} \tag{3.146}$$

This considerably simplifies the expression of the numerator of the transfer function. For example, the above problem, figure 3.11, for the case of an input force at mass 1 and ignoring motion of the ground, results in a simultaneous equation that can be expressed in matrix form

$$\begin{bmatrix} -m_1\omega^2 + i\omega b_1 + k_1 & -(i\omega b_1 + k_1) \\ -(i\omega b_1 + k_1) & -m_2\omega^2 + i\omega(b_1 + b_2) + (k_1 + k_2) \end{bmatrix} \begin{Bmatrix} H_{11} \\ H_{12} \end{Bmatrix} = \begin{Bmatrix} 1 \\ 0 \end{Bmatrix}$$

or

$$[A]\{H\} = \{1\} \tag{3.147}$$

Therefore, to solve for the numerator of H_{11} it is necessary to compute the minor determinant

$$\begin{vmatrix} 1 & -(i\omega b_1 + k_1) \\ 0 & -m_2\omega^2 + i\omega(b_1 + b_2) + (k_1 + k_2) \end{vmatrix} = \left(-m_2\omega^2 + i\omega(b_1 + b_2) + (k_1 + k_2)\right) \quad (3.148)$$

which is, of course, the same as the numerator of equation (3.138) only this time more easily derived.

Similarly, the numerator for H_{21}, also in accord with (3.139), is simply

$$\begin{vmatrix} -m_1\omega^2 + i\omega b_1 + k_1 & 1 \\ -(i\omega b_1 + k_1) & 0 \end{vmatrix} = (i\omega b_1 + k_1) \quad (3.149)$$

Because we are considering linear systems, it is possible to derive the general system response through consideration of individual responses to a single input at each coordinate. A rather striking feature of equation (3.144) is the fact that the denominator will be the same for all of the frequency responses. As a consequence it is possible to state the important lemma that

all denominators of the response function of a linear system are the same and can be obtained from the determinant of the response matrix [A].

Clearly, this enables the derivation of a generalized approach to frequency response calculation. Also it is apparent that the frequency responses computed at any point to forces applied at arbitrary positions around the system will have common poles. However, the number and frequency of zeros will vary at different coordinates of the system.

3.9.1 Rayleigh's approach to the problem of computing generalized frequency response functions

The use of determinants to derive a generalized form of the frequency response that is both mathematically robust and readily computed did not escape Rayleigh's notice. However, it must be acknowledged that, prior to Rayleigh's exposition, a more generalized approach had been thoroughly investigated by his "coach" E. J. Routh. To determine the frequency response to a generalized dynamic system, it is necessary to return to the original rigid body approach to dynamics. Assuming that we can split our system into discrete masses, springs and viscous damping elements, through appropriate choice of the generalized coordinates, q_i, for small displacements, it is possible to write the energy components of the Lagrangian in the form

$$T=\sum_{i=1}^{n}a_i\dot{q}_i^2$$
$$D=\sum_{i=1}^{n}\sum_{j=1}^{n}b_{ij}\dot{q}_i\dot{q}_j \qquad (3.150)$$
$$V=\sum_{i=1}^{n}\sum_{j=1}^{n}c_{ij}q_iq_j$$

It should be noted that

$$b_{ij}=b_{ji}$$
$$c_{ij}=c_{ji} \qquad (3.151)$$

Substitution of these into Lagrange's equation immediately yields

$$a_1\ddot{q}_1+b_{11}\dot{q}_1+b_{12}\dot{q}_2...+c_{11}q_1+c_{12}q_2+...=Q_1$$
$$a_2\ddot{q}_2+b_{21}\dot{q}_1+b_{22}\dot{q}_2...+c_{21}q_1+c_{22}q_2+...=Q_2 \qquad (3.152)$$

and so on for all other coordinates.

Assuming harmonic applied forces, and subsequent responses, the fluxions in (3.152) can be replaced with $i\omega$. Collecting all terms common to each coordinate we are left with the simultaneous linear equation

$$e_{11}q_1+e_{12}q_2+...=Q_1$$
$$e_{21}q_1+e_{22}q_2+...=Q_2 \ etc. \qquad (3.153)$$

where

$$e_{rs}=a_r(i\omega)^2+b_{rs}(i\omega)+c_{rs} \qquad (3.154)$$

Clearly, equation (3.153) can be expressed in matrix form

$$\begin{bmatrix} e_{11} & e_{12} & ... \\ e_{21} & e_{22} & ... \\ ... & ... & ... \end{bmatrix}\begin{Bmatrix} q_1 \\ . \\ q_n \end{Bmatrix}=[\mathbf{e}]\{q\}=\{Q\} \qquad (3.155)$$

In the absence of applied forces, for a non-trivial solution it is necessary that the determinant of **e** vanishes. It remains to link (3.155) with the general form of a linear simultaneous equation given in equation (3.143). Denoting ∇ as the determinant of **e**, its minor is given by

$$\frac{\partial \nabla}{\partial e_{rs}} \tag{3.156}$$

A general solution to the simultaneous equation (3.155) is given by

$$\begin{aligned}
\nabla q_1 &= \frac{\partial \nabla}{\partial e_{11}} Q_1 + \frac{\partial \nabla}{\partial e_{21}} Q_2 + \frac{\partial \nabla}{\partial e_{31}} Q_3 + \ldots \\
\nabla q_2 &= \frac{\partial \nabla}{\partial e_{12}} Q_1 + \frac{\partial \nabla}{\partial e_{22}} Q_2 + \frac{\partial \nabla}{\partial e_{32}} Q_3 + \ldots \\
&\textit{etc}
\end{aligned} \tag{3.157}$$

Setting all of the generalized forces to zero except for Q_s say, the response of the ith coordinate can be found from the equation

$$\nabla q_i = \frac{\partial \nabla}{\partial e_{is}} Q_s \tag{3.158}$$

Rearranging the above, the frequency response function can be readily recognized as

$$H_{is}(i\omega) = \frac{q_i}{Q_s} = \frac{1}{\nabla} \frac{\partial \nabla}{\partial e_{is}} \tag{3.159}$$

For those of a mathematical disposition, it is noteworthy that, by equation (3.151), a principle of reciprocity exists for dynamic systems i.e.

$$H_{is}(i\omega) = H_{si}(i\omega) \tag{3.160}$$

The reader is left to show, once again, that equation (3.159) can be used to reproduce the frequency response functions given by (3.140) and (3.141).

The utility of equation (3.159) is readily apparent for systems of two or more degrees of freedom. Once the equations of motion of a system have been determined (using Lagranges equation, of course) it is no more than a collecting of terms and programming a computer to evaluate the determinants for individual frequencies of interest. From such a computation, the gain and phase can be computed directly. An example of a six-degree of freedom system is presented in section 3.11.5.

3.9.1.1 Example of a two degree of freedom system

For the two-degree of freedom system considered above, (3.147) gives

$$\nabla = \det\begin{bmatrix} -m_1\omega^2 + i\omega b_1 + k_1 & -(i\omega b_1 + k_1) \\ -(i\omega b_1 + k_1) & -m_2\omega^2 + i\omega(b_1 + b_2) + (k_1 + k_2) \end{bmatrix}$$

$$= \begin{vmatrix} -m_1\omega^2 + i\omega b_1 + k_1 & -(i\omega b_1 + k_1) \\ -(i\omega b_1 + k_1) & -m_2\omega^2 + i\omega(b_1 + b_2) + (k_1 + k_2) \end{vmatrix}$$

$$= \begin{vmatrix} e_{11} & e_{12} \\ e_{21} & e_{22} \end{vmatrix}$$

From which the frequency responses can be computed from

$$H_{11} = \frac{\begin{vmatrix} 1 & e_{12} \\ 0 & e_{22} \end{vmatrix}}{\begin{vmatrix} e_{11} & e_{12} \\ e_{21} & e_{22} \end{vmatrix}}$$

$$H_{12} = \frac{\begin{vmatrix} e_{11} & 1 \\ e_{21} & 0 \end{vmatrix}}{\begin{vmatrix} e_{11} & e_{12} \\ e_{21} & e_{22} \end{vmatrix}} = H_{21}$$

$$H_{22} = \frac{\begin{vmatrix} e_{11} & 0 \\ e_{21} & 1 \end{vmatrix}}{\begin{vmatrix} e_{11} & e_{12} \\ e_{21} & e_{22} \end{vmatrix}} \; etc.$$

3.10 Eigen analysis

Fortunately, it has been the preoccupation of many mathematicians to be able to express the determinant in the form of a series of simple products or, alternatively, the sum of simple polynomials. Before looking at this technique, let us back-track a little and return to a simple transfer function of the form

$$H(i\omega) = \frac{1}{m}\left(\frac{1}{(\lambda_1 - \lambda_2)(i\omega - \lambda_1)} + \frac{1}{(\lambda_2 - \lambda_1)(i\omega - \lambda_2)}\right) \tag{3.161}$$

This can be factored to give the alternative form

$$H(i\omega) = \frac{1}{m}\left(\frac{1}{(i\omega - \lambda_1)(i\omega - \lambda_2)}\right) \tag{3.162}$$

Clearly, the two values for λ correspond to the roots or characteristic values of the denominator of the frequency response. If we substitute the two eigenvalues for a simple single degree of freedom spring/mass/damper system given by

$$\lambda_{1,2} = -\xi\omega_n \pm i\omega_n\sqrt{1-\xi^2} \tag{3.163}$$

we derive the frequency response function

$$H(i\omega) = \frac{1}{m}\left(\frac{1}{\omega_n^2 - \omega^2 + i2\xi\omega\omega_n}\right) \tag{3.164}$$

This is exactly the same as the frequency response derived in equations (3.78) and (3.95). Therefore we have two alternative forms of the frequency response function in terms of the characteristic roots, or *eigenvalues* of the system. This problem has been extensively studied and there now exist a large number of software packages capable of computing the eigenvalues of a matrix. Although beyond the scope of the present book, it is possible to produce the frequency response of any linear system in the form, Newland, 1993,

$$H(i\omega) = \frac{d(i\omega)}{c\prod_{j=1}^{2n}(i\omega - \lambda_j)} \tag{3.166}$$

Derivation of equations for d and c is a complex task and adds little to our understanding of system behavior. There are two characteristics of the frequency response that are of interest in design, these are

1. Values of input frequency at which the numerator becomes small
2. Values of input frequency at which the denominator becomes small

The first of these results in the magnitude of the frequency response tending to zero and, as a consequence, these points are called *zeros*. The second condition results in rapid increases, or spikes, in the frequency response and these are consequently known as *poles*. It is the poles that result in large stresses and deformations and their identification is of some concern to the designer. It is clear from (3.162) or (3.166) that, for purely complex roots, the poles occur at the eigenvalues of the system.

The remaining piece of the jigsaw puzzle for our multi-degree of freedom systems is the computation of the *characteristic*, or, identically, *eigen*, values.

3.10.1 Conservative systems

It can be shown that any linear mechanical system can be described by a governing equation of 'free' motion in the generalized matrix form

$$[m]\{\ddot{x}\}+[b]\{\dot{x}\}+[k]\{x\}=0 \tag{3.167}$$

Clearly, any solution to this equation will represent its transient behavior. Again, a solution is assumed to be of the form

$$\{x(t)\}=\{X_i e^{\lambda t}\} \tag{3.168}$$

Substituting this into the above gives

$$\begin{aligned}&\lambda^2[m]\{X\}+\lambda[b]\{X\}+[k]\{X\}=0\\&=\lambda^2[m]+\lambda[b]+[k]\end{aligned} \tag{3.169}$$

Clearly, there are only certain values for λ which satisfy the above equation and these are the eigenvalues of our system. As a first example, consider a simple three mass system with no damping as shown in figure 3.13. The matrix equations of motion for this system are easily derived and are given by

$$\begin{bmatrix}m_1&0&0\\0&m_2&0\\0&0&m_3\end{bmatrix}\begin{Bmatrix}\ddot{x}_1\\\ddot{x}_2\\\ddot{x}_3\end{Bmatrix}-\begin{bmatrix}-(k_2+k_1)&k_2&0\\k_2&-(k_3+k_2)&k_3\\0&k_3&-(k_4+k_3)\end{bmatrix}\begin{Bmatrix}x_1\\x_2\\x_3\end{Bmatrix}=\{0\} \tag{3.170}$$

which is of the form (3.167) with $[b]=0$.
Substituting (3.168) into (3.167) and canceling out the exponent gives

$$\left(\lambda^2[m]-[k]\right)\{X\}=\{0\} \tag{3.171}$$

Multiplying through by the inverse mass matrix (which is a diagonal of $1/m$ terms if it is a diagonal matrix as for this example) results in an equation of the form

$$\begin{aligned}&\left[\lambda^2[I]-[m]^{-1}[k]\right]\{X\}=0\\&=\left[\lambda^2[I]-[A]\right]\{X\}\end{aligned} \tag{3.172}$$

Apart from the trivial solution $\{X\}=0$, solutions for λ^2 and $\{X\}$ are called the eigenvalues and eigenvectors respectively. Usually, if the system has more than two degrees of freedom it is common to employ computer solutions for the eigenvalues and eigenvectors based on the system matrix $[A]$. For equation (3.168) to be valid, the eigenvalue must be an imaginary number. Intuitively, this is obvious because, in the absence of damping, any transient motion will result in a sinusoidal oscillation that will continue indefinitely. It is clear from the solution (3.168) that this is only true if the eigenvalue is imaginary. Two eigenvalues are possible for each solution corresponding to a positive and negative imaginary number. This leads to an important conclusion that the solution gives two eigenvalues that are complex conjugate pairs. Substituting these into (3.168) and expanding for the ith degree of freedom and the n and n+1 eigenvalues we have

$$x_i(t) = X_{in}e^{\lambda_n t} + X_{i+1n}e^{\lambda_{n+1}t} \tag{3.173}$$

Associating the eigenvalues with the frequency of oscillation

$$\begin{aligned} \lambda_n &= i\omega_n \\ \lambda_{n+1} &= -i\omega_n \end{aligned} \tag{3.174}$$

and expanding (3.173) using Abraham De Moivre's theorem

$$x_i(t) = X_{in}(\cos\omega_n t + i\sin\omega_n t) + X_{i+1}(\cos\omega_n t - i\sin\omega_n t) \tag{3.175}$$

This gives the displacement of the ith mass in terms of imaginary quantities! Clearly, this cannot be representative of the real world and it is easy to show that the corresponding amplitudes of motion, or eigenvectors, must, and do, also occur as complex conjugate pairs for $x(t)$ to be real.

For a non-trivial solution to equation (3.172), the determinant of the matrix must vanish. This will reveal the roots of the characteristic equation which also yield an expression for the denominator of the linear system frequency response function.

As an example, consider the three mass system of figure 3.13 for the parameters given by

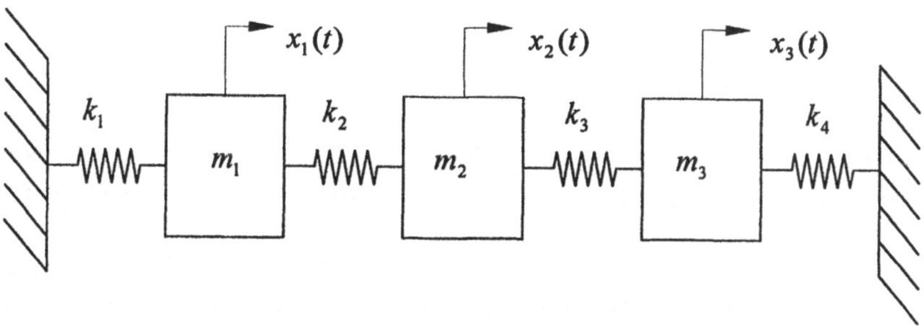

Figure 3.13: A simple three mass linear system without damping

$$m_1 = m_3 = 2m$$
$$m_2 = m$$
$$k_1 = k_4 = 3k$$
$$k_2 = k_3 = k$$

After substituting the assumed solution and rearranging into the form of equation (3.172), the matrix equation of motion for free oscillation becomes

$$\begin{bmatrix} (4k - 2m\lambda^2) & -k & 0 \\ -k & (2k - m\lambda^2) & -k \\ 0 & -k & (4k - 2m\lambda^2) \end{bmatrix} \left\{ X \right\} = \left\{ 0 \right\}$$

Setting the determinant of the first term to zero will correspond to the condition for a resonant frequency. In the absence of damping, this is the same as the natural frequency. Expanding the determinant of the above matrix gives

$$(4k - 2m\lambda^2)^2 (2k - m\lambda^2) - 2k^2 (4k - 2m\lambda^2) = 0$$

Expanding this further and dividing by $4m^3$ yields

$$4m^3 \left(6\left(\frac{k}{m}\right)^3 - 11\left(\frac{k}{m}\right)^2 \lambda^2 + 6\left(\frac{k}{m}\right)\lambda^4 - \lambda^6 \right) = 0$$

This can be factorized to give an expression for the determinant in the form

$$4m^3 \left(\frac{k}{m} - \lambda^2 \right)\left(2\frac{k}{m} - \lambda^2 \right)\left(3\frac{k}{m} - \lambda^2 \right) = 0$$

$$= 4m^3 \left(\sqrt{\frac{k}{m}} - i\lambda \right)\left(\sqrt{\frac{k}{m}} + i\lambda \right)\left(\sqrt{2\frac{k}{m}} - i\lambda \right)\left(\sqrt{2\frac{k}{m}} + i\lambda \right)\left(\sqrt{3\frac{k}{m}} - i\lambda \right)\left(\sqrt{3\frac{k}{m}} + i\lambda \right)$$

(3.176)

Comparison of equation (3.176) with (3.162) indicates that the characteristic roots can be found from the eigenvalues of the free vibration matrix $[A]$ of equation (3.172). In this particular instance, the constant c is simply the product of inertial terms. As can also be seen, the eigenvalues occur in complex conjugate pairs and in the absence of damping are purely imaginary. This system could be represented by plotting the eigenvalues (or characteristic roots) on an Argand diagram whereby it is obvious that they will all be situated on the imaginary axis with no real parts. The values for the eigenvalues that satisfy equation (3.176) are the characteristic values of the system and these are given by

$$\lambda_{1,2} = \pm i\sqrt{\frac{k}{m}}$$
$$\lambda_{3,4} = \pm i\sqrt{2\frac{k}{m}} \qquad (3.177)$$
$$\lambda_{5,6} = \pm i\sqrt{3\frac{k}{m}}$$

The frequency response of any mass to any steady state sinusoidal input can be obtained from equations (3.162) and (3.177)

$$H(i\omega) = \frac{|A_i|}{4m^3(i\omega-\lambda_1)(i\omega+\lambda_2)(i\omega-\lambda_3)(i\omega+\lambda_4)(i\omega-\lambda_5)(i\omega+\lambda_6)}$$
$$= \frac{|A_i|}{4m^3(\omega^2-\lambda_1^2)(\omega^2-\lambda_3^2)(\omega^2-\lambda_5^2)} \qquad (3.178)$$
$$= \frac{|A_i|}{4m^3\left(\frac{k}{m}-\omega^2\right)\left(2\frac{k}{m}-\omega^2\right)\left(3\frac{k}{m}-\omega^2\right)}$$

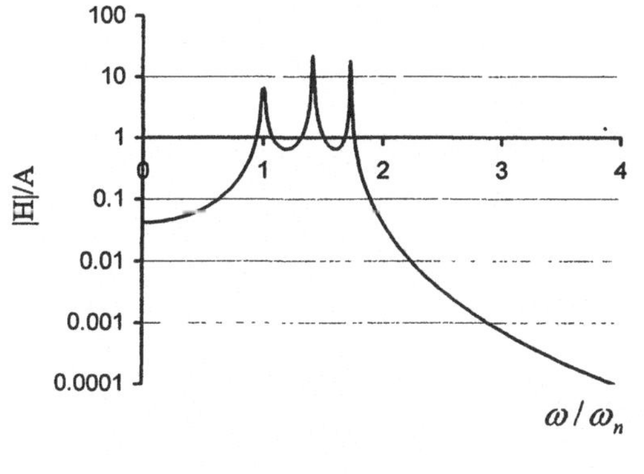

Figure 3.14: Graph of the frequency response of the three mass system of figure 3.13 (note that the resonant amplitudes have been suppressed)

It should be noted that the phase shifts by π every time the system goes through a resonance and therefore it is more informative to plot the magnitude of $1/|A|$ which for $k = m = 1$ is shown in figure 3.14. As would be expected, the magnitudes tend to infinity at the three natural frequencies. At frequencies higher than the largest resonance, the amplitude drops off at rate approaching 18 dB per octave corresponding to the reciprocal sixth power law with frequency.

Determination of the numerator of equation (3.178) is left for the following example.

3.10.2 Systems with damping

In the previous eigenvalue analysis system damping has been ignored in order to obtain the undamped natural frequencies of our system. These undamped natural frequencies are also the eigenvalues and were found to be purely imaginary. In reality there will always be some damping present and we must therefore assume a system equation of the general form

$$[m]\{\ddot{x}\} + [b]\{\dot{x}\} + [k]\{x\} = 0$$

Again assuming a solution of the form (3.168), we seek a solution to the equation

$$\lambda^2[m]\{X\} + \lambda[b]\{X\} + [k]\{X\} = 0 \tag{3.179}$$

This can be rearranged to give

$$\lambda[[m]\{V\} + [b]\{V\} + [k]\{X\} = 0 \tag{3.180}$$

where

$$\{V\} = \lambda\{X\} \tag{3.181}$$

which, by examination of equation (3.168), obviously corresponds to the velocity eigenvector.

Equations (3.180) and (3.181) can be rearranged into the equivalent eigenvalue problem

$$\lambda\begin{Bmatrix} X \\ V \end{Bmatrix} = \begin{bmatrix} 0 & I \\ -[m]^{-1}[k] & -[m]^{-1}[b] \end{bmatrix}\begin{Bmatrix} X \\ V \end{Bmatrix} \tag{3.182}$$

By defining

$$\{Z\} = \begin{Bmatrix} X \\ V \end{Bmatrix} \tag{3.183}$$

and

$$[A] = \begin{bmatrix} 0 & I \\ -[m]^{-1}[k] & -[m]^{-1}[b] \end{bmatrix} \tag{3.184}$$

equation (3.182) can be written in the standard eigenvalue form

$$[A]\{Z\} = \lambda\{Z\} \tag{3.185}$$

There are many software packages that will supply the eigenvalues and eigenvectors for the [A] and {Z} matrices respectively. However, in this case both the eigenvalues and eigenvectors occur in complex conjugate pairs. To see why this is so, consider the following interpretation of these values.

3.10.3 Interpretation of complex eigenvalues and eigenvectors

From equation (3.168), it can be seen that if both λ and X are real, then, provided that λ is negative, the solution for $x(t)$ is an exponential decay. If the real part of the eigenvalue (or characteristic root) is positive, the output will be exponentially increasing with time which corresponds to an unstable system. This does not occur unless there is some form of feedback or self-induced effects. When the eigenvalues and eigenvectors are complex they will occur in conjugate pairs. Therefore considering the ith mode transient response of the kth coordinate of our system we have the solution

$$\begin{aligned} x_k(t) &= c_k X_k e^{\lambda_i t} + c_{k+1} X_{k+1} e^{\lambda_{i+1} t} \\ &= c_k X_k e^{\lambda_i t} + c_k^* X_k^* e^{\lambda_i^* t} \end{aligned} \tag{3.186}$$

The constant c values depend upon the boundary conditions and, for the purposes of illustration, these can be set at arbitrary values and unity is the obvious choice. The eigen value can be described by a constant complex number of the form

$$\lambda_i = -\alpha + i\omega \tag{3.187}$$

where α and ω are real positive constants. Inserting (3.187) into (3.186) and expanding yields

$$x_k(t) = e^{-\alpha t}\left(X_k e^{i\omega t} + X_k^* e^{-i\omega t}\right) \tag{3.188}$$

Making $X = A + iB$

$$\begin{aligned} x_k(t) &= e^{-\alpha t}(2A\cos(\omega t) + 2B\sin(\omega t)) \\ &= e^{-\alpha t}\left(2\sqrt{A^2+B^2}\cos(\omega t + \tan^{-1}(B/A)\right) \\ &= 2e^{-\alpha t}|X_k|\cos(\omega t + \angle X_k) \end{aligned} \tag{3.189}$$

Consequently, the displacement of x is purely real as would be expected. This can be represented on a phasor diagram as the sum of two equal vectors rotating at equal velocity in opposite directions and *each* is shortening exponentially at a rate

$$e^{-\alpha t} \qquad (3.190)$$

The reader is recommended to draw this. Plot the first vector rotating clockwise with its tail at the origin of an Argand diagram (another name for a real/imaginary set of Cartesian axes). The second vector should be connected to the tip of the first and rotating at the same rate in the opposite direction. Its tip should be always touching the real axis and at the point given by equation (3.189). This corresponds to the transient solution for the kth coordinate of the freely oscillating multi-degree of freedom mechanism.

3.10.4 Summary of primary steps in the derivation the response function of a linear multi-degree of freedom system

The mathematical arguments result in a reasonably simple approach to the computation of response functions. In general the response can be obtained through the following steps

1. Derive the equations of motion using Lagranges equation or Hamilton's principle.
2. Linearize equations by assuming small displacements.

   Set up the equations of motion in the standard matrix form

   $$[\mathbf{m}]\{\ddot{x}\}+[\mathbf{b}]\{\dot{x}\}+[\mathbf{k}]\{x\}=\{f(t)\}$$

3. Substitute in the trial solution

   $$\{x_i(t)\}=\{X_i e^{\lambda t}\}$$

   to derive simultaneous equation in $\{x\}$ in the form

   $$[\mathbf{A}]\{X\}=\{1\}$$

4. Determine the numerator of the frequency response function for the response of the kth coordinate to input at j from the equation

   $$d_{jk}(i\omega)=|A_k|$$

   where $|A_k|$ is the minor determinant of the matrix $[A]$ with the kth column replaced with zero's, apart from a value of 1 at row j.

Compute the response at any frequency using the equation

$$H_{ls}(i\omega) = \frac{q_l}{Q_s} = \frac{1}{\nabla}\frac{\partial \nabla}{\partial e_{ls}}$$

4. Finally, to determine the characteristic roots (eigenvalues) and mode shapes (eigenvectors) of the system, rearrange the solution matrix in the form of an eigenvalue problem of the form

$$\left[\lambda[I] - [A]\right]\{Z\} = 0$$

where

$$[A] = \begin{bmatrix} 0 & I \\ -[m]^{-1}[k] & -[m]^{-1}[b] \end{bmatrix}$$

Use any computational mathematics package to determine the eigenvalues and vectors.

3.10.5 Example: A series, six mass vibration isolator

Before moving on to continuous systems, an example of a multi-degree of freedom vibration isolator is considered. The global interest in the detection of the extremely low perturbations induced in solid bodies (the 'antennae') through the effects of gravity waves propagated by the creaking together of distant massive objects might provide some more clues to the nature of the universe to which we are bound. Requiring the measurement of induced strains which could be as low as 10^{-20} (which would correspond to an extension of about one atoms length in a bar that that is about the length of 1,700 earth's strung together), it is probably an understatement to say that this effect is small. Clearly, it is not acceptable to have our 'antennae' mounted on a platform that may be vibrating due to ground and air-borne vibrations. Adding more spring/mass/damper combinations to our isolator leads to successively higher and higher 'roll-off' rates, or attenuation, in response to perturbations at frequencies well above the highest resonance. It is for this reason that multi-stage isolators have received considerable attention over the last decade. With this in mind, let us consider the six-stage vibration isolation mount shown in figure 3.15. In our conceptual example, we will assume that the final mass number 1 is the antennae and that all inputs originate from the ground in the form of seismic disturbances. To further simplify the analysis it will be assumed that each of the masses, springs and damping elements are nominally identical.

Step1

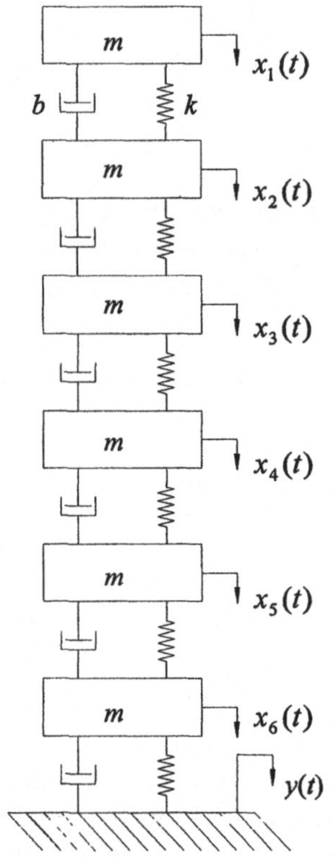

Figure 3.15: A six-stage vibration isolation mount

Clearly, for this simple mechanism, the system matrices are given by

$$[m]=\begin{bmatrix} m & 0 & 0 & 0 & 0 & 0 \\ 0 & m & 0 & 0 & 0 & 0 \\ 0 & 0 & m & 0 & 0 & 0 \\ 0 & 0 & 0 & m & 0 & 0 \\ 0 & 0 & 0 & 0 & m & 0 \\ 0 & 0 & 0 & 0 & 0 & m \end{bmatrix}$$

$$[k]=\begin{bmatrix} k & -k & 0 & 0 & 0 & 0 \\ -k & 2k & -k & 0 & 0 & 0 \\ 0 & -k & 2k & -k & 0 & 0 \\ 0 & 0 & -k & 2k & -k & 0 \\ 0 & 0 & 0 & -k & 2k & -k \\ 0 & 0 & 0 & 0 & -k & 2k \end{bmatrix}$$

$$[b]=\begin{bmatrix} b & -b & 0 & 0 & 0 & 0 \\ -b & 2b & -b & 0 & 0 & 0 \\ 0 & -b & 2b & -b & 0 & 0 \\ 0 & 0 & -b & 2b & -b & 0 \\ 0 & 0 & 0 & -b & 2b & -b \\ 0 & 0 & 0 & 0 & -b & 2b \end{bmatrix}$$

$$\{F\}=\begin{Bmatrix} 0 \\ 0 \\ 0 \\ 0 \\ 0 \\ Y_s \end{Bmatrix}$$

where Y_s corresponds to the seismic motion of the ground.

Step 2
The numerator of the frequency response can be found from the determinant of the following matrix

$|A_y| =$

$$\begin{bmatrix} 0 & -(i\omega b+k) & 0 & 0 & 0 & 0 \\ 0 & (-m\omega^2+i\omega b+k) & -(i\omega b+k) & 0 & 0 & 0 \\ 0 & -(i\omega b+k) & \ldots & -(i\omega b+k) & 0 & 0 \\ 0 & 0 & -(i\omega b+k) & \ldots & -(i\omega b+k) & 0 \\ 0 & 0 & 0 & -(i\omega b+k) & \ldots & -(i\omega b+k) \\ Y_0(i\omega b+k) & 0 & 0 & 0 & -(i\omega b+k) & (-m\omega^2+i\omega b+k) \end{bmatrix}$$

Step 3

The determinant of this matrix is simply

$$\begin{aligned} |A_y| &= Y_0(i\omega b+k)^6(-1)^5 \\ &= -Y_0(-(\omega b)^6 + 6i(\omega b)^5 k + 15(\omega b)^4 k^2 - 20i(\omega b)^3 k^3 - 15(\omega b)^2 k^4 + 6i(\omega b)k^5 + k^6) \end{aligned}$$

Step 4

This requires the reformulation of the matrix [A] for computation of the eigenvalues and eigen vectors. This was carried out using MATLAB™ for $m = k = 1$ and two values of b = 0.0, 0.4 and 0.1 respectively. The corresponding eigenvalues were

For $b = 0$

$$\lambda(b=0) = \begin{Bmatrix} 1.9419 \\ 1.7709 \\ 1.4970 \\ 1.1361 \\ 0.7092 \\ 0.2410 \end{Bmatrix}, \lambda(b=0.1) = \begin{Bmatrix} -0.1885 \pm i1.9327 \\ -0.1568 \pm i1.7640 \\ -0.1121 \pm i1.4928 \\ -0.0645 \pm i1.1343 \\ -0.0251 \pm i0.7088 \\ -0.0029 \pm i0.2411 \end{Bmatrix}$$

$$\lambda(b=0.4) = \begin{Bmatrix} -0.7542 \pm i1.7890 \\ -0.6272 \pm i1.6561 \\ -0.4482 \pm i1.4283 \\ -0.2582 \pm i1.1064 \\ -0.1006 \pm i0.7020 \\ -0.0116 \pm i0.2408 \end{Bmatrix}$$

It is worth checking that the product of squares of the eigenvalues yields the appropriate value for the static response. In this case the output of the top mass will be equal to the displacement of the base. It is therefore clear that the product of squares of the eigenvalues should be equal to the ratio of products of stiffness and mass which is equal to 1. A quick calculation will verify this.

Step 5

The above values plus the product of masses (=1) have been inserted into equation (3.178) and numerically computed using a spreadsheet that provides the complex coefficients of the frequency response function.

3.10.5.1 Results

For zero damping the response function has six frequency's at which the response will tend to infinity with a corresponding instantaneous phase shift of π (i.e. a change in sign) as the frequency passes through each resonance. Interestingly the eigenvalues and eigenvectors are useful for the identification and visualization of characteristic mode shapes. The eigenvectors are

$$\{X\} = \left\{ \begin{Bmatrix} 055 \\ 0.52 \\ 0.46 \\ 0.37 \\ 0.26 \\ 0.13 \end{Bmatrix}, \begin{Bmatrix} -0.52 \\ -0.26 \\ 0.13 \\ 0.46 \\ 0.55 \\ 0.37 \end{Bmatrix}, \begin{Bmatrix} -0.46 \\ 0.13 \\ 0.55 \\ 0.26 \\ -0.37 \\ -0.52 \end{Bmatrix}, \begin{Bmatrix} 0.37 \\ -0.46 \\ -0.26 \\ 0.52 \\ 0.13 \\ -0.55 \end{Bmatrix}, \begin{Bmatrix} -0.26 \\ 0.55 \\ -0.37 \\ -0.13 \\ 0.52 \\ -0.46 \end{Bmatrix}, \begin{Bmatrix} -0.13 \\ 0.37 \\ -0.52 \\ 0.55 \\ -0.46 \\ 0.26 \end{Bmatrix} \right\}$$

The first three modes are plotted in figure 3.16. One can see the characteristic shapes for each. The zero crossings are stationary over time and are the characteristic nodes of the system. The number of nodes is directly related to the frequency mode m. In this case, for the mth mode there will be $(m\text{-}1)$ nodes.

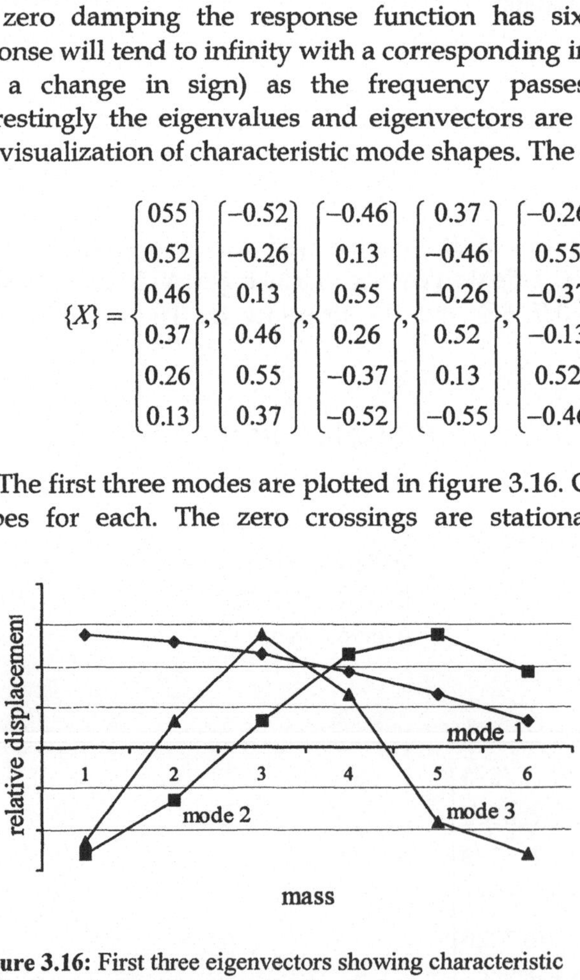

Figure 3.16: First three eigenvectors showing characteristic mode shape for the six-mass vibration isolator

For a relatively low damping coefficient corresponding to $b = 0.1$ the frequency response function for the upper mass is shown in figure 3.17. The relatively high Q value for the lower

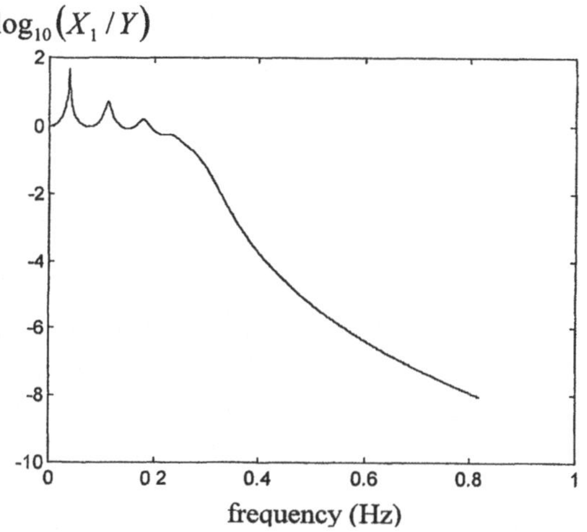

Figure 3.17: Frequency response of the six mass isolator for $b = 0.1$

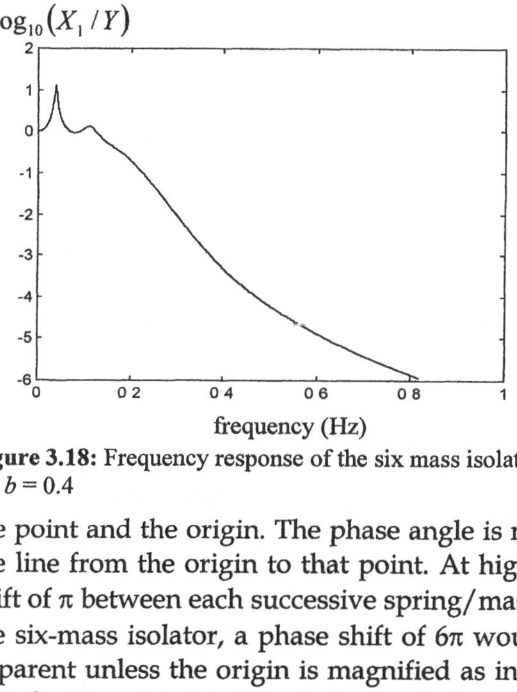

Figure 3.18: Frequency response of the six mass isolator for $b = 0.4$

modes is readily apparent although at the higher modes there is so much attenuation that the final two peaks are difficult to discern even when a logarithmic scale is used for the magnitude of the response. It is also interesting that the attenuation occurs at a relatively high frequency of a little over 1.5 radians per second (0.24 Hz).

For even higher damping corresponding to a damping factor of $b = 0.4$, the frequency response still shows peaks at the first two fundamental frequencies, but they are almost impossible to discern visually at the high mode resonance's, figure 3.18. One method of visualization is that of the polar, or Nyquist, plot in which the real and imaginary components of the response function are plotted on an Argand diagram, figure 3.19. Interpretation of this is simple, the magnitude of the gain at any frequency on the plot is the length of a straight line drawn between the point and the origin. The phase angle is measured between the real axis and the line from the origin to that point. At high frequencies, there is also a phase shift of π between each successive spring/mass/damper stage. Consequently, for the six-mass isolator, a phase shift of 6π would be expected. This is not readily apparent unless the origin is magnified as in figure 3.20 in which it can be seen that the Nyquist plot does indeed spiral around the origin the requisite number of times.

It is interesting to assess the effects of adding damping to any system. Choosing a six mass system where each mass is 10 kg, and with stiffness values of 10,000, the eigenvalues for zero damping are

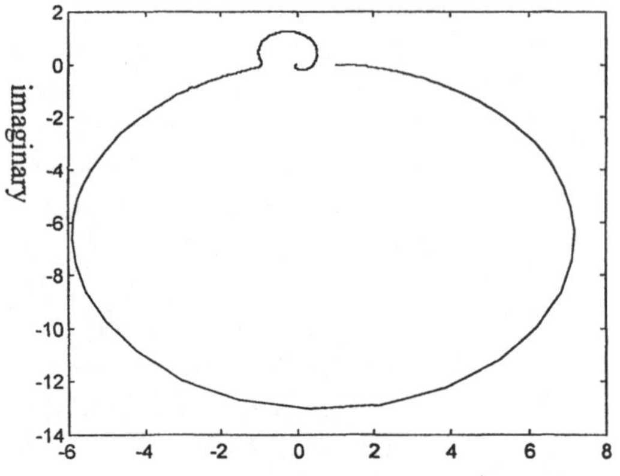

Figure 3.19: Polar plot of the frequency response of figure 3.18

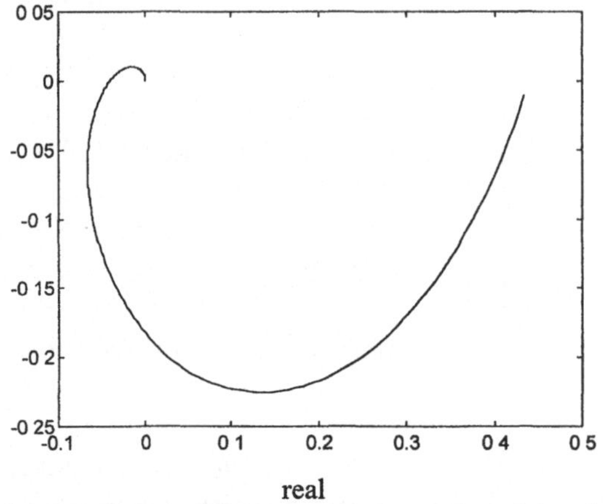

Figure 3.20: Amplified portion around the origin of the polar plot of figure 3.19

4.450, 8.678, 12.470, 15.637, 18.019, and 19.499 (rad s^{-1})

Starting at zero damping the eigenvalues all lay on the imaginary axis. Increasing the damping results in the individual eigenvalues having increasingly negative real parts with the imaginary component reducing, see figure 3.21. Eventually the eigenvalue will become purely real at which point the system is critically damped. Interestingly, each characteristic root traces out a locus that is similar in form to a single degree of freedom system. For obvious reasons, this plot of often referred to as a *roots locus*.

3.11 Vibrations and natural frequencies of continuous systems

In this section, equations for computing the natural frequencies of continuous systems such as wires, beams and plates that are common to many flexures will be presented. Invariably, Lagrange's equation and Hamilton's principle are used to produce 'exact' solutions to these problems. Rayleigh's method, although only approximate, produces reasonable estimates and often results in an equation that can be computed on the 'back of an envelope'. In all of the following, it will be assumed that distortions and, therefore stresses, are small. As a consequence, all will behave in the manner of a linear system and therefore the principle of superposition applies. Almost all of the following can be found in Rayleigh's treatise, in most cases in a more compact and complete form.

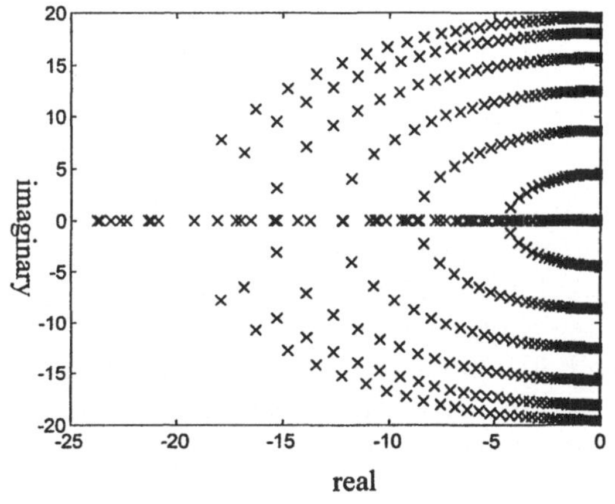

Figure 3.21: Roots locus

3.11.1 Strings

Strings are probably the simplest vibratory systems to analyze. In most texts, this is discussed with reference to plucking of the string and, in some cases, analysis is extended to include the response to forced vibrations. Such discussions are not relevant to this text and the curious reader is referred to the excellent discussion given by Rayleigh.. Imagine a string subject to a tensile force of magnitude T_o. In the absence of gravity forces, the string, when stationary will be straight. If it is then set into vibratory motion by a force applied perpendicular to its axis, it will attain additional strain energy due to the extension of the string in this deformed state. Assuming that the tension does not change significantly the strain energy will be equal to the work done. Choosing an element of length dx, the work done on the element as it increases in length to ds is

$$WD = T_o(dx - ds) \tag{3.191}$$

For small slopes this can be reduced to

$$\begin{aligned} WD &= T_o\left(dx - \sqrt{(dx)^2 + (dy)^2} \right) \\ &\approx \frac{T_o}{2}\left(\frac{dy}{dx}\right)^2 dx \end{aligned} \tag{3.192}$$

The strain energy throughout the length of the wire can be obtained by integration. Applying the variational operator to this integral gives

$$\delta V = [y]_0^l - \int_0^l \frac{\partial^2 y}{\partial x^2}\delta y dx \tag{3.193}$$

For a string of mass per unit length ρ', the kinetic energy of the string, T, is given by

$$T = \frac{\rho'}{2}\int_0^l \dot{y}^2 dx \tag{3.194}$$

From Hamilton's principle defined by equations (3.28) and (3.35), the equation of motion to be satisfied at all points along the string is given by

$$\frac{\partial^2 y}{\partial t^2} = \frac{T_o}{\rho'}\frac{\partial^2 y}{\partial x^2} = c^2 \frac{\partial^2 y}{\partial x^2} \tag{3.195}$$

Equation (3.195) is the one-dimensional wave equation with c being the characteristic wave velocity. There is, of course, another independent wave equation governing the vibration of the string in the z axis direction. For a uniform string, this will be identical, with identical solutions. However, it should be emphasized that the motion of a point on a string undergoing oscillations in both directions will trace out an ellipse perpendicular to its axis. The general solution to (3.183) is

$$y(t) = f(x - ct) + g(x + ct) \tag{3.196}$$

The above solution represents two arbitrary shapes travelling in opposite directions with a velocity c. Shapes of the waves are determined by the arbitrary functions f and g. For example, if $z = x - ct$ then any function of z will satisfy equation (3.195). Examples that the reader might try could be z^n, $e^z \cos z$, $\log z$, etc. If the two waves are of identical shape and amplitude, the subsequent superposition of the two will result in a *standing* wave of the same shape.

A solution to equation (3.195) of the form (3.196) is given by

$$y(t) = C\cos(m(x - ct)) + D\sin(m(x + ct)) \tag{3.197}$$

This can be rearranged to yield

$$y(t) = (A\sin(mct) + B\cos(mct))(E\cos(mx) + F\sin(mx)) \tag{3.198}$$

For a string held fast at each end, it is clear that $E = 0$.

Assuming a boundary condition of zero displacement at each end at all times it is relatively easy to derive the condition

$$\sin ml = 0 \tag{3.199}$$

From which

$$m = \frac{n\pi}{l} \qquad n = 1, 2.. \infty \tag{3.200}$$

The characteristic frequencies of the string can be easily deduced as

$$\omega_n = \frac{n\pi}{l}\sqrt{\frac{T_o}{\rho'}} \qquad n = 1, 2.. \infty \tag{3.201}$$

Clearly, there are an infinite number of solutions for the free vibration of the string. Equation (3.197) can be rewritten with separate time and displacement terms.

$$y(t) = \sum_{n=1}^{\infty} \sin\left(\frac{n\pi x}{l}\right)\left(A_n \cos\left(\frac{n\pi ct}{l}\right) + B_n \sin\left(\frac{n\pi ct}{l}\right)\right) \tag{3.202}$$

Because the shape of the deflection curve is independent of time, as expected, this represents a standing wave. Equation (3.202) constitutes a Fourier series for an odd function that is periodic in l. Consequently, for *any* initial displacement of the string, the constant coefficients may be determined and subsequent motion of the string over time computed. Because the Fourier series can be used to describe arbitrary distortions of the string, equation (3.202) corresponds to the most general solution of this problem. Following Fourier's original approach, the coefficients can be determined based on the initial displacement and velocity along the string at time $t = 0$ given by

$$\begin{aligned} y_0 &= \sum_{n=1}^{\infty} A_n \sin\left(\frac{n\pi x}{l}\right) \\ \dot{y}_0 &= \frac{\pi c}{l}\sum_{n=1}^{\infty} nB_n \sin\left(\frac{n\pi x}{l}\right) \end{aligned} \tag{3.203}$$

Exploiting the orthogonal properties of equation's (3.203), the coefficients can be determined by multiplying through by $\sin(n\pi x/l)$ and integrating over the length of the string to give

$$\begin{aligned} \int_0^l y_0 \sin\left(\frac{m\pi x}{l}\right)dx &= \int_0^l \sum_{n=1}^{\infty} A_n \sin\left(\frac{n\pi x}{l}\right)\sin\left(\frac{m\pi x}{l}\right)dx \\ &= \frac{lA_n}{2} \\ \int_0^l \dot{y}_0 \sin\left(\frac{m\pi x}{l}\right)dx &= \frac{\pi c}{l}\sum_{n=1}^{\infty} nB_n \sin\left(\frac{n\pi x}{l}\right)\sin\left(\frac{m\pi x}{l}\right)dx \\ &= \frac{\pi cn}{2}B_n \end{aligned} \tag{3.204}$$

From which

$$A_n = \int_0^l y_0 \sin\left(\frac{n\pi x}{l}\right) dx$$
$$B_n = \frac{2}{\pi c n} \int_0^l \dot{y}_0 \sin\left(\frac{n\pi x}{l}\right) dx \tag{3.205}$$

For example, for a string plucked a distance γ at a position measuring b from one end, it is can be shown that

$$B_n = 0$$
$$A_n = \frac{2\gamma l^2}{\pi^2 s^2 b(l-b)} \sin\left(\frac{n\pi b}{l}\right) \tag{3.206}$$

Note that both coefficients can be zero if the point at which the string is plucked corresponds to a node. Hence, if the string is plucked at a node, that particular mode will not be excited. For example, a guitar string plucked in the middle will produce a sound devoid of all even harmonics, a less pleasing sound.

The above analysis can be applied to all systems governed by a wave equation.

3.11.2 Longitudinal vibrations of a rod

Instead of the Hamiltonian approach of previous sections, use of Lagrange's equation can sometimes provide insight into the nature of vibrations of continuous systems. In the case of longitudinal vibrations of a homogeneous, prismatic beam, it is possible to split the beam into longitudinal elements of length dx. Each element will have a stiffness and mass given by

$$k = \frac{EA}{dx}$$
$$m = \rho A dx \tag{3.207}$$

where A is the cross sectional area of the rod [m^2], ρ is the density of the material [kg m^{-3}], and x is a distance measured along the bar. During longitudinal vibration, each element will undergo a displacement u measured relative to an inertial frame. For an arbitrary motion, the potential and kinetic energy in the bar is given by

$$V = \frac{1}{2}\sum_{i=1}^{\infty} k(u_i - u_{i-1})^2$$
$$T = \frac{1}{2}\sum_{i=1}^{\infty} \rho A \dot{u}_i^2 dx \tag{3.208}$$

Substituting equation's (3.208) into Lagrange's equation (noting that there are two terms involving u_i in the expansion for the potential energy), the equation governing free motion of the bar is

$$\begin{aligned} &\rho A\ddot{u}dx + k(2u_i - u_{i+1} - u_{i-1}) = 0 \\ &\rho A\ddot{u}dx + \frac{EA}{dx}(2u_i - u_{i+1} - u_{i-1}) = \\ &\ddot{u} + \frac{E}{\rho}\frac{(2u_i - u_{i+1} - u_{i-1})}{(dx)^2} = 0 \end{aligned} \tag{3.209}$$

In the limit as dx tends to zero (3.209) becomes the familiar wave equation

$$\ddot{u} = \frac{E}{\rho}\frac{\partial^2 u}{\partial x^2} = c^2\frac{\partial^2 u}{\partial x^2} \tag{3.210}$$

Here $c = (E/\rho)^{1/2}$ is the velocity of sound along the thin bar.

Assuming that the bar is undergoing free oscillation, in the absence of damping, it will assume a continuous vibration of constant shape $f(x)$ varying harmonically with time. With this assumption, it is possible to immediately write the solution

$$u = C_1 f(x)\cos(\omega_n t + \varphi) \tag{3.211}$$

Substituting (3.211) into (3.210) and canceling $\cos(\omega_n t)$ on both sides gives

$$\frac{d^2 f(x)}{dx^2} + \omega_n^2 \frac{\rho}{E} f(x) = 0 \tag{3.212}$$

For which the solution for $f(x)$ is

$$\begin{aligned} &f(x) = C_2 \sin\left(\omega_n\sqrt{\frac{\rho}{E}}x + \beta\right) \\ &\therefore \\ &u = R\sin\left(\omega_n\sqrt{\frac{\rho}{E}}x + \beta\right)\cos(\omega_n t + \varphi) \end{aligned} \tag{3.213}$$

To determine the natural frequencies corresponding to longitudinal vibrations of a bar, it is necessary that the constants in (3.213) be chosen to match the boundary conditions. Three examples that apply to some flexure designs are given below.

3.11.2.1 Longitudinal vibrations of a clamped-free rod

The boundary conditions require that, at all times, there is zero displacement at the fixed end while the free end will be strain free i.e.

$$u = 0 @ \dot{x} = 0$$

$$\frac{\partial u}{\partial x} = 0 @ x = l$$

From the first of these conditions $\beta = 0$, while from the second we have

$$\cos\left(\omega_n \sqrt{\frac{\rho}{E}} l\right) = 0$$

$$\therefore \qquad n = 1, 2 \,..\, \infty \qquad (3.214)$$

$$\omega_n = \frac{2n-1}{2l}\pi\sqrt{\frac{E}{\rho}}$$

Note that the period of the fundamental frequency corresponds to the time required for a pulse to travel the bar four times.

3.11.2.2 Longitudinal vibrations of a free-free or fixed-fixed rod

For a bar in space, each end will be correspondingly free from strain giving the boundary conditions

$$\frac{\partial u}{\partial x} = 0 @ x = l, x = 0 \qquad (3.215)$$

Substituting these into the second of (3.213) gives $\beta = \pi/2$ and also

$$\sin\left(\omega_n \sqrt{\frac{\rho}{E}} l\right) = 0 \qquad (3.216)$$

From this the natural frequencies of the bar are

$$\omega_n = \frac{n\pi}{l}\sqrt{\frac{E}{\rho}} \qquad n = 0, 1, 2 \,..\, \infty \qquad (3.217)$$

A general solution for the free vibration of the bar is

$$u(t) = \sum_{n=0}^{\infty} \cos\left(\frac{n\pi x}{l}\right)\left(A_n \cos\left(\frac{n\pi ct}{l}\right) + B_n \sin\left(\frac{n\pi ct}{l}\right)\right) \qquad (3.218)$$

Comparing this with equation (3.202), it is apparent that Fourier decomposition can be used to deduce the magnitudes of the coefficients A and B. Note the change in summation limits in which a value of zero for n corresponds to the position of the bar.

When both ends of the bar are fixed, the boundary conditions require that the displacement u be zero at both ends. Substitution of these conditions yields the same equation (3.217) for the natural frequencies of the bar. Rayleigh (section 154), points out that the general equation for displacement is given by (with symbols changed for consistency with the present text)

$$u(t) = \varepsilon_0 \frac{x}{l} + \sum_{n=1}^{\infty} \sin\left(\frac{n\pi x}{l}\right)\left(A_n \cos\left(\frac{n\pi ct}{l}\right) + B_n \sin\left(\frac{n\pi ct}{l}\right)\right) \tag{3.219}$$

Where ε_0 is the residual strain in the rod, which is, after all, over-constrained.

3.11.2.3 Longitudinal vibrations of a clamped-free rod with a rigid mass at the free end

For a rigid mass, M, attached at the free end of the cantilever, the force due to the product of stress and area must provide the requisite acceleration to the mass. Mathematically

$$AE\left(\frac{\partial u}{\partial x}\right)_{x=l} = M\left(\frac{\partial^2 u}{\partial t^2}\right)_{x=l} \tag{3.220}$$

Substituting this into (3.200) gives

$$-AER\omega_n\sqrt{\frac{\rho}{E}}\cos\left(\omega_n\sqrt{\frac{\rho}{E}}l\right) = -MR\omega_n^2 \sin\left(\omega_n\sqrt{\frac{\rho}{E}}l\right) \tag{3.221}$$

Noting that ρAl is equal to the total mass of the rod m, equation (3.221) can be rewritten in the form

$$\begin{aligned}\frac{m}{M} &= \left(\omega_n\sqrt{\frac{\rho}{E}}l\right)\tan\left(\omega_n\sqrt{\frac{\rho}{E}}l\right) \\ &= C\tan C\end{aligned} \tag{3.222}$$

Clearly, for any given ratio of masses there will be an infinite number of solutions to equation (3.222). The first two solutions for a number of different mass ratios are tabulated overleaf

m/M	0.1	1	10	100	∞ (M = 0)
C_1	0.311	0.860	1.429	1.555	$\pi/2$
C_2	3.173	3.426	4.306	4.666	$3\pi/2$
$\left(\sqrt{M/m+1/3}\right)^{-1}$	0.311	0.866	1.519	1.707	1.732
$\pi/\left(2\sqrt{1+M/m}\right)$	0.473	1.111	1.498	1.563	1.5707

Assuming that, during vibration at the fundamental frequency, the rod extends linearly to a distance δ at the free end, it is possible to derive equations for the kinetic and potential energy

$$\begin{aligned} T &= \frac{1}{2}M\omega_n^2\delta^2 + \frac{\rho A\omega_n^2\delta^2}{2}\int_0^l\left(\frac{x}{l}\right)^2 dx \\ &= \frac{\omega_n^2\delta^2}{2}(M+m/3) \\ V &= \frac{EA}{2l}\delta^2 = \frac{k\delta^2}{2} \end{aligned} \tag{3.223}$$

Using Rayleigh's method these two can be equated to yield a approximation for the fundamental mode natural frequency given by

$$\omega_1 = \sqrt{\frac{k}{M+m/3}} \tag{3.224}$$

Using this simplified equation a slightly high frequency is predicted as can be seen by comparing rows 2 and 4 of the above table. For appreciable masses attached to the bar, the difference is, for most purposes, negligible.

For small masses attached at any point x along the bar, it is reasonable to assume that the mode shapes will remain unaltered. Under these conditions, the kinetic energy for any mode is given by, Rayleigh, 1894,

$$\begin{aligned} T &= \frac{1}{2}\int_0^l \rho A\sin^2\left(\frac{(2n-1)\pi x}{4l}\right)dx + \frac{1}{2}M\sin^2\left(\frac{(2n-1)\pi x}{4l}\right) \\ &= \frac{m}{4}\left(1+\frac{2M}{m}\sin^2\left(\frac{(2n-1)\pi x}{4l}\right)\right) \end{aligned} \tag{3.225}$$

Consequently, it can be seen that each natural frequency will reduce in proportion

$$1:\left(1+\frac{M}{m}\sin^2\left(\frac{(2n-1)\pi x}{4l}\right)\right)^{-1/2} \tag{3.226}$$

For a mass attached to the free end this becomes

$$1:\left(1+\frac{M}{m}\right)^{-1/2} \tag{3.228}$$

Compare rows 5 and 2 of the above table to assess the relative error of this assumption on the first mode natural frequency as a function of the mass at the free end.

3.11.3 Lateral vibration of a bar

The equation governing motion for lateral vibrations of a bar has already been derived in section 3.3 using Hamilton's principle and is reproduced here for convenience

$$EI\frac{\partial^4 y}{\partial x^4} = -\rho A\frac{\partial^2 y}{\partial t^2} \tag{3.229}$$

Again, assuming that the vibration shapes are independent of time, it is possible to write the solution to (3.229) in the form

$$y(t) = f(x)\sin(\omega_n t + \varphi) \tag{3.230}$$

Substituting this into (3.229) yields

$$\frac{d^4 f(x)}{dx^4} = \frac{m\omega_n^2}{lEI} f(x) = \alpha^4 f(x) \tag{3.231}$$

where m is the *total* mass of the bar and

$$\omega_n = (\alpha_n l)^2\sqrt{\frac{EI}{ml^3}} \tag{3.232}$$

Assuming a general solution of the form

$$f(x) = Ce^{\lambda t} \tag{3.234}$$

The roots of equation (3.229) are

$$\lambda^4 = \alpha^4 \tag{3.235}$$

Consequently, the four solutions to (3.235) are

$$\lambda = \alpha\sqrt[4]{1+i0} = \pm\alpha, \pm i\alpha \tag{3.236}$$

Substituting (3.236) into (3.234) and rearranging, the solution to (3.229) can be expressed in the form

$$f(x) = C_1 \sin \alpha x + C_2 \cos \alpha x + C_3 \sinh \alpha x + C_4 \cosh \alpha x \tag{3.237}$$

Clearly, the constants C must be determined from the boundary conditions.

In general, the case of a beam being either free or hinged or clamped at the ends will be considered. In the first of the following examples, all steps in a general solution will be indicated. Thereafter, for brevity, only major components of a solution will be presented. The readers are once again referred to Rayleigh to see how it is rigorously pursued and Timoshenko and Young, 1955, for detailed discussions.

3.11.3.1 Hinged-hinged beam.

For the hinged beam, at all times the deflection and bending moment must be zero at each end i.e.

$$f(x) = 0 \;\&\; \frac{\partial f(x)}{\partial x} @ x = 0, x = l \tag{3.238}$$

For $x = 0$

$$\begin{aligned} f(x) &= 0 = C_2 + C_4 \\ \frac{\partial f(x)}{\partial x} &= 0 = -C_2 + C_4 \end{aligned} \tag{3.239}$$

This is only satisfied if both constants are equal to zero. For $x = l$

$$\begin{aligned} f(x) &= 0 = C_1 \sin \alpha l + C_3 \sinh \alpha l \\ \frac{\partial f(x)}{\partial x} &= 0 = -C_1 \sin \alpha l + C_3 \sinh \alpha l \end{aligned} \tag{3.240}$$

Because $\sinh \alpha l$ cannot be zero, then C_3 must be zero. The only non-trivial solution that satisfies *all* boundary conditions is

$$\sin \alpha l = 0 \quad \text{or} \quad \alpha l = n\pi \qquad n = 0, 1 \ldots \infty \tag{3.241}$$

Consequently, the natural frequencies of the hinged beam given by (3.232) are

$$\omega_n = (n\pi)^2 \sqrt{\frac{EI}{ml^3}} \tag{3.242}$$

From the above, it can be deduced that the mode shapes for this bar, being successive integral sine waves are the same as those previously derived for a string. However, the frequencies, or tones, for the bar increase by the *square* of n as opposed to linearly in the case of a string.

A general solution for the bar is

$$y(t) = \sum_{n=1}^{\infty} B_n \sin\left(\frac{n\pi x}{l}\right)\cos\left((n\pi)^2 \sqrt{\frac{EI}{ml^3}}.t\right) \tag{3.243}$$

Again, exploitation of the orthogonality of this equation will enable the computation of the coefficients for B_n based on the initial distortion of the bar.

3.11.3.2 The free-free or clamped-clamped bar

For a bar 'floating in space', the conditions there will be an absence of both bending moments and shear forces at each end. Mathematically

$$\frac{\partial^2 f(x)}{\partial x^2} = \frac{\partial^3 f(x)}{\partial x^3} = 0 @ x = 0, x = l \tag{3.244}$$

The second and third differential equations for $f(x)$ are

$$\frac{\partial^2 f(x)}{\partial x^2} = -C_1\alpha^2 \sin\alpha x - C_2\alpha^2 \cos\alpha x + C_3\alpha^2 \sinh\alpha x + C_4\alpha^2 \cosh\alpha x$$

$$\frac{\partial^2 f(x)}{\partial x^2} = -C_1\alpha^3 \cos\alpha x + C_2\alpha^3 \cos\alpha x + C_3\alpha^3 \sinh\alpha x + C_4\alpha^3 \cosh\alpha x$$

From the first of (3.244) we have

$$-C_2 + C_4 = 0$$
$$-C_1 + C_3 = 0$$

The second condition gives

$$\begin{aligned} &C_1(\sinh\alpha l - \sin\alpha l) + C_2(\cosh\alpha l - \cos\alpha l) = 0 \\ &C_1(\cosh\alpha l - \cos\alpha l) + C_2(\sin\alpha l + \sinh\alpha l) = 0 \end{aligned} \tag{3.245}$$

Having two equations and two unknowns, it is possible to solve for the two constants in (3.245) after which these can be substituted back into the general solution for $f(x)$.

From equations (3.245) it is possible to express the ratio of the constants by two equations

$$-\frac{C_2}{C_1} = \frac{\sinh \alpha l - \sin \alpha l}{\cosh \alpha l - \cos \alpha l} = \frac{\cosh \alpha l - \cos \alpha l}{\sinh \alpha l + \sin \alpha l} \tag{3.246}$$

Rearranging (3.246), compatibility requires that

$$(\cosh \alpha l - \cos \alpha l)^2 = \sinh^2 \alpha l - \sin^2 \alpha l$$

$$\text{or} \tag{3.247}$$

$$\cos \alpha l \cosh \alpha l = 1$$

Equation (3.246) can be solved to provide an infinite number of roots, the first two of which are

$$\begin{aligned} \alpha_1 l &= 4.7300408 \\ \alpha_2 l &= 7.8532046 \end{aligned} \tag{3.248}$$

Values for each natural frequency can then be calculated from equation (3.232)

Based on the above analysis, the general solution for the free-free beam is given by

$$y(t) = \sum_{n=1}^{\infty} \begin{pmatrix} (\sin \alpha_n l - \sinh \alpha_n l)\{\cos \alpha_n x + \cosh \alpha_n x\} + \\ ..(\cosh \alpha_n l - \cos \alpha_n l)\{\sin \alpha_n x + \sinh \alpha_n x\} \end{pmatrix} A_n \cos(\omega_n t) \tag{3.249}$$

In the case of a clamped bar it is necessary that the deflection and slope are zero at both ends. Substitution of these boundary conditions results in the same compatibility equation (3.247) as that necessary for a free-free bar. As a rather surprising consequence (to me anyway) the natural frequencies in each case are of the same value. However, it is obvious that the mode shape corresponding to each frequency will differ. Again, because (3.249) corresponds to a series of orthogonal functions, it is possible, although rather tedious, to determine the constant coefficients of (3.249) using the techniques of Fourier series, see for example Timoshenko and Young, 1955.

Case 4: The cantilever or clamped-free bar

For a clamped-free bar the compatibility equation is given by

$$\cos \alpha l \cosh \alpha l + 1 = 0 \tag{3.250}$$

To determine the natural frequencies for any of the above configurations it is necessary to calculate the coefficients of $\alpha_i l$. For convenience these are tabulated below.

Tabulated values for the coefficients $\alpha_i l$

i	Clamped-clamped Free-free	Clamped-free	Hinge-hinge
1	4.7300408	1.875104	3.141593
2	7.8532046	4.694091	6.283185
3	10.9956078	7.854757	9.424778
4	14.1371655	10.995541	12.566371
5	17.2787596	14.137168	15.707963

Finally, for want of a better place, the effective mass of a beam, when it vibrates with an 's' shaped deflection curve of the form given by equation (4.76) is assessed using Rayleigh's method.

Case 5: Effective mass of a bar vibrating with an 's' shaped deflection curve

Many flexures consist of two leaf-type spring connected at each end by two rigid platforms. Application of an appropriate force will result in the linear motion between the two platforms. Under the correct conditions, and for small distortions, the platforms will move parallel to each other in a near rectilinear path, see chapter 4. Each of the leaf springs will undergo an 's' shaped deflection with zero slope at each end. From the beam bending equation (see also section 4.1.5), the deflection perpendicular to the axis of a beam is given by

$$y = \delta\left(\frac{2x^3}{l^3} - \frac{3x^2}{l^2}\right) \tag{3.251}$$

Assuming that this represents the fundamental frequency mode shape, the kinetic energy of an element along the beam is

$$dT = \frac{1}{2}dm\dot{y}^2 = \frac{1}{2}\rho A\omega_n^2\partial^2\left(\frac{2x^3}{l^3} - \frac{3x^2}{l^2}\right)^2 \tag{3.252}$$

Integrating equation (3.252), the total kinetic energy for a single leaf spring is

$$T = \frac{1}{2}\rho A\omega_n^2\partial^2\int_0^l\left(\frac{4x^6}{l^6} - \frac{12x^5}{l^5} + \frac{9x^4}{l^4}\right)dx$$
$$= \frac{1}{2}\rho A\omega_n^2\partial^2 l\left[\frac{4}{7} - 2 + \frac{9}{5}\right] = \frac{1}{2}\rho A\omega_n^2\partial^2 l\left(\frac{13}{35}\right) \quad (3.253)$$
$$= \frac{1}{2}m\omega_n^2\partial^2\left(\frac{13}{35}\right)$$

Equating maximum potential and kinetic energies for this system, the first mode natural frequency for s springs can be approximated by

$$\omega_n^2 = \frac{k}{M + \sum_s 0.37m} \quad (3.254)$$

where k is the stiffness of the complete platform.

3.11.4 Lateral vibration of bars with a rigid mass attached

Although extension of the above analyses can be used to determine equations for the lateral vibrations of bars having rigid masses attached, these are rather cumbersome, see the comprehensive paper of Maltbaek, 1961. In contrast, for these examples Rayleigh's method produces simple expressions for which the errors usually deviate less than 1% from the 'true' value. These and many more examples can be found in Rayleigh, 1877, or Timoshenko and Young, 1955.

3.11.4.1 Cantilever beam with a rigid mass attached at the free end

As a reasonable guess, the deflection curve for a massless cantilever beam with a concentrated mass, M, on the free end will satisfy the appropriate boundary conditions. From simple bending theory, the deflection along the beam can be written in the form

$$y(x) = \frac{\delta}{2}\left(\frac{3x^2}{l^2} - \frac{x^3}{l^3}\right) \quad (3.255)$$

The maximum kinetic energy of an element along the beam is given by

$$dT = \frac{1}{2}\rho A\dot{y}^2 dx = \rho A\frac{\delta^2}{8}\left(\frac{3x^2}{l^2} - \frac{x^3}{l^3}\right)^2 dx$$
$$= \rho A\frac{\delta^2}{8}\left(\frac{9x^4}{l^4} - \frac{6x^5}{l^5} + \frac{x^6}{l^6}\right)dx \quad (3.256)$$

Integrating over the length of the beam yields

$$T=\int_0^l \rho A\omega_n^2\frac{\delta^2}{8}\left(\frac{9x^4}{l^4}-\frac{6x^5}{l^5}+\frac{x^6}{l^6}\right)dx$$

$$=\rho A\omega_n^2\frac{\delta^2}{8}\left[\frac{9x^5}{5l^4}-\frac{x^6}{l^5}+\frac{x^7}{7l^6}\right]_0^l$$

$$=\frac{1}{2}\rho Al\left(\frac{33}{140}\right)\omega_n^2\delta^2=\frac{1}{2}m\left(\frac{33}{140}\right)\omega_n^2\delta^2$$

Consequently, the total kinetic energy for the beam and attached mass is

$$T=\frac{1}{2}\left(M+\left(\frac{33}{140}\right)m\right)(\omega_n\delta)^2 \tag{3.257}$$

The maximum potential energy is readily calculated

$$V=\frac{1}{2}k\delta^2=\frac{1}{2}\frac{3EI}{l^3}\delta^2 \tag{3.258}$$

Equating equations (3.258) and (3.257) an estimate for the fundamental mode natural frequency is given by the equation

$$\omega_n^2=\frac{k}{M+0.2357m} \tag{3.259}$$

Interestingly, it can be shown, Maltbaek, 1961, that the error between this approximate equation and the 'true' value is never greater than 1%. For almost all engineering applications this is more than adequate.

3.11.4.2 Hinged beam with a central mass, M

Choosing the static deflection curve for a massless beam with a central force of value Mg, the kinetic energy of an element dx at a distance x from one end is

$$dT=\frac{1}{2}\rho A\omega_n^2\delta^2\left(\frac{3x}{l}-\frac{4x^3}{l^3}\right)^2dx \tag{3.260}$$

Integrating over the length of the beam yields

$$\begin{aligned}T&=\frac{1}{2}\rho A\omega_n^2\delta^2\int_0^l\left(\frac{9x^2}{l^2}-\frac{24x^4}{l^4}+\frac{16x^6}{l^6}\right)dx\\&=\frac{1}{2}\rho A\omega_n^2\delta l^2\left(\frac{17}{35}\right)\end{aligned} \tag{3.261}$$

Following the previous analysis, an estimate of the fundamental mode natural frequency is given by

$$\omega_n^2 = \frac{k}{M + 0.4857m} \tag{3.262}$$

Again, irrespective of the relative values of these masses, (3.262) is sufficiently close to the 'true' value for most engineering applications.

3.11.5 Vibration of plates

As might be expected, when departing from two to three-dimensional problems, the mathematical difficulties are correspondingly enhanced. In fact, it appears that the solution for vibrations of a circular plate was a harsh testing ground for some of the best mathematicians of the early 1800's, the full solution finally yielding to Kirchoff in 1850. Of course, Rayleigh's succinct discussion of the vibration of membranes and plates remains one of the more complete accounts of the theory to this day. Solutions to the vibrations of a rectangular plate remain intransigent and will be discussed only in brief.

3.11.5.1 Vibrations of a circular plate

From the theory of plates, the potential of an element of a distorted flat plate can be expressed in terms of the two principle curvatures, r

$$dV = \frac{D}{2}\left\{\left(\frac{1}{r_1} + \frac{1}{r_2}\right)^2 - \frac{2(1-\upsilon)}{r_1 r_2}\right\}dxdy \tag{3.263}$$

D is the flexural rigidity given by

$$D = \frac{Et^3}{12(1-\upsilon^2)} \tag{3.264}$$

and t is the total thickness of the plate.

For small displacements, u, in the z-axis direction, the total potential energy in the thin plate can be obtained by integrating equation (3.263) over the surface giving

$$V = \frac{D}{2}\iint\left\{(\nabla^2 u)^2 - 2(1-\upsilon)\left(\frac{\partial^2 u}{\partial x^2}\frac{\partial^2 u}{\partial y^2} - \left(\frac{\partial^2 u}{\partial x \partial y}\right)^2\right)\right\}dxdy \tag{3.265}$$

where ∇ represents the two-dimensional Laplacian

$$\nabla^2 = \frac{\partial^2}{\partial x^2} + \frac{\partial^2}{\partial y^2} \tag{3.266}$$

The total kinetic energy of the plate is readily obtained from the integral

$$T = \iint_s \rho t \dot{u}^2 dxdy \tag{3.267}$$

Application of Hamilton's principle (eventually) yields the equation governing free motion at all points on the plate

$$\frac{D}{\rho t}\nabla^4 u - Z + \ddot{u} = 0 \tag{3.268}$$

where Z represents body forces on the plate. Assuming that these are gravitational and, therefore, constant, this will not affect the vibration characteristics of the plate and will be dropped henceforth.

Again, for convenience, the constant in (3.268) can be reassigned as

$$c^4 = \frac{D}{\rho t} \tag{3.269}$$

Assuming a solution to (3.268) of the form

$$u(r,\theta,t) = f(r,\theta)e^{i\omega_{mn}t} \tag{3.270}$$

Substituting (3.270) into (3.268) gives

$$\nabla^4 f = k_{mn}^4 f \tag{3.271}$$

where

$$k_{nm}^4 = \frac{\omega_{mn}^2}{c^4} \tag{3.272}$$

3.11.5.1.1 Free vibration of a circular plate clamped at the perimeter

To analyze the vibration response of a circular plate, it is convenient to express equation (3.268) in polar form (the r and θ in (3.270)) in which the Laplacian becomes

$$\nabla^2 = \frac{\partial^2}{\partial r^2} + \frac{1}{r}\frac{\partial}{\partial r} + \frac{1}{r^2}\frac{\partial^2}{\partial \theta^2} \tag{3.273}$$

Assuming a separable solution for the radial and tangential terms in (3.270) the solution for f will be of the form

$$f(r,\theta) = g(r)h(\theta) \tag{3.274}$$

Substituting (3.274) and (3.273) into (3.271) gives

$$\left(\frac{\partial^2}{\partial r^2} + \frac{1}{r}\frac{\partial}{\partial r} + \frac{1}{r^2}\frac{\partial^2}{\partial \theta^2} + k_{mn}^2\right)\left(\frac{\partial^2}{\partial r^2} + \frac{1}{r}\frac{\partial}{\partial r} + \frac{1}{r^2}\frac{\partial^2}{\partial \theta^2} - k_{mn}^2\right)g = -\frac{1}{h}\frac{\partial^2 h}{\partial \theta^2} \tag{3.275}$$

Because r and θ are independent of each other, equation (3.275) can only be satisfied if both are equal to a similar integer constant, n^2 say. From the right hand side of (3.275) we have

$$\frac{\partial^2 h}{\partial \theta^2} + n^2 h = 0 \tag{3.276}$$

for which the solution is readily obtained as

$$h = A\cos(n\theta - \varphi) \tag{3.277}$$

A and φ are constants that depend on initial conditions. For symmetry, it is necessary, and convenient, that n is an integer. From the left hand side of (3.275) the solution for the function g can be obtained from

$$\left(\frac{\partial^2}{\partial r^2} + \frac{1}{r}\frac{\partial}{\partial r} - \frac{n^2}{r^2} \pm k_{nm}^2\right)g = 0 \tag{3.278}$$

From which

$$\begin{aligned} g &= BJ_n(k_{mn}r) + CJ_n(ik_{mn}r) \\ &= BJ_n(k_{mn}r) + CI_n(k_{mn}r) \end{aligned} \tag{3.279}$$

J and I represent Bessel functions of order n while the constants B and C can be determined from the boundary conditions. In addition, there will be an infinite number of admissible values for k the determination of which provides the natural frequency for that mode.

From equations (3.270), (3.277) and (3.279) the complete solution for displacement of the plate is given by

$$u(t) = A_{nm}\cos(n\theta - \varphi_{nm})(J_n(k_{mn}r) + \psi_n I_n(k_{mn}r))\cos(\omega_{nm}t - \varepsilon_{nm}) \tag{3.280}$$

where A, ε, ψ and φ are constants.

The two integers, n and m correspond to the number of radial and circumferential nodal lines on the plate respectively. Based on the value of n there are an infinite number of discrete frequencies determined from the Bessel functions in the above. For flexure applications, the circular plate represents a diaphragm. In most cases the plate will be clamped at the perimeter for which the boundary conditions are

$$u\big|_{r=R} = \frac{\partial u}{\partial r}\bigg|_{r=R} = 0 \tag{3.281}$$

Because these conditions must be satisfied around the complete periphery of the plate over all time, dropping the subscripts on k, substitution of (3.281) into (3.280) gives

$$J_n(kR)\frac{dI_n(kR)}{dr} - I_n(kR)\frac{dJ_n(kR)}{dr} = 0$$

or (3.282)

$$\frac{J_n(kR)}{J_n'(kR)} - \frac{I_n(kR)}{I_n'(kR)} = 0$$

It can also be seen that the constant ψ is given by

$$\psi = -\frac{J_n(kR)}{I_n(kR)} \tag{3.283}$$

Equation (3.282) provides an infinite number of roots, $(kR)_{mn}$ corresponding to each value of n. From (3.269) and (3.272), the natural frequencies for each mode are given by

$$\omega_{nm} = \frac{(kR)^2_{mn}}{R^2}\sqrt{\frac{D}{\rho t}} \tag{3.284}$$

The first 16 roots of equation (3.282) are listed in the table below

Values of the roots $(kR)_{mn}$ corresponding to equation (3.272)

m/n	0	1	2	3
1	3.196	4.611	5.906	7.143
2	6.306	7.799	9.197	10.537
3	9.440	10.958	12.577	14.108
4	12.577	14.108	15.579	17.005

The complete solution for vibration of the disc is given by

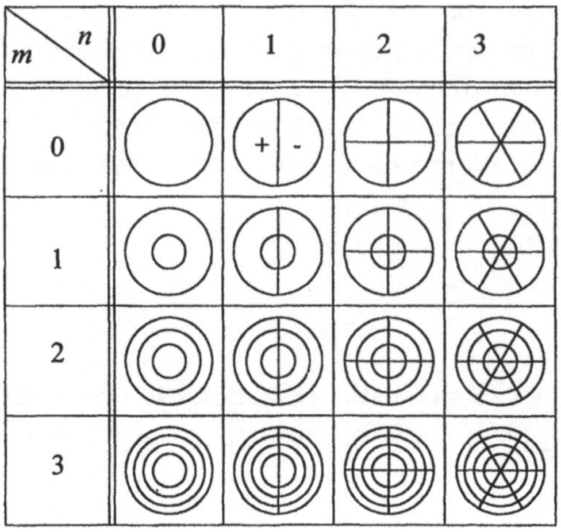

Figure 3.22: Nodal lines corresponding to different mode shapes for free vibration of a circular plate

$$u(t)=\sum_{n=0}^{\infty}\sum_{m=1}^{\infty}A_{mn}\cos(n\theta-\varphi)\left(J_n(k_{mn}r)+\psi_{mn}I_n(k_{mn}r)\right)\cos\left(\omega_{nm}t-\varepsilon_{mn}\right) \tag{3.285}$$

The constant coefficients A_{mn} are dependent on the initial deflection $u(0)$. Again, because (3.285) represents an orthogonal function, it is possible to determine values for each coefficient using the methods of Fourier series. Normalization coefficients for the series expansion to determine A_{mn}, were first recognized by Bessel. For a symmetric mode deformation both n and I are zero and the Fourier-Bessel series of equation (3.285) is, Hildebrand, 1976,

$$A_m=\frac{2}{R^2\left[J_1(k_{mn}R)\right]^2}\int_0^R ru(0)J_0(k_{mn}r)dr \tag{3.286}$$

To visualize each of these modes, figure 3.22 shows the nodal lines. Figures 23 show isometric views of the first four modes represented in figure 3.22.

3.11.5.2 Rayleigh's method applied to a circular plate with a central mass

Determination of the natural frequencies for different plate configurations involves solving complex differential equations. For many flexures, it is the fundamental frequency that limits performance. As a consequence, Rayleigh's method can be used to derive simple approximate formulae. In this section, the clamped plate with and without a central point mass is analyzed.

Case 1: Circular plate clamped at the perimeter

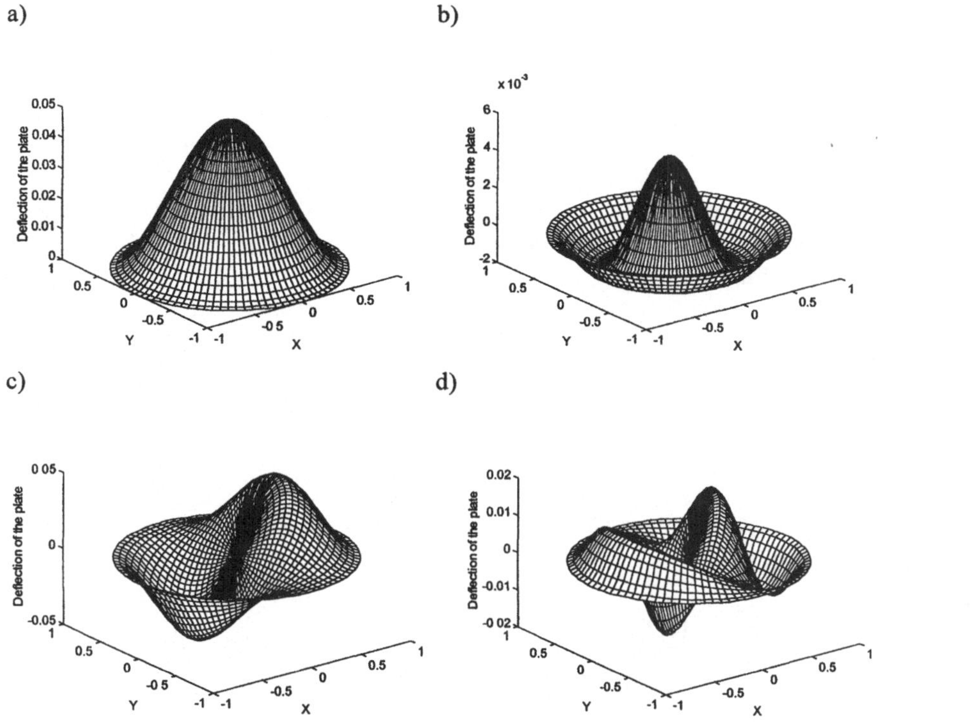

Figure 3.23: Mode shapes for a clamped circular plate a) $m=0$, $n=0$, b) $m=1$, $n=0$, c) $m=0$, $n=1$, d) $m=1$, $n=1$

As a reasonable guess for the mode shape, the static deflection due to a uniformly distributed load is

$$\begin{aligned} u &= \frac{q}{64Dt}\left(R^2 - r^2\right)^2 \\ &= \delta\left(R^2 - r^2\right)^2 \end{aligned} \tag{3.287}$$

where q is load per unit area and δ is a constant.

Converting equation (3.265) to polar coordinates, the potential energy of a circular plate with deflections symmetric about the center (i.e. contour of constant height being circles) is given by

$$V = \pi D \int_0^R \left\{ \left(\frac{\partial^2 u}{\partial r^2} + \frac{1}{r}\frac{\partial u}{\partial r} \right)^2 - 2(1-\upsilon)\frac{\partial^2 u}{\partial r^2}\frac{1}{r}\frac{\partial u}{\partial r} \right\} r\,dr \tag{3.288}$$

Substituting (3.287) into (3.288) gives

$$V = 2^5 \pi D \delta^2 \int_0^R \left(\left(R^4 r - 4R^2 r^3 + 5r^5 \right) + \upsilon \left(R^4 r - 4R^2 r^3 + 3r^5 \right) \right) dr$$
$$= \frac{\pi D 2^5 \delta^2 R^6}{3} \tag{3.289}$$

As would be expected, the term involving Poisson's ratio corresponding to energy of twisting vanishes. The total kinetic energy in the plate can be determined from the integral

$$T = \pi \rho t \omega_n^2 \delta^2 \int_0^R \left(R^2 - r^2 \right)^2 r dr$$
$$= \frac{\pi \rho t \omega_n^2 \delta^2 R^{10}}{10} \tag{3.290}$$

Equating (3.289) and (3.290), an estimate of the fundamental mode natural frequency is given by

$$\omega_n = \frac{8\sqrt{5}}{R^2 \sqrt{3}} \sqrt{\frac{D}{\rho t}} = \frac{10.328}{R^2} \sqrt{\frac{D}{\rho t}} \tag{3.291}$$

Compared to the 'true' value of 10.21, the approximate method is a little over 1% higher.

Case 2: A clamped circular plate with a central mass

Assuming that the dominant static forces on a plate are those due to a central point mass, *M*, the subsequent deflection curve can be represented by the equation

$$u = \frac{F}{8\pi D} \left(r^2 \ln\left(\frac{r}{R} \right) + \frac{\left(R^2 - r^2 \right)}{2} \right)$$
$$= \delta \left(r^2 \ln\left(\frac{r}{R} \right) + \frac{\left(R^2 - r^2 \right)}{2} \right) \tag{3.292}$$

Substituting this into (3.288) and ignoring the term involving Poisson's ratio, the integral can be rewritten in the form

$$V = \pi D 4 R^2 \delta^2 \int_{-\infty}^{0} \left(e \xi^2 e^{2\xi} + \xi^{2\xi} + e^{2\xi} \right) d\xi$$
$$= 2\pi D R^2 \delta^2 \tag{3.293}$$

where

$$\xi = \ln\left(\frac{r}{R}\right)$$

Alternatively, the potential can be deduced from

$$V = \frac{1}{2}ku(0)^2 = \frac{1}{2}\frac{16\pi D}{R^2}\frac{\delta^2 R^4}{4} = 2\pi D R^2 \delta^2$$

Similarly, the kinetic energy can be computed from the integral

$$\begin{aligned} T &= \pi\rho t\omega_n^2 \int_0^R u^2 r dr + \frac{M\omega_n^2 u(0)}{2} \\ &= \frac{7m\omega_n^2\delta^2 R^4}{2^4.3^3} + \frac{M\omega_n^2\delta^2 R^4}{8} \end{aligned} \tag{3.294}$$

Equating the maximum values for potential and kinetic energy, an estimate for the lowest mode natural frequency is given by

$$\begin{aligned} \omega_n &= \frac{4\sqrt{\pi}}{R}\sqrt{\frac{D}{M + \frac{7}{2.3^3}m}} = \sqrt{\frac{k}{M + \frac{7}{2.3^3}m}} \\ &= \sqrt{\frac{k}{M + 0.13m}} \end{aligned} \tag{3.295}$$

At low values for the central mass, M, the frequency equation becomes

$$\omega_n^2 = \frac{12}{R^2}\sqrt{\frac{6}{7}}\sqrt{\frac{D}{\rho t}} = \frac{11.1}{R^2}\sqrt{\frac{D}{\rho t}} \tag{3.296}$$

In view of the initial assumption that the mass of the plate is insignificant in comparison, the error is still less than 10%. In fact the boundary condition of zero slope at the center of the disc is also violated.

If the central mass is relative small with respect to that of the plate, it may be assumed that the shape is relatively unchanged from that assumed for the plate of case 1 above. Consequently, the potential energy given by (3.289) will remain unchanged while the kinetic energy becomes

$$\begin{aligned} T &= \frac{\pi\rho t R^2\omega_n^2\delta^2 R^8}{2.5} + \frac{1}{2}M\delta^2 R^8\omega_n^2 \\ &= \frac{m\omega_n^2\delta^2 R^8}{2.5} + \frac{1}{2}M\delta^2 R^8\omega_n^2 \end{aligned} \tag{3.297}$$

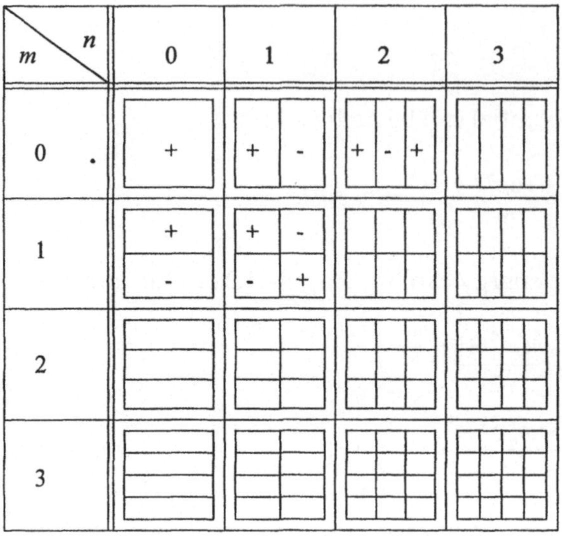

Figure 3.24: Nodal lines corresponding to different mode shapes for free vibration of a rectangular plate

Consequently, for small values of the central mass, *M*, the fundamental frequency is given by

$$\omega_n = \frac{\sqrt{5\pi}.8}{R\sqrt{3}}\sqrt{\frac{D}{m+5M}} \tag{3.298}$$

It can be readily verified that (3.297) become equal to (3.291) when the central mass becomes vanishingly small.

If the radius of the central mass cannot be neglected, an approximate formula can be obtained for the case where the mass of the disk is considered negligible. In this case, the stiffness, k, of the disk is, Den Hartog, 1952,

$$k \approx \frac{Et^3}{R^2\left(-\alpha^2/435+\alpha/22-1/18\right)} \tag{3.299}$$

where $\alpha = R/r$ and is accurate to within 10% for values ranging from 2 to 5. In this case, the fundamental mode natural frequency can be computed from

$$\omega_n = \sqrt{\frac{k}{M}} \tag{3.300}$$

3.11.6 Vibrations of a rectangular plate

Exact solutions for the free vibrations of rectangular plates require some mathematical rigor. Fortunately, there is one exception.

3.11.6.1 Free vibrations of a rectangular plate simply supported at the edges

By substituting back into equation (3.271), it can be readily verified that a solution for the vibration of a rectangular plate is given by

$$u(t) = \sum_{m=1}^{\infty}\sum_{n=1}^{\infty} A_{mn} \sin\left(\frac{m\pi x}{a}\right)\sin\left(\frac{n\pi y}{b}\right)\cos(\omega_{mn}t) \tag{3.301}$$

where a and b are the lengths of the sides and, yet again, the constants A can be obtained thorough Fourier decomposition of the initial conditions. Correspondingly, the natural frequencies are readily shown to be

$$\omega_{mn} = \pi^2\left(\frac{m^2}{a^2}+\frac{n^2}{b^2}\right)\sqrt{\frac{D}{\rho t}} \tag{3.302}$$

A representation of these various mode shapes is shown in figure 3.24. To visualize the schematic representations of figure 3.24, isometric views of the first four modes are shown in figure 3.25.

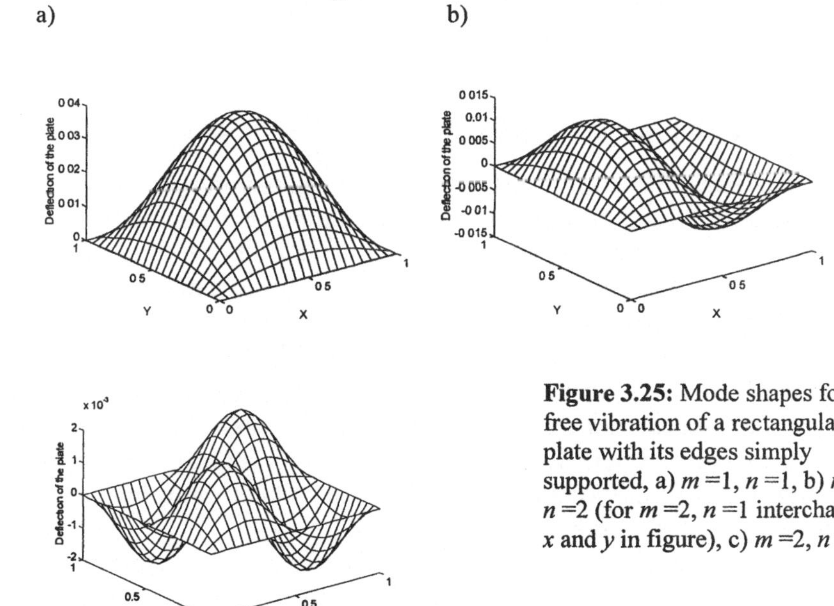

Figure 3.25: Mode shapes for free vibration of a rectangular plate with its edges simply supported, a) $m=1$, $n=1$, b) $m=1$, $n=2$ (for $m=2$, $n=1$ interchange x and y in figure), c) $m=2$, $n=2$

3.11.6.2 Fundamental frequency's of rectangular plates with other boundary conditions

Investigations of plates having other boundary condition are best approached through the successive refinement of Rayleigh's method commonly called the Ritz method. Using this technique, equations for predicting natural frequencies for many mode shapes and boundary conditions have been investigated, Blevins, 1985, Shockley, 1995. Considering equation (3.257)

$$\omega_{mn} = k^2 c^2 = \frac{k^2}{a^2}\sqrt{\frac{D}{\rho t}} \tag{3.303}$$

Values for k^2 corresponding to a variety of boundary conditions are tabulated below

Table of values for the coefficient k^2 in (3.303) for a square plate

Boundary conditions	k^2
Clamped at one edge free elsewhere (cantilevered plate)	3.494
Clamped all around	35.99
Clamped at two edges free elsewhere	6.958

3.12 Case study I: A simple, two degree of freedom flexure mechanism

This case study has been chosen to illustrate some of the techniques introduced in this chapter, most of which will be repeated throughout the rest of the book. Figure 3.26 shows a notch type planar flexure that consists of two, parallel, simple four-bar links. A simple planar mobility analysis indicates a total of 2 freedoms (=3(7-1)-2(8)) which are assigned the coordinates q_1 & q_2. It is readily apparent that, for small motions, these correspond to linear displacements of the two platforms. It is desired that, through suitable selection of the notch hinge thickness, *t*, the fundamental mode undamped natural frequency be greater than 30 Hz. Identical viscous dampers are to be applied to each platform as shown. Determine the damping force necessary to produce a critical damping ratio of value 0.2 at the fundamental mode natural frequency.

As a first step in any dynamic analysis, it is necessary to determine the equations governing motion for the system. To do so it is necessary to compute the kinetic and potential energies and the Rayleigh dissipation function in terms of the independent coordinates of the mechanism. Using local coordinates for each rigid body, the kinetic energy is given by

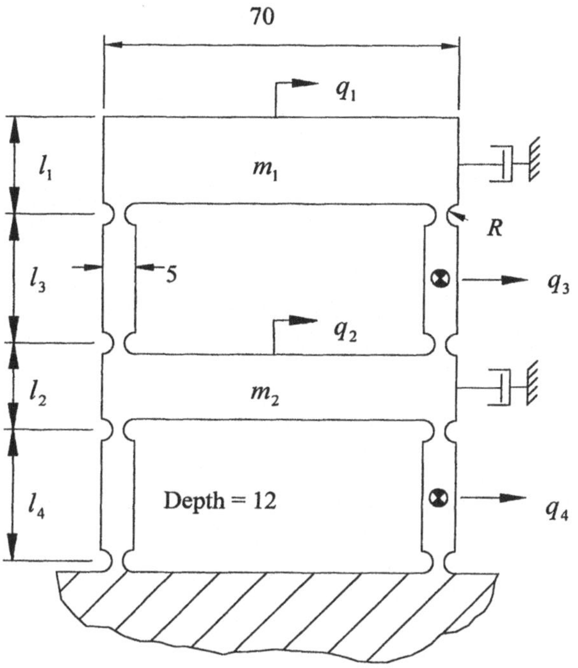

Figure 3.26: A stacked, simple linear spring providing two, parallel linear freedoms $l_1 = 22.5$, $l_2 = 25$, $l_3 = l_4 = 45$, (all dimensions in mm)

$$T = \frac{1}{2}\left[m_1\dot{q}_1^2 + m_2\dot{q}_2^2 + 2I_3\dot{\theta}_3^2 + 2m_3\left(\frac{\dot{q}_1 + \dot{q}_2}{2}\right)^2 + 2I_4\dot{\theta}_4^2 + 2m_4\left(\frac{\dot{q}_2}{2}\right)^2 \right]$$

However, the two angular coordinates are not independent and, noting that $l_3 = l_4$, can be expressed in terms of the linear coordinates by the equations

$$\theta_3 = \frac{q_1 - q_2}{l_3}$$

$$\theta_4 = \frac{q_2}{l_4} = \frac{q_2}{l_3}$$

Also, assuming that the support legs can be considered as long thin rods, the values for their second moment of mass about the centriods can be computed from the equations

$$I = \frac{ml^2}{12}$$

Ignoring the mass of the notches, the kinetic energy can be rewritten to give

$$T = \frac{1}{2}\left[m_1\dot{q}_1^2 + m_2\dot{q}_2^2 + \frac{m_3}{6}(\dot{q}_1 - \dot{q}_2)^2 + 2m_3\left(\frac{\dot{q}_1 + \dot{q}_2}{2}\right)^2 + \frac{m_4}{6}\dot{q}_2^2 + 2m_4\left(\frac{\dot{q}_2}{2}\right)^2 \right]$$

Expanding terms in parentheses, rearranging and noting that, in this case, $m_3 = m_4$ finally yields

$$T = \frac{1}{2}\left[\left(m_1 + \frac{2m_3}{3}\right)\dot{q}_1^2 + \left(\frac{2m_3}{3}\right)\dot{q}_1\dot{q}_2 + \left(m_2 + \frac{4m_3}{3}\right)\dot{q}_2^2\right]$$

The potential energy can be obtained from the equation

$$V = 4\frac{1}{2}k\left(\theta_3^2 + \theta_4^2\right)$$

Again, expressing this in terms of the linear coordinates gives

$$\begin{aligned} V &= 2k\left(\left(\frac{q_1 - q_2}{l_3}\right)^2 + \left(\frac{q_2}{l_3}\right)^2\right) \\ &= 2\frac{k}{l_3^2}\left(q_1^2 - 2q_1q_2 + 2q_2^2\right) \end{aligned}$$

Finally, the dissipation function can be readily seen to be

$$D = \frac{b}{2}\left(\dot{q}_1^2 + \dot{q}_2^2\right)$$

Substituting the above into Lagrange's equation gives the two equations governing motion

$$\begin{aligned} &\left(m_1 + \frac{2m_3}{3}\right)\ddot{q}_1 + \frac{2m_3}{3}\ddot{q}_2 + b\dot{q}_1 + \frac{4k}{l_3^2}q_1 - \frac{4k}{l_3^2}q_2 = Q_1 \\ &\frac{2m_3}{3}\ddot{q}_1 + \left(m_2 + \frac{4m_3}{3}\right)\ddot{q}_2 + b\dot{q}_2 - \frac{4k}{l_3^2}q_1 + \frac{8k}{l_3^2}q_2 = Q_2 \end{aligned} \qquad \text{(a)}$$

For a notch hinge, the angular stiffness k is given by equation (4.91), reproduced here

$$k = \frac{2Ebt^{5/2}}{9\pi R^{1/2}} \qquad \text{(b)}$$

In the absence of damping and externally applied forces, the equations of motion in matrix form become

$$\begin{bmatrix} \left(m_1 + \dfrac{2m_3}{3}\right) & \dfrac{2m_3}{3} \\ \dfrac{2m_3}{3} & \left(m_2 + \dfrac{4m_3}{3}\right) \end{bmatrix} \begin{Bmatrix} \ddot{q}_1 \\ \ddot{q}_2 \end{Bmatrix} + \begin{bmatrix} \dfrac{4k}{l_3^2} & -\dfrac{4k}{l_3^2} \\ -\dfrac{4k}{l_3^2} & \dfrac{8k}{l_3^2} \end{bmatrix} \begin{Bmatrix} q_1 \\ q_2 \end{Bmatrix} = \begin{Bmatrix} 0 \\ 0 \end{Bmatrix} \qquad \text{(c)}$$

Harmonic motions of each of the coordinates can be assumed of the form

$$\begin{aligned} q_1(t) &= \mathrm{Re}\{A_1 e^{i\omega_n t}\} \\ q_2(t) &= \mathrm{Re}\{A_2 e^{i\omega_n t}\} \end{aligned} \qquad \text{(d)}$$

Where the constant coefficients of A correspond to the eigenvectors. Substituting (d) into (c) and rearranging gives

$$\begin{bmatrix} \dfrac{4k}{l_3^2} - \omega_n^2\left(m_1 + \dfrac{2m_3}{3}\right) & -\dfrac{4k}{l_3^2} - \omega_n^2 \dfrac{2m_3}{3} \\ -\dfrac{4k}{l_3^2} - \omega_n^2 \dfrac{2m_3}{3} & \dfrac{8k}{l_3^2} - \omega_n^2\left(m_2 + \dfrac{4m_3}{3}\right) \end{bmatrix} \begin{Bmatrix} A_1 \\ A_2 \end{Bmatrix} = \begin{Bmatrix} 0 \\ 0 \end{Bmatrix} \qquad \text{(e)}$$

The undamped natural frequencies can be determined from the roots of the determinant

$$\begin{vmatrix} \dfrac{4k}{l_3^2} - \omega_n^2\left(m_1 + \dfrac{2m_3}{3}\right) & -\dfrac{4k}{l_3^2} - \omega_n^2 \dfrac{2m_3}{3} \\ -\dfrac{4k}{l_3^2} - \omega_n^2 \dfrac{2m_3}{3} & \dfrac{8k}{l_3^2} - \omega_n^2\left(m_2 + \dfrac{4m_3}{3}\right) \end{vmatrix} = 0 \qquad \text{(f)}$$

Expanding the terms in this determinant gives a quadratic in α $(=\omega_n^2)$ given by

$$\left(\frac{4k}{l_3^2} - \alpha\left(m_1 + \frac{2m_3}{3}\right)\right)\left(\frac{8k}{l_3^2} - \alpha\left(m_2 + \frac{4m_3}{3}\right)\right) - \left(\frac{4k}{l_3^2} + \alpha\frac{2m_3}{3}\right)\left(\frac{4k}{l_3^2} + \alpha\frac{2m_3}{3}\right)$$
$$= 0$$

For the dimensions shown, the mass of the platforms is 7 times greater than that of the individual support legs. This can be used to further reduce the above equation to

$$\left(\frac{4k}{l_3^2}-\alpha\left(\frac{23m_3}{3}\right)\right)\left(\frac{8k}{l_3^2}-\alpha\left(\frac{25m_3}{3}\right)\right)-\left(\frac{4k}{l_3^2}+\alpha\frac{2m_3}{3}\right)\left(\frac{4k}{l_3^2}+\alpha\frac{2m_3}{3}\right)$$
$$=\alpha^2\left(\frac{571m_3^2}{9}\right)-\alpha\left(\frac{278m_3k}{3l_3^2}\right)+\frac{16k^2}{l_3^4}=0 \tag{g}$$

Dividing through by the first term in the above gives

$$\alpha^2-\alpha\left(\frac{834k}{571l_3^2m_3}\right)+\frac{144k^2}{571l_3^4m_3^2}=0$$

From which, the lowest value for α is given by

$$\alpha=\frac{k}{l_3^2m_3}\frac{\left(\frac{834}{571}\right)-\sqrt{\left(\frac{834}{571}\right)^2-\frac{576}{571}}}{2} \tag{h}$$
$$\approx\frac{k}{5l_3^2m_3}=\omega_n^2=355530$$

From equation (b), the thickness of the notch hinge is given by

$$t=\left(\frac{45l_3^2\pi m_3R^{1/2}\alpha}{2Eb}\right)=0.00304\text{ m} \qquad \text{(QED)} \tag{i}$$

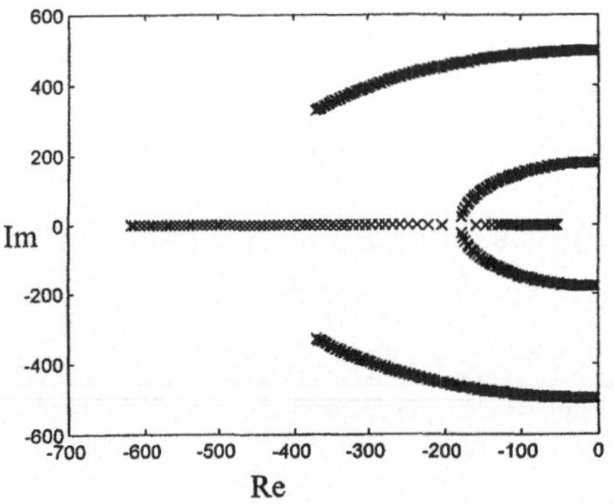

Figure 3.27: Roots locus for the two degree of freedom flexure

Substituting equation (i) into (b), the stiffness, k, of each individual notch and mass of each support leg, m_3, is

$k=1.91$ [Nm rad^{-1}]

$m_3=0.00672$ [kg] (j)

Computation of the damping coefficients is not as straightforward and requires a computational approach. Modeling the first mode natural frequency in terms of a simple second order system, the optimal damping coefficient can be obtained from a roots locus of the corresponding eigenvalue. Consequently, it is necessary to develop the complete system equation of free motion given by

$$\begin{bmatrix} \frac{23m_3}{3} & \frac{2m_3}{3} \\ \frac{2m_3}{3} & \frac{25m_3}{3} \end{bmatrix} \begin{Bmatrix} \ddot{q}_1 \\ \ddot{q}_2 \end{Bmatrix} + \begin{bmatrix} b & 0 \\ 0 & b \end{bmatrix} \begin{Bmatrix} \dot{q}_1 \\ \dot{q}_2 \end{Bmatrix} + \begin{bmatrix} \frac{4k}{l_3^2} & -\frac{4k}{l_3^2} \\ -\frac{4k}{l_3^2} & \frac{8k}{l_3^2} \end{bmatrix} \begin{Bmatrix} q_1 \\ q_2 \end{Bmatrix} \tag{k}$$

$$= \mathbf{M}\ddot{\mathbf{q}} + \mathbf{b}\dot{\mathbf{q}} + \mathbf{k}\mathbf{q} = \mathbf{0}$$

From section 3.10.2, this can be written in the form of an eigenvalue equation with the eigen matrix given by

$$[A] = \mathbf{A} = \begin{bmatrix} \mathbf{0} & \mathbf{I} \\ -\mathbf{M}^{-1}\mathbf{k} & -\mathbf{M}^{-1}\mathbf{b} \end{bmatrix} \tag{l}$$

Eigenvalues of this matrix will comprise two complex conjugate pairs, the lowest (or gravest) mode natural frequency corresponding to the pair with the lowest complex part. Modeling this as a single degree of freedom second order system, the ratio of the real to imaginary parts of the eigenvalue can be represented by the equation

$$\frac{\text{Re}(\lambda)}{\text{Im}(\lambda)} = \frac{\xi}{\sqrt{1-\xi^2}} = 0.204 \tag{m}$$

Figure 3.27 shows a plot of the eigenvalues as a function of the damping force. Equation (m) is satisfied at a value of

$$b \approx 10 \qquad [\text{Ns m}^{-1}] \qquad \text{(QED)}$$

If desired, the frequency response functions for this system can be computed using the techniques outlined in section 3.9.

References

Blevins R.D., 1985, *Formulas for Natural Frequency and Mode Shapes*, Krieger Publishing Co., FL

Den Hartog J.P., 1952, *Advanced Strength of Materials*, McGraw-Hill, NY, p131

Feynman R.P., R.B. Leighton and M. Sands, 1964, *The Feynman Lectures on Physics*, vols I - III, Addison Wesley, Reading, Ma.

Goldstein H., 1980, *Classical Mechanics*, Addison-Wesley. An advanced level text, reading this text is like playing chess. Likewise; L. A. Pars, 1961, *Treatise on Analytical Dynamics*.

Hildebrand F.B., 1976, *Advanced Calculus for Applications*, Prentice-Hall, Inc., NJ.

Lanczos C., 1970, *The Variational Principles of Mechanics*, Dover publications Inc., NY. Chpt's 1 -5.

Longair M.S., 1984, *Theoretical concepts in physics*, Cambridge University Press.

Maltbaek J.C., 1961, The influence of a concentrated mass on the free vibrations of a uniform beam, *J. Mech. Sci.*, **3**, 197-218

Newland D.E., 1989, *Mechanical vibration analysis and computation*, J. Wiley and Sons Inc., NY.

Rayleigh J.W.S., 1894, *Theory of Sound*, vol. I, Dover Publications. Quote is to be found in section 96 on page 130 of this edition. I must confess to a rather biased view of this treatise. Rarely a month goes by in which I do not consult this text to see how it is really done. Unfortunately, in the discussion of the vibration of plates the symbol h is used to represent the *half* thickness. Great care must be applied when translating this to an equation in which the total thickness is used.

Shockley W.F., Vibration of systems having distributed mass and elasticity, in *Shock and Vibration Handbook*, 4th edition, *ed.* Harris C.M., McGraw-Hill, 1995

Timoshenko S.P. and Young D.H., 1955, *Advanced Dynamics*, McGraw-Hill, NY.

Timoshenko S.P. and Young D.H., 1955, *Vibration Problems in Engineering*, Van Nostrand Co. Inc

4 Flexure elements

4.0 Overview

Flexure devices consist of compliant elements that connect 'rigid' bodies to form a mechanism. Fortunately, flexure elements almost invariably consist of either leaf type springs, notch hinges or cross strip pivots (with a variant known as a cartwheel pivot). Ultimately, flexures will be integrated into instrumentation and machine mechanisms to provide defined motions upon application of a force. Consequently, analysis of the behavior of the simple leaf spring under the influence of combined forces is discussed. From this, it is found that there is the possibility of buckling under compressive axial loads and stiffening in the presence of tensile loads. While the former of these represents an unwanted instability, the latter can be tolerated in most applications. Design strategies that favor the latter are presented. Another common element is the notch type hinge. This can be relatively easily manufactured by machining away circular sections either side of a solid body to produce two separate bodies connected by the thin, compliant web between the circular cut-outs. Approximate equations for computing the compliance of this type of flexure element were first presented by Paros and Weisbord in 1965 and are still used to this day. Both the leaf and notch type hinges are singular manifestations of the more general elliptic hinge for which the equations of Paros and Weisbord have been expanded. Cross-strip and cartwheel hinges are also common flexure elements for which a relatively comprehensive analysis is provided. All of the above are often used for single degree of freedom flexures. Elements having two degrees of freedom or greater have not been subject to such scrutiny. Some of the two axis mechanisms amenable to analysis are presented at the end of this chapter.

4.1 Leaf type springs

While there appears no natural hierarchical order to the three different flexure elements presented in this chapter, it is mainly for historical reasons that the leaf spring is discussed first. It is possible that the reason this element has been in use for so long is its ease of manufacture and assembly. Primarily, leaf springs can be either fabricated by clamping a thin plate between two rigid bodies or, with more difficulty, machined out of a monolithic block of material. In the former manifestation, leaf springs have been used in fine instrumentation for physical apparatus since the onset of the industrial revolution and probably before.

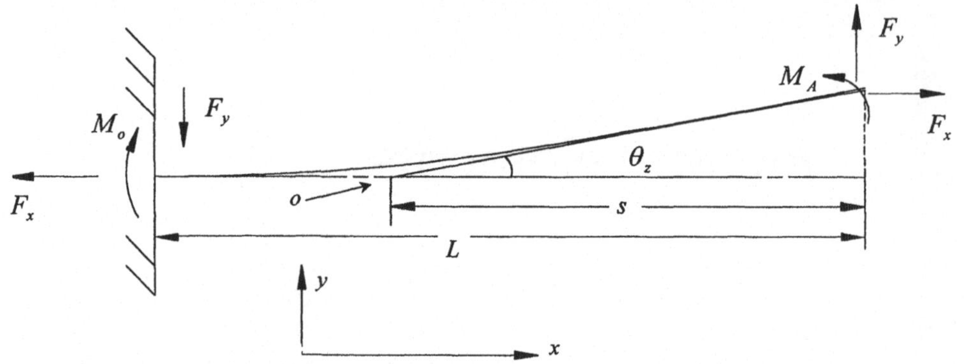

Figure 4.1: A cantilever beam with an arbitrary, in-plane force plus bending moment at the free end

4.1.1 The cantilever as a rotary hinge

Possibly, the simplest flexure mechanism is the cantilever beam. For the beam of figure 4.1, the bending moment due only to a force, F_y, is

$$EI\frac{d^2y}{dx^2} = F_y(L-x) \tag{4.1}$$

For the purpose of computing the maximum stress, the maximum bending moment, $M_{\max} = F_y L$, can be directly substituted into the bending equation, see chapter 2.

Integrating equation (4.1) twice

$$\begin{aligned} EI\frac{dy}{dx} &= F_y\left(Lx - \frac{x^2}{2}\right) + C_1 \\ EIy &= F_y\left(L\frac{x^2}{2} - \frac{x^3}{6}\right) + C_1x + C_2 \end{aligned} \tag{4.2}$$

Substituting the boundary conditions that both slope and deflection are zero at the clamped end ($x = 0$), equations (4.2) can be solved to give

$$\begin{aligned} EI\frac{dy}{dx} &= F_y\left(Lx - \frac{x^2}{2}\right) \\ EIy &= F_y\left(L\frac{x^2}{2} - \frac{x^3}{6}\right) \end{aligned} \tag{4.3}$$

For relatively small displacements, the slope θ and deflection δ at the free end of the beam are given by

$$\theta_{x=L} = \frac{F_y L^2}{2EI}$$
$$\delta_{x=L} = \frac{F_y L^3}{3EI} \qquad (4.4)$$

From the above, the angular and linear stiffness of the cantilever are

$$k_{\theta_z F_y} = \frac{F_y}{\theta_z} = \frac{2EI}{L^2}$$
$$k_{\delta_y F_y} = \frac{F_y}{\delta_y} = \frac{3EI}{L^3} \qquad (4.5)$$

Clearly, if the cantilever is to be used as a hinge, it is necessary to know where the pivot point is. Although the pivot point will vary with deflection, it is apparent that, if the deflections are small, so too will the variations of the pivot point. Defining this as the intersection of the tangent to the free end of the cantilever with the unloaded beam axis, it is clear that the pivot point, *o*, in figure 4.1 is located from the free end of the undistorted beam a distance, *s*, given by

$$s = \frac{\delta_y}{\tan\theta_z} \approx \frac{2}{3}L \qquad (4.6)$$

Equation (4.6) implies a stationary pivot point located a distance of one third of the way along the axis from the fixed end of the cantilever. Before moving on to more complex load conditions, it will be informative to look at a few applications of the leaf type hinge. Four different flexure mechanisms are shown in figure 4.2. In all figures, the stationary link of the flexure mechanism is shaded. The first two mechanisms are variants on the simple linear spring to be discussed in chapter 5. Application of a force to the moving platform will produce distortions resulting in a parallelogram displacement of the four bar linkage. For mechanisms in which the direction of applied forces may vary, it is often desirable to use more than one hinge as illustrated in figure 4.2 c). In this diagram, vertical and horizontal forces are each carried by the corresponding hinge. Commonly, such a requirement is typical for a flexure that is required to support varying weights, for example, a spring balance or truck stop weigh platform. Construction of the crossed-strip hinge provides a challenge to both designer and manufacturer. A simpler construction is shown in the hinge of figure 4.2 d). The crossed strip hinge is discussed towards the end of this chapter.

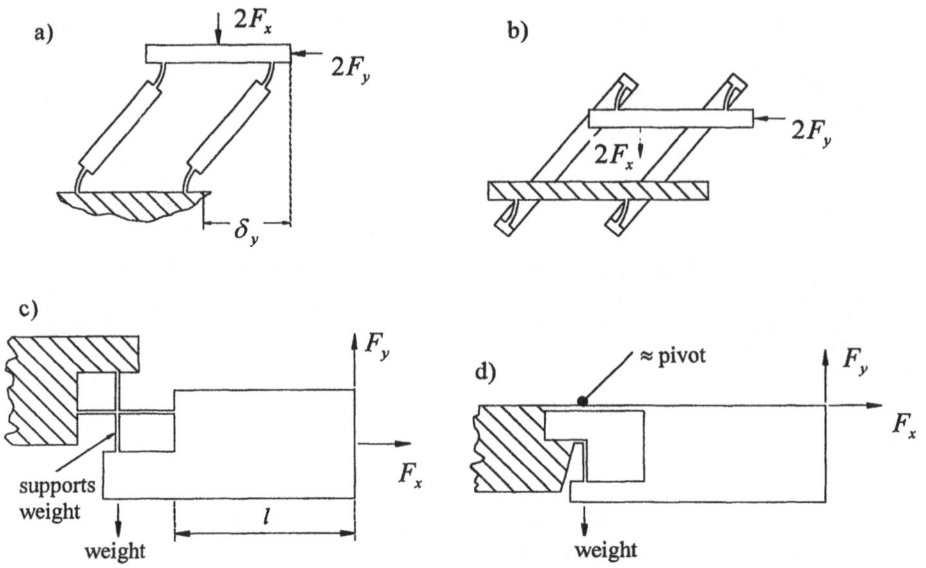

Figure 4.2: Four flexure mechanisms utilizing the leaf type hinge, a) four-bar parallelogram flexure with hinges under compressive axial load, b) four bar flexure with hinges under tensile axial load, c) two-hinge pivot with elements crossed (the cartwheel hinge), d) separated two hinge pivot (reproduced from Thorpe, 1953)

Figure 4.2 also illustrates another feature common to flexures, that is, the forces are often applied to a rigid body remote from the hinge. As a consequence it is often necessary to assess the effects of this on the subsequent distortion of the hinge. Fortunately, such an analysis has been carried out by Thorpe, 1953, in which various cases of hinges subject combinations of z-axis moments and axial

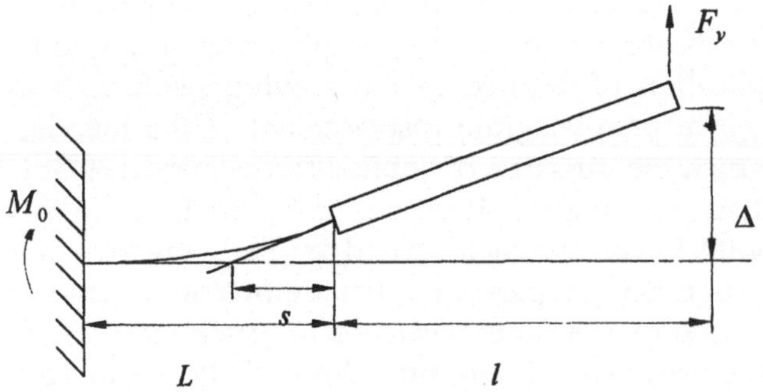

Figure 4.3: Mathematical model for a cantilever hinge with a rigid body attached to the free end subject to a tangential force

and tangential forces are presented. For the case of a single tangential force applied to a rigid body attached to the free end of a cantilever hinge, figure 4.3, the deflection Δ is given by

$$\Delta_y = F_y \frac{L\left(l^2 + Ll + L^2/3\right)}{EI} \tag{4.7}$$

The distance from the 'free' end of the hinge to the effective pivot point is

$$s = \frac{lL + 2L^2/3}{L + 2l} \tag{4.8}$$

Finally, the slope and deflection at any position from the stationary body to the end of the rigid body attached to the free end of the cantilever can be determined from the equations

$$\begin{aligned} y &= \frac{F_y}{EI}\left(\frac{L+l}{2}x^2 - \frac{x^3}{6}\right) \\ \frac{dy}{dx} &= \frac{F_y}{EI}\left((L+l)x - \frac{x^2}{2}\right) \end{aligned} \tag{4.9}$$

At the free end of the rigid body (i.e. $x = L + l$), equations (4.9) become

$$\begin{aligned} y &= \frac{F_y}{3EI}(L+l)^3 \\ \frac{dy}{dx} &= \frac{F_y}{2EI}(L+l)^2 \end{aligned} \tag{4.10}$$

Setting the length of the rigid body l to zero, it can be readily verified that equations (4.7) to (4.10) reduce to the equations for the hinge with a tangential force applied at the free end, equations (4.2) to (4.6).

Clearly, for many applications, bending moments about the free end as well as axial and tangential forces will be present. Therefore, it is necessary to assess the effects of these combined forces on the stiffness of the hinge. In the following analysis, only bending moments about the z-axis will be considered. It is also noted that, to determine the angle of twist and shear stress due to a moment about the x-axis, equation (2.102) and (2.104) of section 2.4.9.1 can be added to the stress due to the other loads. In all cases it will be assumed that the principle of superposition is applicable.

4.1.2 The cantilever hinge subject to an axial compressive force

Fred Eastman investigated the effects of axial loads on the stiffness of a cantilever beam in 1937 for application to the design of fine balances. In particular, it was observed that

> *With the aid of ordinary beam theory, it is a simple matter to compute the stiffness of a flexure if it is subjected to a little or no load, but if it is heavily loaded in compression, its stiffness will be greatly reduced and may be even zero or negative. Negative stiffness indicates instability resulting from column action* (i.e. buckling). *On the other hand, if the flexure is heavily loaded in tension, its stiffness increases and may become many times it's no load magnitude.*

In fact, the balance design presented by this author represents an attempt to use this 'tunable' stiffness to create a free hinge to compete with the knife-edge hinges currently in use. The victor in this competition between the flexure and knife-edge pivot remains unresolved.

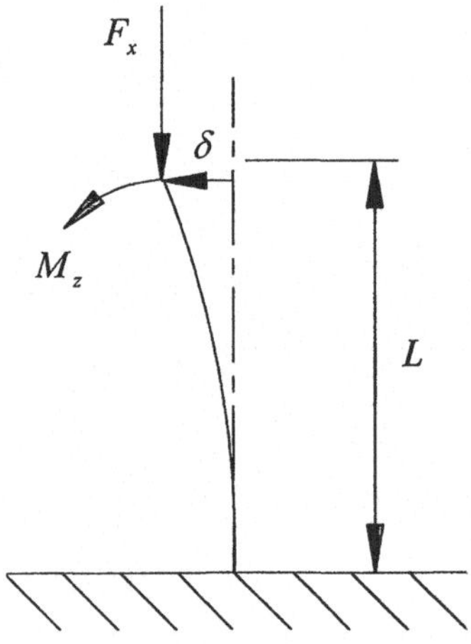

Figure 4.4: A cantilever beam subject to an axial compressive force plus bending moment

For small axial loads, the compression or extension of a straight beam can be readily computed from

$$\delta_x = \frac{F_x L}{EA} \tag{4.11}$$

where A is the cross sectional area of the beam

For larger loads there will be a significant bending moment due to the deflection of the beam itself from the central axis. For a compressive load as shown in figure 4.4, Eastman has shown that the stiffness expressed as the ratio of the bending moment, M_z, required to induce an angular variation, θ_z is given by

$$\begin{aligned} K_{\theta_z M_z} &= \frac{M_z}{\theta_z} = EI\sqrt{\frac{F_x}{EI}}\cot\left(L\sqrt{\frac{F_x}{EI}}\right) = \frac{EI}{L}\omega\cot(\omega) \\ &= \frac{EI}{L}\omega\coth(\omega) \end{aligned} \tag{4.12}$$

Where the first equation should be used if the loads are compressive while the second equation corresponds to a tensile axial load. From the first equation in (4.12) the stiffness reduces to zero at a compressive load given by

$$\omega = L\sqrt{\frac{F_x}{EI}} = 2\gamma = \pi/2$$

or (4.13)

$$F_x = \frac{\pi^2 EI}{4L^2}$$

This corresponds to the familiar Euler buckling condition. For zero axial load, it can be shown that both of equations (4.12) reduce to

$$K_{\theta_z M_z} = \frac{M_z}{\theta_z} = \frac{EI}{L} \tag{4.14}$$

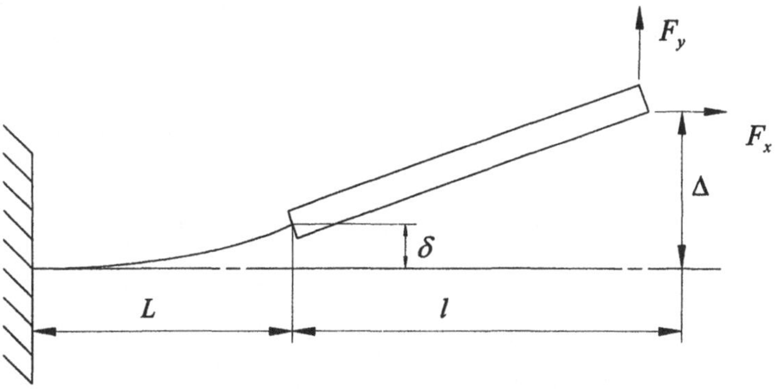

Figure 4.5: A cantilever beam subject to combined tangential and tensile forces at the free end

4.1.3 Combined axial and tangential loads

Consider a cantilever hinge with a rigid rod attached at the free end. At the end of this rigid rod, both axial and tangential forces are applied, figure 4.5. Different solutions result if the axial load is tensile or compressive.

Case 1: Cantilever beam subject to combined tangential and tensile axial load

For this case, the deflection y at an arbitrary position x is, Thorpe, 1953,

$$y = \Delta\left\{C_1 \cosh\left(x\sqrt{\frac{F_x}{EI}}\right) - \frac{1+C_1}{(L+l)\sqrt{\frac{F_x}{EI}}}\sinh\left(x\sqrt{\frac{F_x}{EI}}\right) + \left(\frac{1+C_1}{L+l}\right)x - C_1\right\} \tag{4.15}$$

As in the previous section a dimensionless axial load rating is defined by

$$\omega = L\sqrt{\frac{F_x}{EI}} \tag{4.16}$$

at which equation (4.16) becomes

$$y = \Delta\left\{C_1 \cosh\left(\omega \frac{x}{L}\right) - \frac{1+C_1}{\omega(1+l/L)}\sinh\left(\omega \frac{x}{L}\right) + \left(\frac{1+C_1}{L+l}\right)x - C_1\right\} \tag{4.17}$$

where

$$C_1 = \frac{M_z}{F_x\Delta} = \frac{\omega \coth(\omega) + \frac{L}{l}}{(\omega)^2\left(1 + \frac{l}{L}\right) + L/l\,(\omega\coth(\omega) - 1)} \tag{4.18}$$

The displacement at the free end of the attached rigid body can be determined from the equation

$$\Delta = \frac{F_y}{F_x}\frac{L+l}{1+C_1} \tag{4.19}$$

The distance from the 'free' end of the hinge to the effective pivot point is

$$s = \frac{1}{C_1 \frac{\omega}{L}\sinh(\omega) - \left(\frac{1+C_1}{L+l}\right)(\cosh(\omega) - 1)} - l \tag{4.20}$$

Differentiating (4.17), the slope at any position x is

$$\frac{dy}{dx} = \Delta\left\{\frac{\omega}{L}C_1 \sinh\left(\omega\frac{x}{L}\right) - \frac{1+C_1}{L+l}\cosh\left(\omega\frac{x}{L}\right) + \left(\frac{1+C_1}{L+l}\right)\right\} \tag{4.21}$$

Setting the length of the attached rigid body to zero (i.e. $l \Rightarrow 0$) corresponds to the case where the forces are applied to the free end of the cantilever. In this case, the constant C_1 becomes

$$C_1 = \frac{M_z}{F_x\delta} = \frac{1}{(\omega\coth(\omega) - 1)} \tag{4.22}$$

After which, the deflection along the cantilever is given by

$$y = \delta\left\{\frac{\tanh(\omega)\cosh\left(\omega\frac{x}{L}\right) - \sinh\left(\omega\frac{x}{L}\right) + \omega\frac{x}{L} - \tanh(\omega)}{(\omega - \tanh(\omega))}\right\} \tag{4.23}$$

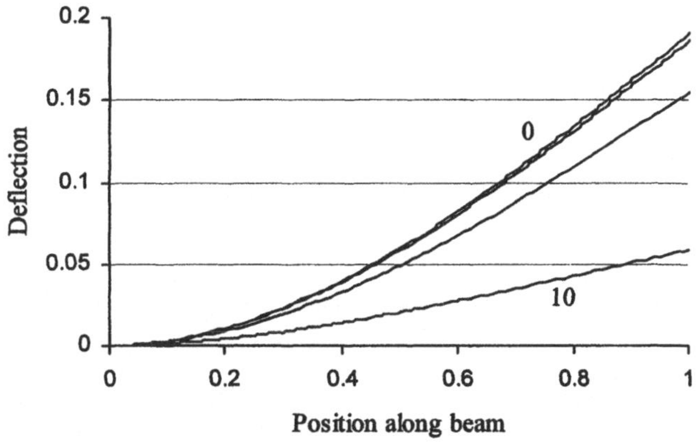

Figure 4.6: A cantilever beam with tangential and tensile, axial forces applied to the free end $F_x = 0, 0.1, 1$ & 10 N. Dimensions are; width = 0.1 m, length = 1 m, thickness = 0.01, tangential force = 1 N.

and

$$\delta = \frac{F_y}{F_x}\frac{(\omega - \tanh(\omega))L}{\omega} \tag{4.24}$$

Substituting (4.24) into (4.23) and rearranging yields

$$y = \frac{LF_y}{\omega F_x}\left(\tanh(\omega)\cosh\left(\omega\frac{x}{L}\right) - \sinh\left(\omega\frac{x}{L}\right) + \omega\frac{x}{L} - \tanh(\omega)\right) \tag{4.25}$$

Plots of equation (4.25) for increasing axial load are shown in figure 4.6. It can be seen that, as the axial force increases, the deflection reduces. For comparison, equation (4.3) has been used to plot the deflection curve corresponding to a beam with no axial load.

Differentiating equation (4.24) with respect to the tangential load, the stiffness of the cantilever is given by

$$k_{\delta_y F_y} = \frac{dF_y}{d\delta_y} = \frac{F_x\omega}{L(\omega - \tanh(\omega))} \tag{4.26}$$

From figure 4.7, it can be seen that the stiffness varies almost with the 3/2 power of the applied axial load. For the purposes of computing the maximum stresses in the cantilever, the maximum bending moment occurs at the base of the cantilever and can be readily obtained. This is not so for the following case.

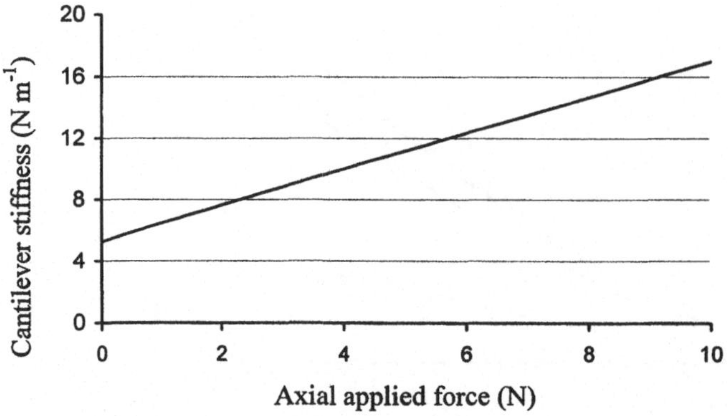

Figure 4.7: Plot of the axial stiffness of a cantilever beam as a function of tensile axial force. All other parameters are as given in caption for figure 4.6.

Case 2: Cantilever beam subject to combined, tangential and compressive, axial load

When the axial load is compressive, the situation is complicated by the possibility of buckling. In this case, the sign of both loads has been reversed so that the tangential force is acting in the negative y direction. The equations corresponding to those used in *case 1* above are

$$y = \Delta\left\{\left(\frac{1-C_2}{L+l}\right)x + C_2 - C_2\cos\left(\omega\frac{x}{L}\right) - \frac{L(1-C_2)}{\omega(L+l)}\sin\left(\omega\frac{x}{L}\right)\right\} \tag{4.27}$$

$$C_2 = \frac{M_z}{F_x\Delta} = \frac{\omega\cot(\omega) + \dfrac{L}{l}}{(\omega)^2\left(1+\frac{l}{L}\right) + L/l\,(1-\omega\cot(\omega))} \tag{4.28}$$

$$\Delta = \frac{F_y}{F_x}\frac{L+l}{1-C_2} \tag{4.29}$$

$$s = \frac{1}{C_2\dfrac{\omega}{L}\sin(\omega) + \left(\dfrac{1-C_2}{L+l}\right)(1-\cos(\omega))} - l \tag{4.30}$$

$$\frac{dy}{dx} = \Delta\left\{\frac{\omega}{L}C_2\sin\left(\omega\frac{x}{L}\right) - \frac{1-C_2}{L+l}\cos\left(\omega\frac{x}{L}\right) + \left(\frac{1-C_2}{L+l}\right)\right\} \tag{4.31}$$

In this particular case, the maximum bending moment does not occur at the root of the cantilever, but at a position along from the rigid support given by the relationship

$$\left(\frac{x}{L}\right)_{M_{max}} = \frac{1}{\omega}\tan^{-1}\left(\frac{L(1-C_2)}{C_2\omega(L+l)}\right) \tag{4.32}$$

The magnitude of the moment can be obtained from

$$\frac{M_{max}}{F_x\Delta} = \frac{\sqrt{(\omega)^2+\left(\frac{L}{l}\right)^2}}{\sin(\omega)\left\{(\omega)^2\left(1+\frac{l}{L}\right)+\frac{L}{l}(1-\cos(\omega))\right\}} \tag{4.33}$$

Again, it can be shown that, as the rigid member attached to the free end of the cantilever reduces to zero length, the above equation becomes

$$C_2 = \frac{M_z}{F_x\delta} = \frac{1}{(1-\omega\cot(\omega))} \tag{4.34}$$

$$\delta_y = \frac{F_y}{F_x}\frac{L(\tan(\omega)-\omega)}{\omega} \tag{4.35}$$

$$y = \frac{LF_y}{\omega F_x}\left(\tan(\omega)\cos\left(\omega\frac{x}{L}\right)-\sin\left(\omega\frac{x}{L}\right)+\omega\frac{x}{L}-\tan(\omega)\right) \tag{4.36}$$

$$k_{\delta_y F_y} = \frac{dF_y}{d\delta_y} = \frac{F_x\omega}{L(\tan(\omega)-\omega)} \tag{4.37}$$

In this latter case, the maximum bending moment occurs at the fixed end of the cantilever.

Equation's (4.35) to (4.37) contain some initially unusual characteristics. It will be recalled that the tangential force is applied in a direction opposite to the direction of deflection. Clearly, if the axial force is zero, the deflection of the beam will proceed in a negative direction. This can be seen in figure 4.8 when *axial* loads are other than 10 N. However for an axial load of 10 N there is a deflection in the opposite direction, the magnitude of which is *less* than that for zero axial load! This can be explained if it is assumed that the beam is initially deflected in the positive y direction. After this, the axial load is applied and the

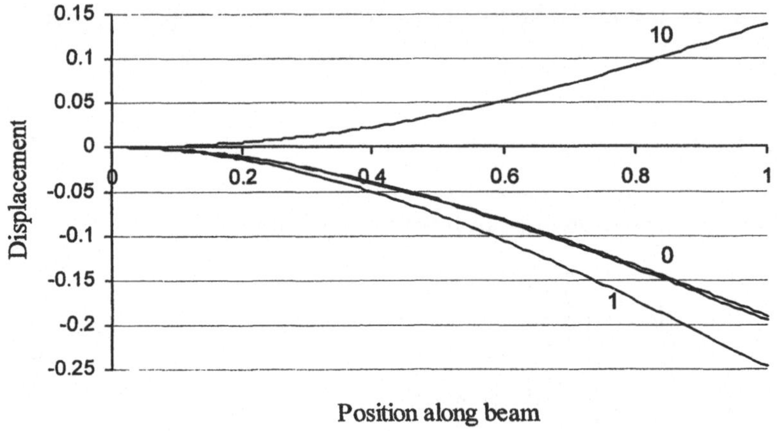

Figure 4.8: Deflection along a cantilever beam with tangential and compressive, axial forces applied to the free end $F_x = 0, 0.1, 1$ and 10 N. Dimensions are width = 0.1 m, length = 1 m, thickness = 0.01, tangential force = 1 N.

original deflection forces reduced to zero. Clearly, under these conditions, there will be a residual bending of the beam in the positive y direction. Now, applying the tangential force, it is obvious that the beam will be restored towards the original unstressed shape. If, however, either the tangential force is large or the axial force is small, this restraining force will push the beam past the origin ($y = 0$) and continue to distort the beam in the opposite direction. Under these circumstances the bending moments due to both of the applied forces will be of the same sign and the deflections due to each will add. If, however, the beam were originally deflected in the opposite direction, the deflections due to each force would always add. Consequently, it is not possible to determine the deflection of a cantilever subject to such loads without first knowing its deflection history. This state of affairs is further exacerbated if the beam is first of all subject to an axial force of sufficient magnitude to induce buckling. The critical load for buckling is obtained from equation (4.35) corresponding to the point at which the deflection becomes infinite or from (4.37) at zero stiffness i.e.

$$\omega = \pi/2$$

or (4.38)

$$F_x = \frac{\pi^2 EI}{4L^2}$$

Under these conditions, at this axial load, the cantilever is in a symmetric state of instability. Consequently, it cannot be decided in which direction the

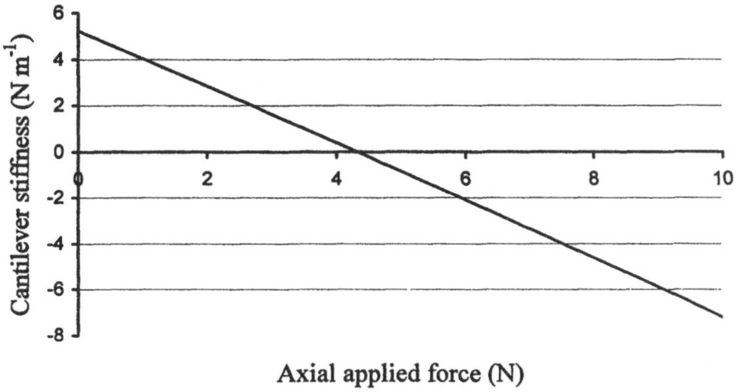

Figure 4.9: The tangential stiffness of a cantilever beam as a function of the compressive axial force. All other parameters are as given in the caption for figure 4.6

beam will deflect. Because of this, cantilever flexures subject to significant axial loads are rarely driven using applied forces. Instead, positive drives are used and the forces in the flexure are then determined from the prescribed deflections. Unfortunately, to produce a positive drive implies adding an element of infinite stiffness in the drive direction. Methods for achieving an approximation to such a drive coupling will be touched upon in chapters 7 and 8.

Figure 4.9 shows the variation in stiffness as a function of the applied load using the same parameters for the previous plots. The point at which the flexure has zero stiffness may be of great utility for some design purposes.

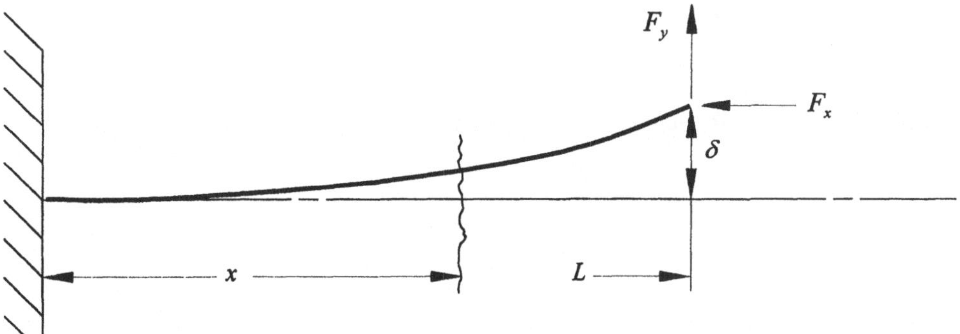

Figure 4.10: Mathematical model for a cantilever bean subject to combined tangential and compressive axial loading

To illustrate the situation when the tangential force is first applied, consider the simple cantilever beam subject to the applied loads shown in figure 4.10. These applied loads are the same as in the previous example with the sign of the tangential load being reversed. Because this solution is not provided in the paper of Thorpe, a more complete derivation will be presented here. The bending equation to be satisfied is given by

$$EI\frac{d^2y}{dx^2} = F_y(l-x) + F_x(\delta_y - y) \tag{4.39}$$

For convenience, this can be written in the form

$$y'' + \frac{\omega^2}{L^2}y = \chi^2 L + \frac{\omega^2}{L^2}\delta_y - \chi^2 x \tag{4.40}$$

With A - D being constants, the general solution to this equation can immediately be written as

$$y = A\cos\left(\omega\frac{x}{L}\right) + B\sin\left(\omega\frac{x}{L}\right) + Cx + D \tag{4.41}$$

Based on the boundary conditions of zero slope and deflection at the free end, equation (4.41) becomes

$$y = \frac{\chi^2 L^3 + \omega^2\delta}{\omega^2}\cos\left(\omega\frac{x}{L}\right) + \frac{L^3\chi^2}{\omega^3}\sin\left(\omega\frac{x}{L}\right) - \frac{L^2\chi^2}{\omega^2}x \tag{4.42}$$

For there to be zero bending moment at the free end of the cantilever, the second derivative must vanish giving the condition

$$\tan(\omega) = \frac{\omega\left(\chi^2 L^3 + \omega^2\delta_y\right)}{L^3\chi^2} \tag{4.43}$$

Substituting (4.43) into (4.42) and rearranging, the deflection along the beam is

$$y = \frac{LF_y}{F_x\omega}\left(\tan(\omega) - \tan(\omega)\cos\left(\omega\frac{x}{L}\right) + \sin\left(\omega\frac{x}{L}\right) - \omega\frac{x}{L}\right) \tag{4.44}$$

The deflection at the free end is given by

$$\delta_y = \frac{LF_y}{F_x\omega}\left(\tan(\omega) - \omega\right) \tag{4.45}$$

It can be seen from the above equation that for axial loads less than that for Euler buckling, the displacement is always positive. For other cases corresponding to the load conditions for the flexures of the type shown in figure 4.2 (b), the readers are recommended to consult the paper of Thorpe, 1953.

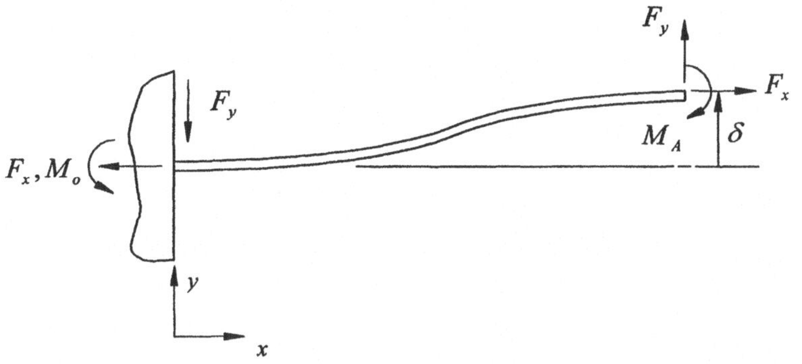

Figure 4.11: Cantilever subject to zero end slope constraint plus axial and tangential loads

4.1.4 Combined axial and tangential loads plus a moment applied to the free end of a simple cantilever

For a simple cantilever beam subject to forces both perpendicular and along the axis of the beam there will be an additional moment due to the deflection of the beam itself, see figure 4.11. Under these conditions, the bending equation can be expressed by

$$EI\frac{d^2y}{dx^2} = -M_0 + F_x y - F_y x \tag{4.46}$$

Following Plainevaux, 1956, it is convenient to introduce the variables

$$\begin{aligned} X &= \frac{x}{L} \\ 4\gamma^2 &= \frac{F_x L^2}{EI} \\ \varphi &= \frac{F_y L^2}{EI} \\ m_0 &= \frac{M_0 L}{EI} \end{aligned} \tag{4.47}$$

It is noted that $\gamma = \omega / 2$ where ω is the axial load variable of the previous section. In all analyses, the value of γ is a direct measure of the influence of the axial load on the stiffness of the beam, for an example see case study 1 of chapter 7. Equation (4.46) can be rearranged to yield

$$\frac{d^2y}{dX^2} - 4\gamma^2 y = -L(m_0 + \varphi X) \tag{4.48}$$

A solution to this problem is given by

$$y = C_1 \cosh 2\gamma X + C_2 \sinh 2\gamma X + \frac{L}{4\gamma^2}(m_0 + \varphi X) \tag{4.49}$$

Differentiating (4.49) gives an expression for the slope along the beam

$$\frac{dy}{dx} = \frac{1}{L}\frac{dy}{dX} = C_1 \frac{2\gamma}{L}\sinh 2\gamma X + C_2 \frac{2\gamma}{L}\cosh 2\gamma X + \frac{\varphi}{4\gamma^2} \tag{4.50}$$

Inserting the boundary conditions of zero slope and deflection at the clamped end of the beam, the general solution for the y-axis deflection is obtained

$$y = \frac{Lm_0}{4\gamma^2}(1 - \cosh(2\gamma X)) + \frac{L\varphi}{8\gamma^3}(2\gamma X - \sinh(2\gamma X)) \tag{4.51}$$

and

$$\begin{aligned} \frac{dy}{dx} &= \frac{\varphi}{4\gamma^2}(1 - \cosh(2\gamma X)) - \frac{m_0}{2\gamma}\sinh(2\gamma X) \\ \frac{d^2y}{dx^2} &= -\frac{m_0}{L}\cosh(2\gamma X) - \frac{\varphi}{2\gamma L}\sinh(2\gamma X) \end{aligned} \tag{4.52}$$

Equations (4.51) and (4.52) can be used to determine the relationships between bending moments and both vertical and horizontal forces for a given axial load and end deflection. Consider the cantilever undergoing an 's' shaped deflection with the boundary condition that the beam is constrained to have zero slope and magnitude δ at the free end. Substituting this into equations (4.51) and (4.52) results in the simultaneous equations

$$\delta = \frac{Lm_0}{4\gamma^2}(1 - \cosh(2\gamma)) + \frac{L\varphi}{8\gamma^3}(2\gamma - \sinh(2\gamma)) \tag{4.53}$$

$$\frac{dy}{dx} = \frac{\varphi}{4\gamma^2}(1 - \cosh(2\gamma)) - \frac{m_0}{2\gamma}\sinh(2\gamma) \tag{4.54}$$

Equation (4.54) can be rearranged in terms of either m_0 or φ to give

$$m_0 = \frac{\varphi}{2\gamma}\frac{\cosh(2\gamma)-1}{\sinh(2\gamma)}$$
$$\varphi = m_0 2\gamma \frac{\sinh(2\gamma)}{\cosh(2\gamma)-1} \tag{4.55}$$

These can be substituted back into (4.53) and solved for each parameter in terms of applied loads and displacements. To see this, using the first of the above equations, (4.53) can be rearranged in the form

$$\begin{aligned}\frac{\delta_y}{L} &= \frac{\varphi}{8\gamma^3}\left[\frac{(\cosh(2\gamma)-1)^2}{\sinh(2\gamma)} - \sinh(2\gamma) + 2\gamma\right] \\ &= \frac{\varphi}{8\gamma^3}\left[\frac{\cosh^2(2\gamma)-2\cosh(2\gamma)+1-\sinh^2(2\gamma)+2\gamma\sinh(2\gamma)}{\sinh(2\gamma)}\right] \\ &= \frac{\varphi}{4\gamma^3}\left[\frac{1-\cosh(2\gamma)+\gamma\sinh(2\gamma)}{\sinh(2\gamma)}\right]\end{aligned} \tag{4.56}$$

Making the substitution

$$\left[\frac{1-\cosh(2\gamma)}{\sinh(2\gamma)}\right] = -\tanh(2\gamma) \tag{4.57}$$

Equation (4.56) can be rearranged to give the relatively simple expression

$$\varphi = \frac{F_y L^2}{EI} = \frac{4\delta_y}{L}\left(\frac{\gamma^3}{\gamma-\tanh(\gamma)}\right) \tag{4.58}$$

This can be further rearranged into the more familiar form

$$\frac{F_y}{\delta_y} = \frac{12EI}{L^3}\left(\frac{\gamma^3}{3(\gamma-\tanh(\gamma))}\right) \tag{4.59}$$

The term in parentheses represents a stiffness factor due to the tensile load applied along the axis of the beam and this has been plotted in figure 4.12. From this graph it can be seen that the stiffness increases slowly for a tensile applied load and vice versa when the axial load is compressive. In the absence of axial loads, the analysis is considerably simplified and is presented in the following section. For the zero load condition, the stiffness is given by the multiplier of the term in parentheses. Consequently, the graph in figure 4.12 represents the deviation in the stiffness as a function of the dimensionless axial load.

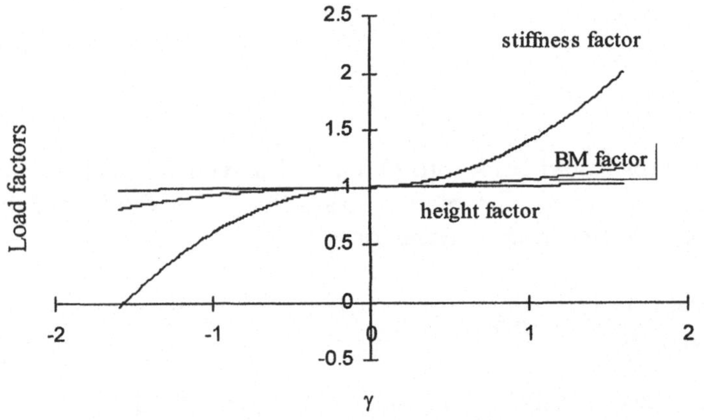

Figure 4.12: Stiffness, bending moment and height factors for a simple cantilever beam subject to an axial load and constrained to have zero slope at each end

Similarly, it can be shown that the bending moment at each end of the beam can be obtained from

$$\frac{M_0 L}{EI} = \frac{\delta_y}{L} \frac{2\gamma^2 \tanh(\gamma)}{\gamma - \tanh(\gamma)} \tag{4.60}$$

Again this can be rearranged into a form readily comparable to that from simple beam bending theory to give

$$\frac{M_0}{\delta_y} = \frac{6EI}{L^2}\left(\frac{\gamma^2 \tanh(\gamma)}{3(\gamma - \tanh(\gamma))}\right) \tag{4.61}$$

Again, the variable in parentheses is the dimensionless bending moment factor, see figure 4.12. This varies less strongly than the stiffness factor.

Finally, to complete our analysis, it is possible to solve for the change in height of the flexure, δ_x for a given deflection in y and axial load given by, Plainevaux, 1956

$$\frac{\delta_x}{L} = \frac{3}{5}\left(\frac{\delta_y}{L}\right)^2 \left(\frac{5\gamma\left[\gamma\left(3 - \tanh^2\gamma\right) - 3\tanh\gamma\right]}{12(\gamma - \tanh\gamma)^2}\right) \tag{4.62}$$

Comparing with equation (4.83), the term in parenthesis on the right hand side is the effective increase in height loss as a function of the tensile axial load and this is also shown in figure 4.12.

If the beam is subject to a compressive load, the bending equation is subject to a change in sign so that it is necessary to seek a solution to the equation

$$\frac{d^2y}{dX^2} + 4\gamma^2 y = L(m_0 + \varphi X) \tag{4.63}$$

This is similar to the previous example only, in this case, the solution involves circular instead of the hyperbolic functions of the previous analysis. Again solving for the applied bending moments, forces and axial compression yields

$$\frac{F_y}{\delta_y} = \frac{12EI}{L^3}\left(\frac{\gamma^3}{3(\tan(\gamma) - \gamma)}\right) \tag{4.64}$$

$$\frac{M_0}{\delta_y} = \frac{6EI}{L^2}\left(\frac{\gamma^2 \tan(\gamma)}{3(\tan(\gamma) - \gamma)}\right) \tag{4.65}$$

$$\frac{\delta_x}{L} = \frac{3}{5}\left(\frac{\delta_y}{L}\right)^2\left(\frac{5\gamma\left[\gamma(3 + \tan^2\gamma) - 3\tan\gamma\right]}{12(\tan(\gamma) - \gamma)^2}\right) \tag{4.66}$$

Again these are plotted in figure 4.12. For this case it can be seen that here is a considerable reduction in the stiffness of the flexure with compressive load and this goes to zero at a value

$$\gamma = \sqrt{\frac{F_x L^2}{4EI}} = 1.57$$

or

$$F_x = 9.859\frac{EI}{L^2} \approx \frac{\pi^2 EI}{L^2} \tag{4.67}$$

This is the familiar Euler load for buckling of beam that is *pinned* at each end. In practice, for application to precision mechanisms, the dimensionless load factor should always be kept as small as possible. It can also be shown that (4.66) tends to negative infinity at $\gamma = \pi$ leading to the Euler buckling condition

$$F \le \frac{\pi^2 EI}{4L^2} \tag{4.68}$$

This is a result more familiar for the prediction of cantilever stability with no applied moment at the free end.

In fact Plainevaux presents solutions to the bending of a cantilever in which the 'free' end is constrained to a tilt angle θ. Such a case corresponds to the free cantilever with a moment applied to the free end. Under these conditions the bending moments at each end of the beam are different, as has been indicated in figure 4.12, and the complete equations for a tensile axial load are

$$\frac{M_A L}{EI} = \tan\theta\left[\gamma\coth(\gamma) + \frac{\gamma^2 \tanh\gamma}{\gamma - \tanh\gamma}\right] - \frac{\delta_y}{L}\frac{2\gamma^2 \tanh\gamma}{\gamma - \tanh\gamma}$$

$$\frac{M_o L}{EI} = \tan\theta\left[\gamma\coth(\gamma) - \frac{\gamma^2 \tanh\gamma}{\gamma - \tanh\gamma}\right] + \frac{\delta_y}{L}\frac{2\gamma^2 \tanh\gamma}{\gamma - \tanh\gamma}$$

$$\frac{F_y L^2}{EI} = \frac{\delta}{L}\frac{4\gamma^3}{\gamma - \tanh\gamma} - \tan\theta\frac{2\gamma^2 \tanh\gamma}{\gamma - \tanh\gamma}$$

$$\frac{\delta_x}{L} = \left(\frac{\delta_y}{L}\right)^2\left(\frac{\gamma\left[\gamma\left(3 - \tanh^2\gamma\right) - 3\tanh\gamma\right]}{4(\gamma - \tanh(\gamma))^2}\right) + \left(\frac{\delta_y}{L}\right)\tan(\theta)\left[\frac{1}{2} - \frac{\gamma\left[\gamma\left(3 - \tanh^2\gamma\right) - 3\tanh\gamma\right]}{4(\gamma - \tanh(\gamma))^2}\right] +$$
$$+ \tan^2(\theta)\left[\frac{\coth\gamma - \gamma\left(1 + \coth^2\gamma\right)}{16\gamma} + \frac{\gamma\left[\gamma\left(3 - \tanh^2\gamma\right) - 3\tanh\gamma\right]}{16(\gamma - \tanh(\gamma))^2}\right]$$

(4.69)

Equations (4.69) can be expanded and expressed in series form

$$\frac{M_A L}{EI} = 4\tan\theta\left[1 + \frac{2\gamma^2}{15} - \frac{11\gamma^4}{1575} + ..\right] - 6\frac{\delta_y}{L}\left[1 + \frac{\gamma^2}{15} - \frac{\gamma^4}{525} + ..\right]$$

$$\frac{M_o L}{EI} = -2\tan\theta\left[1 - \frac{\gamma^2}{15} + \frac{13.\gamma^4}{1575} - ..\right] + 6\frac{\delta_y}{L}\left[1 + \frac{\gamma^2}{15} - \frac{\gamma^4}{525} + ..\right]$$

$$\frac{F_y L^2}{EI} = 12\frac{\delta_y}{L}\left[1 + \frac{2\gamma^2}{5} - \frac{\gamma^4}{525} + ..\right] - 6\tan\theta\left[1 + \frac{\gamma^2}{15} - \frac{\gamma^4}{525} + ..\right]$$

$$\frac{\delta_x}{L} = \frac{3}{5}\left(\frac{\delta_y}{L}\right)^2\left(1 - \frac{\gamma^2}{105} + \frac{\gamma^4}{1575} - ..\right) - \left(\frac{\delta_y}{10L}\right)\tan(\theta)\left[1 - \frac{2\gamma^2}{35} + \frac{2\gamma^4}{525} - ..\right] +$$
$$+ \frac{\tan^2(\theta)}{15}\left[1 - \frac{11\gamma^2}{105} + \frac{\gamma^4}{75} - ..\right]$$

(4.70)

Similarly, for a compressive axial load

$$\frac{M_A L}{EI} = \tan\theta\left[\gamma\cot(\gamma) + \frac{\gamma^2\tan\gamma}{\tan\gamma - \gamma}\right] - \frac{\delta_y}{L}\frac{2\gamma^2\tan\gamma}{\tan\gamma - \gamma}$$

$$\frac{M_o L}{EI} = \tan\theta\left[\gamma\cot(\gamma) - \frac{\gamma^2\tanh\gamma}{\tanh\gamma - \gamma}\right] + \frac{\delta_y}{L}\frac{2\gamma^2\tan\gamma}{\tan\gamma - \gamma}$$

$$\frac{F_y L^2}{EI} = \frac{\delta}{L}\frac{4\gamma^3}{\tan\gamma - \gamma} - \tan\theta\frac{2\gamma^2\tan\gamma}{\tan\gamma - \gamma}$$

$$\frac{\delta_x}{L} = \left(\frac{\delta_y}{L}\right)^2\left(\frac{\gamma\left[\gamma\left(3 + \tan^2\gamma\right) - 3\tan\gamma\right]}{4(\tan(\gamma) - \gamma)^2}\right) + \left(\frac{\delta_y}{L}\right)\tan(\theta)\left[\frac{1}{2} - \frac{\gamma\left[\gamma\left(3 + \tan^2\gamma\right) - 3\tan\gamma\right]}{4(\tan(\gamma) - \gamma)^2}\right] +$$

$$+\tan^2(\theta)\left[\frac{\cot\gamma - \gamma\left(1 + \cot^2\gamma\right)}{16\gamma} + \frac{\gamma\left[\gamma\left(3 + \tan^2\gamma\right) - 3\tan\gamma\right]}{16(\tan(\gamma) - \gamma)^2}\right]$$

(4.71)

Similarly, these can be expanded in series to give

$$\frac{M_A L}{EI} = 4\tan\theta\left[1 - \frac{2\gamma^2}{15} - \frac{11\gamma^4}{1575} - ..\right] - 6\frac{\delta_y}{L}\left[1 - \frac{\gamma^2}{15} - \frac{\gamma^4}{525} - ..\right]$$

$$\frac{M_o L}{EI} = -2\tan\theta\left[1 + \frac{\gamma^2}{15} + \frac{13.\gamma^4}{1575} + ..\right] + 6\frac{\delta_y}{L}\left[1 - \frac{\gamma^2}{15} - \frac{\gamma^4}{525} - ..\right]$$

$$\frac{F_y L^2}{EI} = 12\frac{\delta_y}{L}\left[1 - \frac{2\gamma^2}{5} - \frac{\gamma^4}{525} - ..\right] - 6\tan\theta\left[1 - \frac{\gamma^2}{15} - \frac{\gamma^4}{525} - ..\right] \qquad (4.72)$$

$$\frac{\delta_x}{L} = \frac{3}{5}\left(\frac{\delta_y}{L}\right)^2\left(1 + \frac{\gamma^2}{105} + \frac{\gamma^4}{1575} + ..\right) - \left(\frac{\delta_y}{10L}\right)\tan(\theta)\left[1 + \frac{2\gamma^2}{35} + \frac{2\gamma^4}{525} + ..\right] +$$

$$+\frac{\tan^2(\theta)}{15}\left[1 + \frac{11\gamma^2}{105} + \frac{\gamma^4}{75} + ..\right]$$

It can be easily verified that by setting θ to zero in equations (4.69) - (4.72) reduces them to the simpler versions (4.59) - (4.66) already derived while setting γ to zero results in the original simplified expressions for a beam with no axial load.

4.1.5 Leaf type flexures for parallelogram flexure applications

Because most mechanisms require either a pure rotation about a fixed point or linear translation in a known direction, the arcuate displacement of the cantilever is of little utility. Instead, following the analysis of Smith and Chetwynd, 1992, consider the cantilever of figure 4.13 in which the applied force is transmitted to the beam through a drive bar which, as a consequence of its geometry, imposes

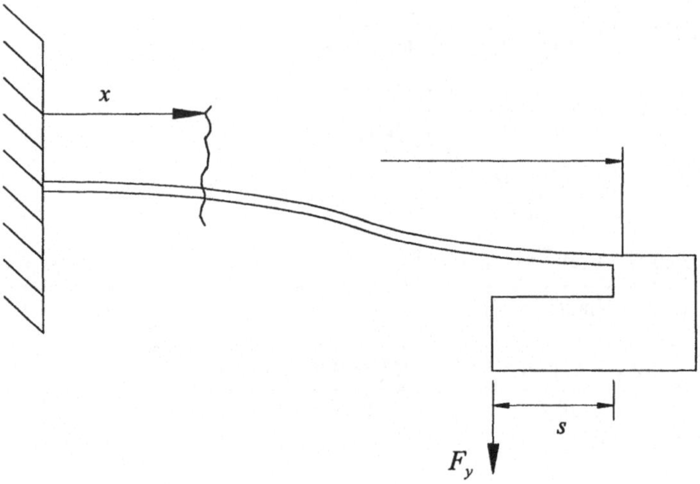

Figure 4.13: Cantilever beam subject to a tangential force plus bending moment at the free end

an additional bending moment. For this particular leaf spring, the bending equation is given by

$$\begin{aligned} EI\frac{d^2y}{dx^2} &= F_y s - F_y(L-x) \\ &= F_y(s+x-L) \end{aligned} \tag{4.73}$$

Again, integrating twice, substituting the same boundary conditions as above and assuming small displacements

$$\begin{aligned} EI\theta &= F_y\left((s-L)x+\frac{x^2}{2}\right) \\ EIy &= F_y\left((s-L)\frac{x^2}{2}+\frac{x^3}{6}\right) \end{aligned} \tag{4.74}$$

It is clear from the first of equations (4.74) that if $s < L$ then it is possible for the slope θ to be zero at a point along the beam other than at the clamped end. Such a condition is governed by the equation

$$s = L - x/2 \tag{4.75}$$

Consequently, the condition for zero slope at the free end is readily found to occur when $s = L/2$. Substituting this condition into equations (4.74), the slope and deflection at the free end of the beam are given by

$$EI\theta_{x=L} = F_y\left(-\frac{Lx}{2}+\frac{x^2}{2}\right) = 0$$
$$EIy_{x=L} = EI\delta = F_y\left(\frac{x^3}{6}-\frac{Lx^2}{4}\right) = -F_y\frac{L^3}{12} \tag{4.76}$$

The second of equations (4.76) can be rearranged into the more familiar form

$$F_y = \frac{12EI}{L^3}\delta_y = K_{\delta_y F_y}\delta_y \tag{4.77}$$

where $K_{\delta_y F_y}$ is the linear stiffness of the cantilever beam when used as a linear spring and the two subscripts represent the displacement and component of generalized force respectively. Each of the subscripts has a subordinate indicating the coordinate in which they are acting. In many cases, the context of the discussion renders such notation unnecessary. Equation (4.77) corresponds to the term on the left of the parentheses for the more complex situation in which there is also an axial load, see equation (4.59).

It is a simple matter to determine the bending moment applied to the end of the beam for a given deflection. Consequently, the bending stiffness is given by

$$K_{\delta_x M_z} = \frac{M_z}{\delta_x} = \frac{F_y L}{\delta_x 2} = \frac{6EI}{L^2} \tag{4.78}$$

Again, equation (4.78) can be compared with equation (4.61) that includes the effects of an axial load.

At first glance, it appears that we have a perfect mechanism producing a pure, rotation free translation. However, it is obvious that for the neutral axis of the beam to maintain it original length, the actual motion of the beam will be an arc with an accompanying displacement along the x-axis. For small deflections the incremental length dL^* of the beam along the neutral axis can be obtained from the integral

$$dL^* = \sqrt{1+\left(\frac{dy}{dx}\right)^2}\,dx$$
$$= \sqrt{1+\left(\frac{F}{EI}\right)^2\left(\frac{x^4}{4}-\frac{Lx^3}{2}+\frac{L^2x^2}{4}\right)}\,dx \tag{4.79}$$

Noting that from equation (4.77)

$$\left(\frac{F}{EI}\right)^2 = \frac{144\delta_y^2}{L^6} \tag{4.80}$$

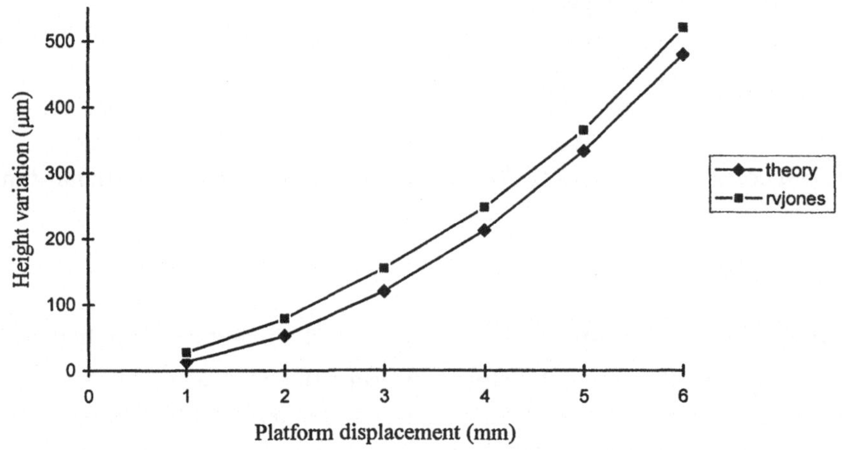

Figure 4.14: Comparison between theoretical model, equation (4.83), and parasitic motion for a simple leaf type hinge measured by R.V. Jones, 1951

the square root term in (4.79) can be expanded giving

$$\sqrt{1+\left(\frac{dy}{dx}\right)^2} = 1+\left(\frac{F}{EI}\right)^2\left(\frac{x^4}{8}-\frac{Lx^3}{4}+\frac{L^2x^2}{8}\right)$$
$$..-\frac{1}{8}\left(\frac{F}{EI}\right)^4\left(\frac{x^4}{4}-\frac{Lx^3}{2}+\frac{L^2x^2}{4}\right)^2.. \tag{4.81}$$
$$\approx 1+\left(\frac{144\delta_y^2}{L^6}\right)^2\left(\frac{x^4}{8}-\frac{Lx^3}{4}+\frac{L^2x^2}{8}\right)$$

Substituting the approximate expression in (4.81) into (4.79) and integrating eventually yields

$$L^* \approx L+\frac{3\delta_y^2}{5L} \tag{4.82}$$

Because of the zero slope at each end, it could be argued that the beam is straight at these points. Consequently, it is reasonable to assume that the length of the neutral axis would be restored to its original value by introducing an x axis displacement, δ_x corresponding to the height deviation, or foreshortening, of the flexure of magnitude

$$\delta_x = -\frac{3\delta_y^2}{5L} \tag{4.83}$$

Typically, the displacement of such a flexure would be less than 1% of its length corresponding to a parasitic displacement of 0.6%. In a study of a range of linear translation flexure designs, Reginald V. Jones studied this, among other, so-called 'parasitic' motion error. Results for a simple flexure of length 45 mm and thickness 0.56 mm subject to rather large deflections of up to 6 mm are presented. This was the more complex simple linear spring mechanism to be discussed in chapter 5. The height error measured in this study is plotted in figure 4.14 with the theoretical predictions also included. Even for these relatively large deflections, the theoretical prediction is reasonably close to the experimental results.

To assess more fully the magnitude of parasitic errors in these simple elements, it is necessary to investigate the influence of forces applied in other directions. However, an arbitrary load applied to one end of the beam can be resolved into three orthogonal forces and torsional moments. Because only small deflections are usually considered, the principle of superposition will apply. Consequently, in many cases distortions and stresses due to each force can be considered in isolation and the effects summed to yield the resultant.

4.2 Notch hinge

Next to the leaf type flexure, the notch hinge is probably the next most popular element. It is for this reason that this element will be discussed in some detail.

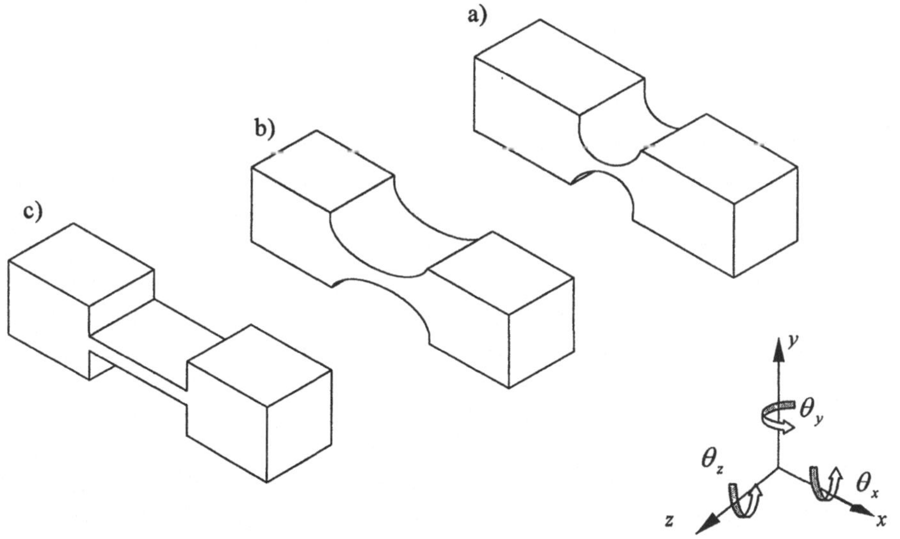

Figure 4.15: Hinges of elliptic geometry, a) the circular hinge (ε =1), b) the elliptic hinge, c) the leaf hinge (ε = ∞)

This section presents closed form equations, based on a modification of those originally derived by Paros and Weisbord in 1965, for the mechanical compliance of a simple monolithic flexure hinge of elliptic shape, the geometry of which is determined by the ratio ε of the major and minor axes. It is shown that these equations converge at ε = 1 to the Paros and Weisbord equations for a hinge of circular section and at $\varepsilon \Rightarrow \infty$ to the equations predicted from simple beam bending theory for the compliance's of a cantilever beam. These equations are then assessed by comparison with results from finite element analysis over a range of geometry's typical of many hinge designs. Based on the finite element analysis, stress concentration factors for the elliptical hinge are also presented. From finite element analysis and experimental data, it has been found that predictions for the compliance of elliptical hinges are likely to be within 12% for a range of geometry's with the ratio β_x (= $t/2a_x$) between 0.06 to 0.2 and for values of ε between 1 and 10, Smith *et al.*, 1997.

Most flexure systems may be divided into two broad categories, notch and leaf type hinges. Making two holes with a small separation between them to form a circular notch, or web, produces notch hinges. It is this thin web which serves as the flexible element (Figure 4.15(a)). Leaf type hinges typically consist of a slender member connected at each end by two rigid bodies to provide a compliant coupling (Figure 4.15(c)). The leaf can be monolithic or fabricated by clamping each end of a thin-strip. Because of its relatively high off-axis stiffness', the notch hinge is correspondingly more immune to parasitic forces. Concentration of the stresses near to the thinnest portion of the notch also results in a localization of strains therefore providing a well-defined axis of rotation. As a direct consequence of this, high local stresses limit the deflection of the notch hinge. The leaf type hinge distributes the deflection over the length of the hinge, thus lowering stress and allowing greater deflection for a given hinge length. If axial forces are present, the effective pivot point is not localized and moves along the leaf hinge as it deflects. While each type of hinge is frequently used, there are many applications where the optimum geometry is likely to be intermediate between the two extremes. Up to the present time, the designer has had only these two extreme options available. The reasons for these two common geometry's may be discerned by looking at them from a manufacturing standpoint. Traditionally, leaf type hinges have been fabricated by clamping a thin strip (the leaf) at its ends or by machining the leaf out of a single piece of material where the leaf thickness permits this. The notch hinge has been traditionally produced by drilling and reaming (or jig-boring) two closely spaced holes to produce the hinge. Thus the circular notches and parallel beam flexures are constraints imposed by the manufacturing process. With the advent of CNC milling machines and, in particular, CNC wire electro-discharge machining (WEDM), hinges of arbitrary shape can now be readily produced.

In this section some of the merits and limitations of monolithic elastic hinges for use as a single rotational degree of freedom mechanism are considered. In particular, a hinge formed from a web that would remain if two elliptical holes were machined from each side of a rectangular bar is assessed, figure 4.15.

Clearly, the hinge geometry is related to the ratio, ε, of the major to minor axes of the ellipse which ranges from 1 for a circular notch to infinity for a leaf type spring. An assessment of the stiffness and induced stresses in such a hinge when subject to a pure bending moment has been performed by comparison with results from continuum mechanics, simple bending theory, finite element analysis and experimental measurements. In reality, hinges are likely to experience combinations of both bending and shear stress, the magnitude of which will depend upon the geometry of the complete flexure mechanism and the nature of the applied loads. Although, in theory, it is possible to produce pure rotation or rectilinear motion by correct application of applied loads, in practice, undesirable or 'parasitic' forces and moments about other axes will always be present. Subsequent off-axis distortions of an elliptical hinge can be calculated using the equations for compliance in the other axes, also presented in this section.

4.2.1 Theoretical considerations

Formula for the stiffness of the leaf spring and circular notch are presented in the next two sections. For comparison between simple bending theory and finite element modeling, an expression for the bending stress in a circular notch hinge has been derived from the continuum mechanics solutions first presented by Ling in 1952.

4.2.1.1 The leaf type flexure reconsidered

From simple bending theory, the angular stiffness of this type of hinge is given by the equation

$$K_{\theta z Mz} = \frac{M}{\theta} = \frac{EI}{2a_x} \tag{4.84}$$

where $2a_x$ is the length of the hinge, E is the elastic modulus, I is the second moment of area about the neutral axis, M the bending moment and θ the angular deflection about the neutral axis. This rather unusual definition of the length of the beam has been chosen for comparison with notch hinges of circular and elliptic geometry.

Ignoring stress concentrations at the outer edges at each end of the flexure, at the onset of yielding, the stress in such a hinge for a given maximum angular displacement, θ_{max}, can be derived from simple bending theory

$$\sigma_y = \frac{Et}{2L}\theta_{max} = \frac{Et}{4a_x}\theta_{max} \tag{4.85}$$

Rearranging equation (4.85), the thickness of such a leaf spring is

$$t = \frac{4a_x}{E\theta_{\max}}\sigma_y \tag{4.86}$$

Note that the above equations are independent of the depth, *b*, of the spring.

At this angular deflection, the maximum stiffness for a given deflection (which is also proportional to the maximum strain energy that can be stored in such a hinge) is given by

$$K_{\theta_z M_z \max} = \frac{bta_x\sigma_y^2}{3E} = \frac{8ba_x^2}{3E^2}\left(\frac{\sigma_y}{\theta_{\max}}\right)^3 \tag{4.87}$$

This illustrates the well-known design rules that for a given material and length of leaf spring, the stiffness scales linearly with its depth.

4.2.2 The circular notch hinge

For the notch hinge of figure 4.15(a), an approximate solution for the angular compliance was first presented by Paros and Weisbord, 1965 (replacing the symbol R used in this paper with a_x), and is given by

$$\frac{1}{K_{\theta z M z}} = \frac{\theta_z}{M_z}$$

$$= \frac{3}{2Eba_x^2}\left[\frac{1}{2\beta+\beta^2}\right]\left\{\begin{aligned}&\left[\frac{1+\beta}{\gamma^2}+\frac{3+2\beta+\beta^2}{\gamma(2\beta+\beta^2)}\right]\sqrt{1-(1+\beta-\gamma)^2}+\\&\left[\frac{6(1+\beta)}{(2\beta+\beta^2)^{3/2}}\right]\tan^{-1}\left(\sqrt{\frac{2+\beta}{\beta}}\frac{(\gamma-\beta)}{\sqrt{1-(1+\beta-\gamma)^2}}\right)\end{aligned}\right\} \tag{4.88}$$

For full semicircular notches that are considered throughout this section, the dimensionless parameter γ is given by

$$\gamma = \frac{D}{2a_x} = \frac{2a_x+t}{2a_x} = 1+\frac{t}{2a_x} = 1+\beta \tag{4.89}$$

whereupon the full expression for the hinge compliance in (4.88) reduces to

$$\frac{1}{K_{\theta zMz}} = \frac{3}{2Eba_x^2}\left[\frac{1}{2\beta+\beta^2}\right]\left\{\begin{array}{l}\left[\dfrac{3+4\beta+2\beta^2}{(1+\beta)(2\beta+\beta^2)}\right]+ \\ \left[\dfrac{6(1+\beta)}{(2\beta+\beta^2)^{3/2}}\right]\tan^{-1}\left(\sqrt{\dfrac{2+\beta}{\beta}}\right)\end{array}\right\} \tag{4.90}$$

$$= \frac{3}{2Eba_x^2} f(\beta)$$

where β is the dimensionless factor representing the hinge geometry and $f(\beta)$ is a dimensionless compliance factor

For small values of β equation (4.90) can be simplified to produce

$$K_{\theta zMz} \approx \frac{2Eba_x^2}{3}\frac{(2\beta)^{5/2}}{3\pi} = \frac{2Ebt^{5/2}}{9\pi a_x^{1/2}} \tag{4.91}$$

This derivation has been obtained by simple integration of the bending equation for elemental strips of the hinge. This is not a continuum mechanics solution and relies on the slightly erroneous assumption that the principal stresses at all points are normal to the *yz* plane. Consequently, this equation cannot be used to calculate the maximum stress that occurs at each of the outer surfaces of the thinnest part of the notch. Using Fourier integral methods and a technique referred to as promotion of rank, Ling, 1952, has obtained a full solution for the stresses in a notch hinge subject to pure bending and this, in turn, has been experimentally validated by the photo-elastic studies of Frocht, 1935, and Goodier, 1941. This solution is reproduced in graphical form in Peterson, 1995, in terms of a stress concentration factor K_t representing a multiplication so that, for an applied bending moment *M*, the true stress, σ, can be calculated from a nominal stress using the equation

$$\sigma = K_t \frac{6M}{t^2 b} \tag{4.92}$$

Over a very wide range of values for β, the stress concentration factor, to within better than 2%, is given by (see figure 4.16)

$$\begin{aligned} K_t &= \left(\frac{2a_x}{D}\right)^{-9/20} \\ &= \left(1+\frac{t}{2a_x}\right)^{9/20} \qquad 0<\beta<2.3 \\ &= (1+\beta)^{9/20} \end{aligned} \tag{4.93}$$

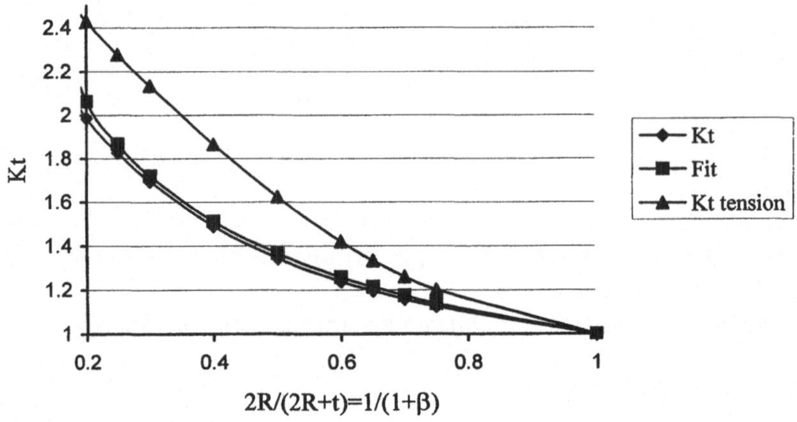

Figure 4.16: Stress concentration in a circular notch hinge as a function of $2a_x/(t+2a_x)$ (from Ling, 1968, Pilkey, 1997). For interest, the stress concentration for a tensile axial load applied to the notch is also included.

Many designs often require a specified angular displacement and therefore the applied bending moment is unknown. In this case, the approximate compliance (4.91) (to be assessed shortly) is used to estimate the bending moment for a given angular displacement θ. Substituting this into equation (4.92) provides an estimate of the stress at the outside edge of the thinnest portion of the hinge given by

$$\sigma = \frac{4Ea_x^2}{f(\beta)t^2}(1+\beta)^{9/20}\theta = \frac{E(1+\beta)^{9/20}}{\beta^2 f(\beta)}\theta \tag{4.94}$$

At a first glance, it would appear that the stress increases with a reduction in thickness, t. However, a reduction in the web thickness also corresponds to a reduction in β. Because the geometric function in the denominator increases more rapidly than the inverse square of β, the stress as given by equation (4.94) reduces with thickness (and therefore β) as would be expected. At the onset of yielding, the maximum thickness of the notch for a given deflection of the hinge can be determined from the equation

$$t^2 = \frac{4Ea_x^2}{f(\beta)}(1+\beta)^{9/20}\frac{\theta}{\sigma_y}$$

or, from equation (4.91)

$$t = \left(\frac{\sigma_y}{\theta}\right)^2 \frac{9\pi^2 a_x}{K_t^2 E^2 16} \tag{4.95}$$

Substituting equation (4.95) into (4.91), the maximum stiffness for such a hinge is given by

$$K_{\theta zMz\max} \approx \frac{b\pi^4 a_x^2}{19K_t^5 E^4}\left(\frac{\sigma_y}{\theta}\right)^5 \tag{4.96}$$

Comparison of the above with equation (4.87) indicates different parametric relationships for the maximum stiffness for the two extremes of elliptic hinge. A generalized approach to the optimal hinge geometry is necessarily dependent upon the design specification in terms of the load capability, displacement range, allowable hinge volume, operating environment and other application specific constraints. The ratio of the maximum hinge stiffness between equations (4.87) and (4.96) is

$$\frac{\text{Maximum stiffness of notch}}{\text{Maximum stiffness of leaf spring}} \approx \frac{3\pi^4}{8.19K_t^5 E^2}\left(\frac{\sigma_y}{\theta}\right)^2$$

$$\approx \frac{1.9}{K_t^5}\left(\frac{\sigma_y}{E\theta}\right)^2$$

Based on a maximum stiffness design criterion, it can be seen that the optimum hinge will depend on the magnitude of the required hinge distortion. Clearly, dependent on the magnitude of the hinge distortion, an optimal geometry is likely to be intermediate between these two hinge types and this provides the motivation for the formulation of equations for predicting stiffness values for an elliptical hinge geometry.

4.2.3 Accuracy of stiffness estimates for a notch type hinge

The difference between stiffness values predicted from the full (this will be referred to as 'exact' in the following discussions) and the 'approximate' theoretical formula given by equations (4.90) and (4.91) as a function of the parameter β is shown in figure 4.17. The approximate equation (4.91) produces a value that is lower than that of the full theoretical formula (4.90) with an error that increases nearly linearly to 8.25% at a value of $\beta = 0.3$. In an effort to assess the validity of these equations, Smith *et al.*, 1997, analyzed a range of notch hinge geometry's using finite element analysis.

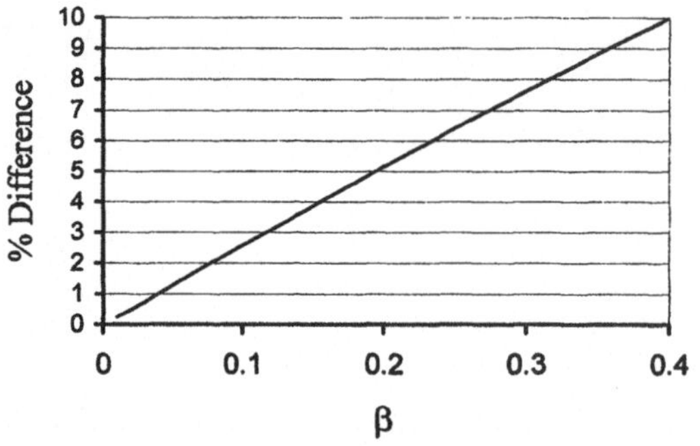

Figure 4.17: The percentage difference between the full stiffness (4.90) and values predicted by the approximate equation (4.91)

The true value for the stress concentration factor is taken to be that produced from continuum analysis. Based on this, it was found that for more than 1000 elements the stress differences were less than 1%. As yet, complete solutions for the prediction of hinge stiffness have not been produced.

Using meshes of between 1,000 and 2,000 elements, the stiffness values for a variety of notch hinges of varying β were computed and compared with the theoretical predictions given by equations (4.90 & 4.91). The results of this analysis are shown in figure 4.18. Each model predicts a different value for the stiffness of a hinge. Assuming that values derived from finite element analysis represent the 'true' stiffness, it is apparent that the approximation to the precise formula is the more accurate predictor for stiffness and that both equations predict a stiffness that is too high. The percentage error between the true stiffness and values predicted by the approximate equation (4.91) is shown in figure 4.19.

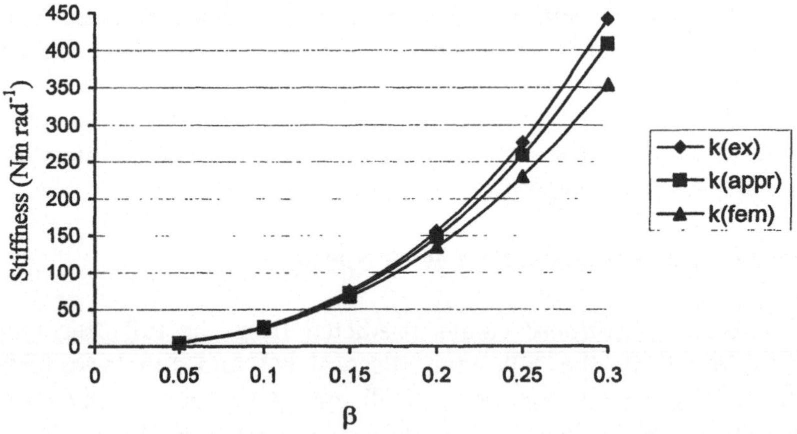

Figure 4.18: Comparison of stiffness calculations for a semi-circular notch hinge of dimensions $R = 10$ mm, $b = 1$ mm, $E = 207$ GPa using equations (4.90), (4.91) and finite element analysis

Finally, the errors between maximum stress calculated from the continuum mechanics solution and the values derived from both finite element analysis and using the bending moment calculated from the approximate theoretical equation

(4.91) due to the angular displacement from FEM results are plotted in figure 4.20. There is agreement between the continuum mechanics and FEM models to better than 2% over the complete range of β from 0.05 to 0.3 while the approximate formula introduces an error that is slightly, but not significantly, larger than the stiffness error of the same analysis.

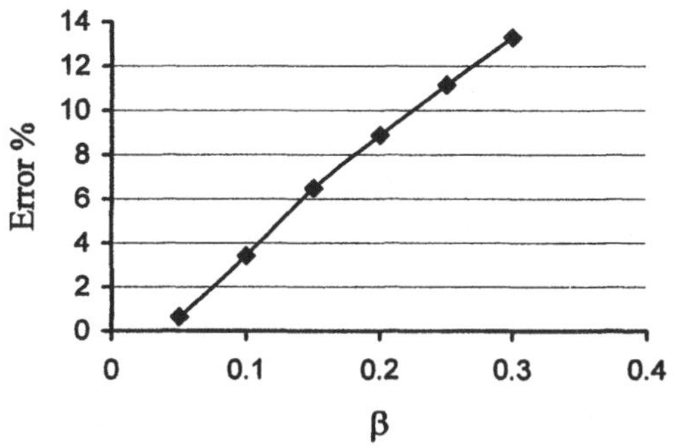

Figure 4.19: Percentage error between stiffness predicted from finite element analysis and the approximate formula (4.91) for the data shown in the previous two figures

4.2.4 The notch hinge of elliptic cross section

Considering the elliptical hinge geometry of figure 4.21, the height y at a position x on the hinge axis is given by

$$y = \left(a_y + t/2\right) - a_y\left[1 - \frac{x^2}{a_x^2}\right]^{1/2} \tag{4.97}$$

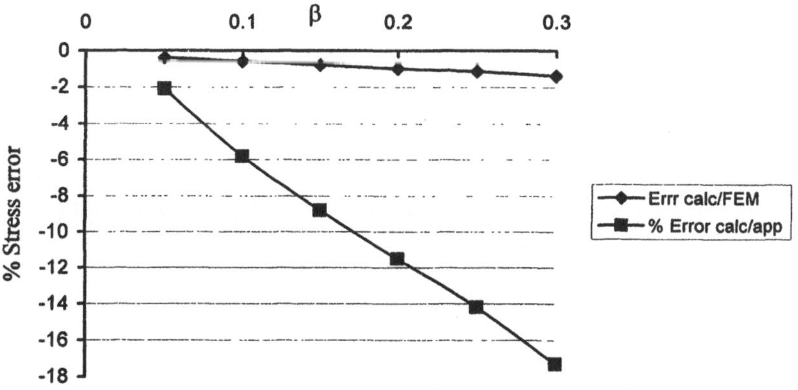

Figure 4.20: Errors between maximum stress calculated from the continuum mechanics solution and the values derived from both finite element analysis and using the bending moment calculated from the approximate theoretical equation (4.91)

Following the analysis of Paros and Weisbord, 1965, an approximate expression for the compliance of the elliptical hinge can be obtained by splitting the hinge into thin vertical strips and integrating the bending equation to give

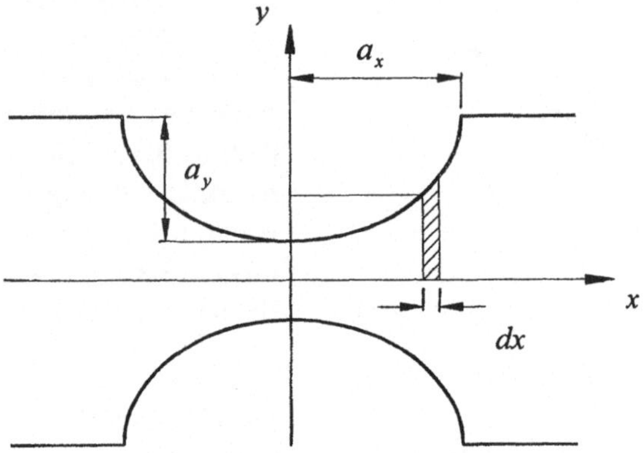

Figure 4.21: Parametric model of the elliptic flexure hinge

$$\begin{aligned}\theta_z &= \int_{-a}^{a} \frac{M_z}{EI_z(x)} dx \\ &= \int_{-a}^{a} \frac{12M_z}{Eb(2y(x))^3} dx\end{aligned} \tag{4.98}$$

Substituting equation (4.97) into the second of equations (4.98) yields

$$\theta_z = \int_{-a}^{a} \frac{3M_z}{2Eb\left[\left(a_y + \frac{t}{2}\right) - a_y\left(1 - \frac{x^2}{a_x^2}\right)^{1/2}\right]^3} dx \tag{4.99}$$

Using the substitution $x = a_x \sin\theta$, equation (4.99) can be rearranged to yield

$$\begin{aligned}\theta_z &= \frac{3M_z a_x}{2Eba_y^3} \int_{-\pi/2}^{\pi/2} \frac{\cos\theta}{\left(\left(1 + \frac{t}{2a_y}\right) - \cos\theta\right)^3} d\theta \\ &= C_1 \int \frac{\cos\theta}{(C_2 - \cos\theta)^3} d\theta\end{aligned} \tag{4.100}$$

The second equation in (4.100) is identical to the integral used in the paper of Paros and Weisbord except for a multiplication factor of the ratio of major to minor axes $\varepsilon = a_x/a_y$, and with a_y replacing the notch radius *R*, and β_y replacing

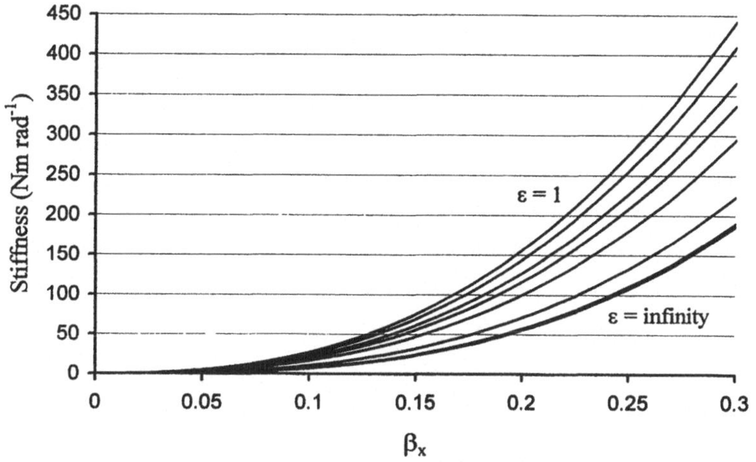

Figure 4.22: Calculated value of stiffness for an elliptical hinge for ratios of major to minor axes of the ellipse of 1 (circular notch), 1.2, 1.6, 2, 3, 10, 100, ∞ (leaf spring). Parameters for calculation are $a_x = 10$ mm, $b = 1$ mm, $E = 207$ GPa.

β. Therefore the compliance equation (4.90) can be replaced by a new expression for an elliptical hinge given by

$$\frac{1}{K_{\theta zMz}} = \frac{3}{2Eba_y^2}\varepsilon f(\beta_y) \tag{4.101}$$

where

$$\beta_y = \frac{t}{2a_y}, \beta_x = \frac{t}{2a_x}$$
$$\varepsilon = \frac{a_x}{a_y} = \frac{\beta_y}{\beta_x} \tag{4.102}$$

Because the length of the hinge, $2a_x$, is of importance in most designs, it is more convenient to express equation (4.101) in the alternative form

$$K_{\theta zMz} = \frac{2Eba_x^2}{3\varepsilon^3 f(\varepsilon\beta_x)} \tag{4.103}$$

Setting ε = 1 the above equation is the same as that for a circular notch hinge. As the major to minor axis ratio is increased, this corresponds to the hinge becoming more elliptic until, as ε tends to infinity, it can be easily shown that the stiffness converges to that for the leaf type given by equation (4.84). If a_x and β_x are maintained at a constant value, then as ε increases, the stiffness value equation (4.103) transforms from a circular notch to a leaf type spring while maintaining a constant thickness at the center. This has been plotted in figure 4.22 for values of ε from 1 to 100 with the leaf spring formula of equation (4.84) included to represent the value of stiffness with ε at infinity. For β_x greater than 0.06 there is less than 10% difference between the stiffness values for ε = 100 and the cantilever formula of equation (4.84). A vertical line through the graph of figure 4.22 represents values of stiffness at constant hinge thickness as the major to minor axis ratio is increased. This is plotted in figure 4.23 for β_x ranging from 0.02 to 0.2.

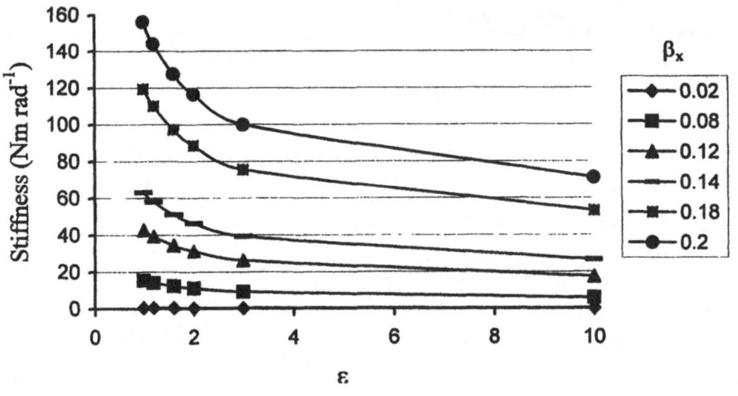

Figure 4.23: Stiffness variation of an elliptic beam of length 20 mm and depth 1 mm for a range of values of β_x

4.2.5 Compliance's of elliptic hinges in other axes

In this section, the formulae for compliance of elliptic hinges in other axes are given in terms of the appropriate modifications to the Paros and Weisbord equations. The rotational compliance about the *y*-axis, see figure 4.15, is given overleaf by

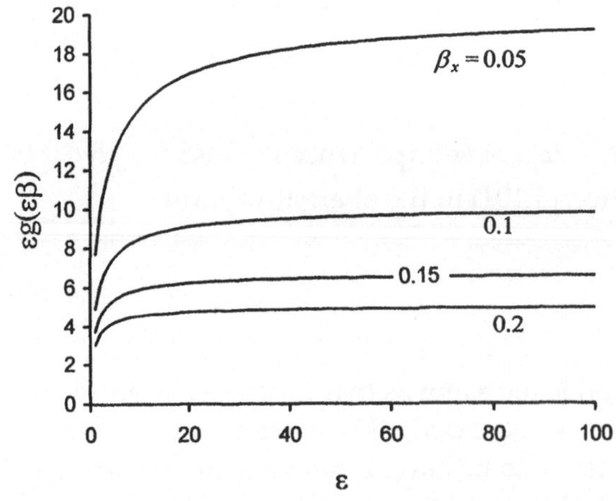

Figure 4.24: Geometric stiffness factor $\varepsilon g(\varepsilon\beta_x)$

$$\frac{1}{K_{\theta yMy}} = \frac{\theta_y}{M_y} = \frac{12\varepsilon}{Eb^3}\left[\frac{2(1+\beta_y)}{\sqrt{2\beta_y+\beta_y^2}}\tan^{-1}\sqrt{\frac{2+\beta_y}{\beta_y}} - \frac{\pi}{2}\right]$$
$$= \frac{12\varepsilon}{Eb^3}g(\beta_y) = \frac{12\varepsilon}{Eb^3}g(\varepsilon\beta_x) \tag{4.104}$$

As the ratio ε tends to infinity, the factor $\varepsilon g(\varepsilon\beta_x)$ converges to β_x, figure 4.24, which corresponds to the stiffness formula for a cantilever as deduced from simple beam theory.

The linear compliance corresponding to tension or compression of the hinge along its axis is

$$\frac{1}{K_{\delta xFx}} = \frac{\delta_x}{F_x} = \frac{\varepsilon}{Eb}g(\varepsilon\beta_x) \tag{4.105}$$

In the z-axis, the bending compliance is given by

$$\frac{1}{K_{\delta zFz}} = a_x^2\frac{\theta_y}{M_y} + \frac{6a_x^2\varepsilon}{Eb^3}\left[\begin{array}{l}2+2\beta_y+\frac{\pi}{2}(1+4\beta_y+2\beta_y^2)-\ldots\\ \ldots4(1+\beta_y)\sqrt{2\beta_y+\beta_y^2}\tan^{-1}\sqrt{\frac{2+\beta_y}{\beta_y}}\end{array}\right]$$
$$= \frac{\delta_z}{F_z} = a_x^2\frac{\theta_y}{M_y} + \frac{6a_x^2\varepsilon}{Eb^3}h(\varepsilon\beta_x) \tag{4.106}$$
$$= \frac{12a_x^2\varepsilon}{Eb^3}\left[g(\varepsilon\beta_x) + \frac{h(\varepsilon\beta_x)}{2}\right]$$

The second function in parentheses in the above equation can also be shown, see figure 4.25, to converge to the correct value of $2\beta_x/3$ as ε tends to infinity, again corresponding to the formula deduced from simple beam theory.

The shear compliance's in the y and z-axes are identical and can be computed from the equation

$$\frac{1}{K_{\delta yFy}} = \frac{1}{K_{\delta zFz}} = \frac{\varepsilon}{Gb}\left[\frac{2(1+\beta_y)}{\sqrt{2\beta_y+\beta_y^2}}\tan^{-1}\sqrt{\frac{2+\beta_y}{\beta_y}} - \frac{\pi}{2}\right]$$
$$= \frac{\varepsilon}{Gb}g(\beta_y) = \frac{\varepsilon}{Gb}g(\varepsilon\beta_x) \tag{4.107}$$

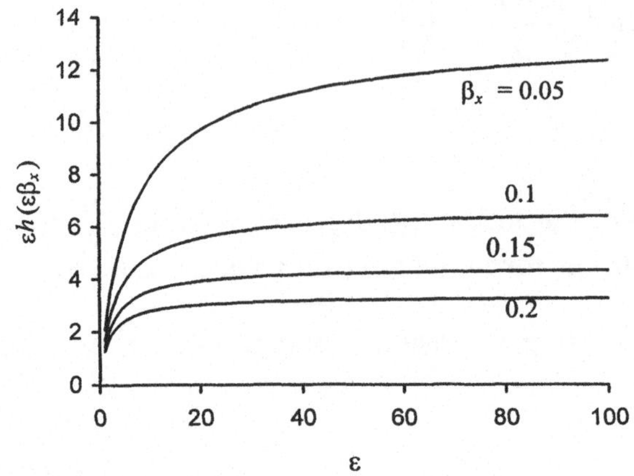

Figure 4.25: Geometric stiffness factor $\varepsilon h(\varepsilon\beta_x)$

Comparison with equation (4.104) verifies that this also produces the formula for a simple shear element as ε tends to infinity.

The above equations for the compliance of an ellipse all converge to the stiffness of a notch hinge for ε = 1 at which value $\beta_x = \beta_y = \beta$ and $a_x = a_y = R$. Under these conditions, it is possible to use the simplified equations of Paros and Weisbord which, for small values of β, are given by the equations

$$K_{\theta zMz} = \frac{M_z}{\theta_z} = \frac{2Ebt^{5/2}}{9\pi R^{1/2}} \tag{4.108}$$

$$K_{\theta zFy} = \frac{F_y}{\theta_z} = \frac{2Ebt^{5/2}}{\sqrt{1-\beta^2}\,9\pi R^{3/2}} \tag{4.109}$$

$$K_{\delta yFy} = \frac{F_y}{\delta_y} = \frac{2Eb}{9\pi\left(1-\beta^2\right)}(2\beta)^{5/2} \tag{4.110}$$

$$\begin{aligned} K_{\theta yFz} = K_{\delta zMy} &= \frac{K_{\theta yMy}}{R} = \frac{K_{\delta xFx}b^2}{12R} \\ &= \frac{M_y}{\theta_y} = \frac{Eb^3\pi}{12R}\left[\left(\frac{1}{2\beta}\right)^{1/2} - \pi\sqrt{\frac{2}{3}}\right]^{-1} \end{aligned} \tag{4.111}$$

$$K_{\delta zFz} = \frac{F_z}{\delta_z} = \frac{Eb^3}{12\pi R^2}\left[\left(\frac{1}{2\beta}\right)^{1/2} - \frac{1}{4}\right]^{-1} \tag{4.112}$$

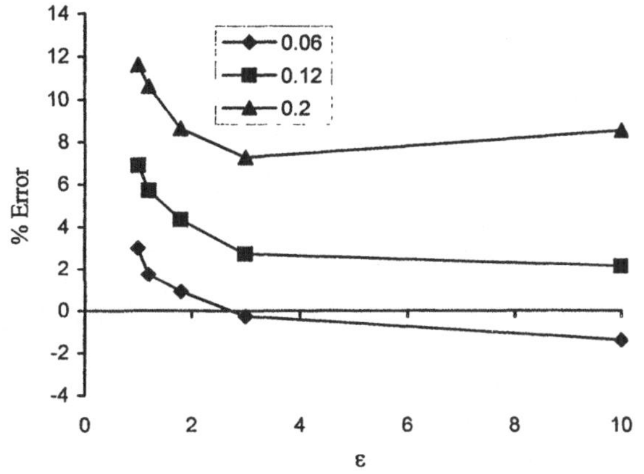

Figure 4.26: Percentage stiffness error for the elliptic hinge as a function of ε for values of β_x = 0.06, 0.12, 0.2

4.2.6 Results

Both finite element analysis and direct experimentation have been used to assess the stiffness predictions of the above equations. Based on the assumption that stresses computed from finite element analysis correspond closely (i.e. to better than 2%) to predictions based on continuum mechanics solutions. An assessment of stress concentration factors for the elliptical hinge is also presented. It should be noted that the figures in the following sections give the percentage errors between theoretical and expected values. Typically these turn out to be within a few percent and represent values that are often adequate for most engineering purposes.

4.2.6.1 Finite element results

Errors between stiffness values predicted from finite element results and the full theoretical equation (4.90) are shown in figure 4.26. In view of possible error mechanisms, for the hinge geometry's considered the theoretical models were shown to produce stiffness values that are within 12 %. Because, as the hinge tends to a leaf spring, the equations converge to those derived from the bending equation, the accuracy of predictions will correspondingly converge. Equivalently, it can be stated that the circular notch hinge represents the worst case error between theoretical and finite element models.

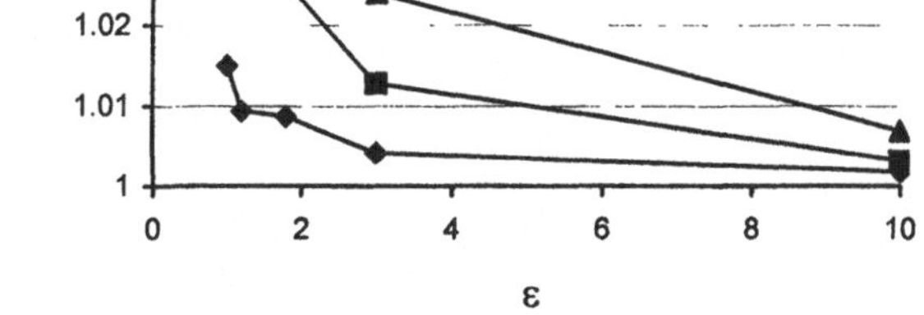

Figure 4.27: Stress concentration factors for the elliptic hinge β_x =0.06, 0.12, 0.2

The stress concentration factors have also been assessed and the results are shown graphically in figure

4.27. In the graph for the concentration factor for $\beta_x = 0.06$ one of the data points appears to be slightly erroneous. However, it is apparent that the difference is only a factor of less than 1 % and could easily be due to errors in analysis and data processing. For the circular notch hinge, this data converges to within 2 % of the 'exact' figure predicted from equation (4.93). In all cases the increase in stress is less than 10 %.

Experiments with aluminum hinges showed the relative errors between measured and theoretical results were less than 10 % and within 12 % of the finite element analysis, Smith *et al.*, 1997. These errors are considered to be within acceptable limits for values of ε between 1 and 10 and represent a reasonable confidence limit for most design purposes.

4.3 Other hinge elements

Although the notch-based hinge is relatively simple to implement and integrate into more complex mechanisms, it suffers from a limited deflection range. Historically, the more compliant cross strip pivot has been used for hinges requiring larger deflections. Such a hinge is relatively simple to construct, consisting of two or more flat strips each attached to a fixed base at one end and the moving platform at the other, figure 4.28. As can be seen in figure 4.28, each strip is mounted at an angle to the other, normally 90 degrees, and the hinge

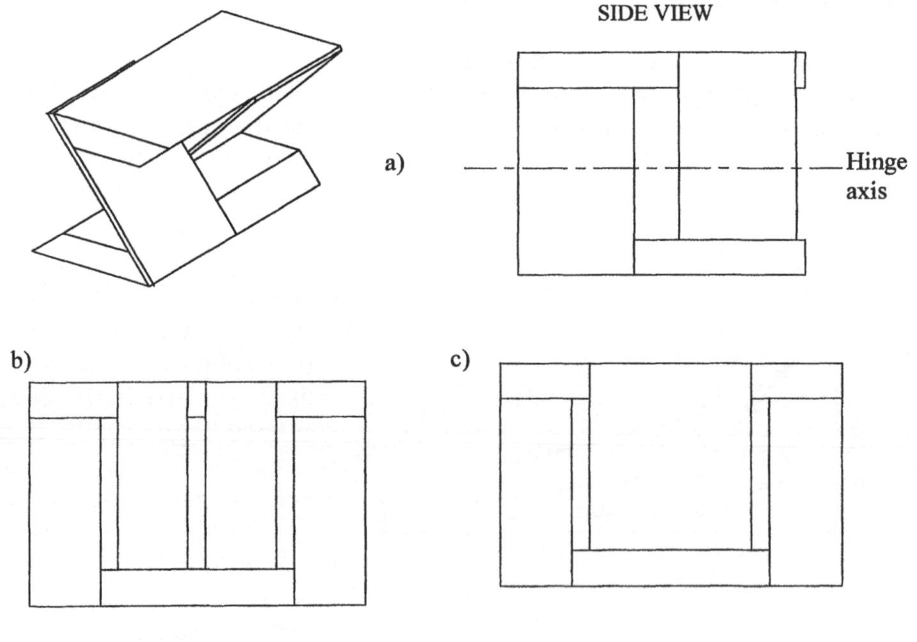

Figure 4.28: The cross strip pivot. The center line of intersection of the flexures represents the hinge axis, a) the simple two strip pivot, b) the symmetric, four leaf pivot, c) the symmetric three leaf pivot (center leaf is twice the width of the strips either side)

pivot is assumed to be about the axis of intersection between the strips. Figure 4.28(a) shows the simplest design in which the hinge is produced by clamping two successive strips of equal width. Although this may be satisfactory for many designs, it has a relatively low torsional stiffness about the vertical axis. Being symmetric about the vertical axis, a more desirable hinge is shown in figure 4.28(b). This consists of two pairs of strips. The strips at each end, being parallel, form the first pair while the two in the center form the second. Each of the two pairs are arranged at an angle as for the simple pivot of figure 4.28(a). Because the deflections of each hinge pair are identical, it makes no difference to the operation of the spring if the central pair is joined thus leading to the equivalent three-strip hinge shown in figure 4.28(c).

Moving the two strips together so that they effectively cross through one another produces a more compact, and symmetric, design. Clearly, to maintain contiguity it is necessary the strips be joined along the hinge axis. For obvious reasons, this is referred to as a 'cartwheel' hinge, Howells, 1995. In the past, there has been less interest in such a design because of the difficulties associated with producing such a shape. However, modern manufacturing techniques such as advanced welding processes and wire electro-discharge machining readily produce such shapes, see chapter 8. The geometric similarities between the simple cross strip and cartwheel hinges result in similar performance characteristics. A comparison indicates the former to offer larger deflections while the latter provides a more stable position of the pivot axis. Each are considered separately in the following.

4.3.1 The cross strip pivot

Although these have been used for many years, there have been relative few studies of this particular mechanism. In particular, Young, 1944, undertook an experimental evaluation of the four-strip pivot, the results of which were used to corroborate the theoretical investigations of Haringx, 1949, upon which the following analysis will be based. At around the same time as Haringx, Wittrick, 1948, 1951, carried out a similar analysis that, where the spring geometry and load conditions were similar, produced exactly the same results. Wittrick's later paper extends this analysis to an investigation of the influence of moving the point at which the pivots cross. In this work, which is not included in this section, it is shown that the center shift of the pivot under load can be reduced by selecting the point at which the strips cross. Although not experimentally verified it is demonstrated that a cross strip pivot arranged so that the strips cross at a point 87.3 percent ($=1/2 + \sqrt{5}/6$) of the distance along each strip will maintain the center of pivot more accurately under the influence of applied loads. Wittrick also demonstrates that a hinge produced from strips that cross at one end will always remain stable in the presence of tensile forces. Such a pivot is presented towards the end of case study 1 in section 7.5.

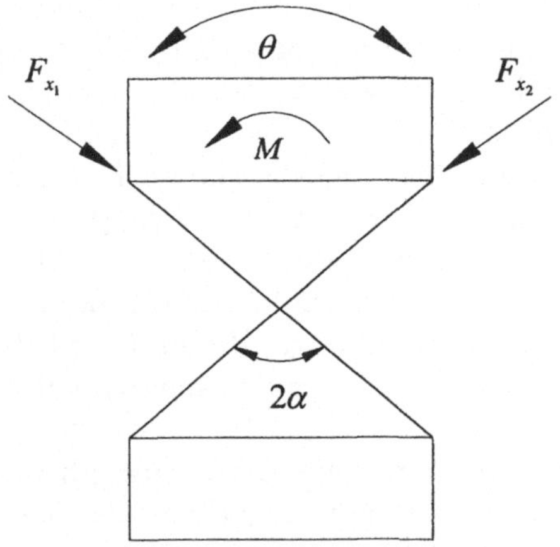

Figure 4.29: Mathematical model for the symmetric, cross-strip pivot

In most applications, the cross strip hinge will be subject to applied forces that can be resolved into an axial load and a pure bending moment as shown in figure 4.29. Applying a pure bending moment to the moving platform will result in the deflections shown in figure 4.30. Because the ends of each beam of the pivot will undergo similar displacements it can be seen from this figure that for small displacements, the angular deflection of the upper platform, *θ*, is given by

$$\theta = \frac{2\delta_y \sin\alpha}{L\sin\alpha} = \frac{2\delta_y}{L} \qquad (4.113)$$

or

$$\delta_y = \frac{L\theta}{2} \qquad (4.114)$$

where *L* is the total length of the cross-strip and δy is the deflection of the strip in a direction perpendicular to its axis.

For a pure applied couple, *M* , and no other applied forces, it is relatively easy to show that the angular stiffness of the cross strip pivot is given by

$$K_{\theta M} = n\frac{EI}{L} = \frac{EI_{tot}}{L} \qquad (4.115)$$

where *n* is the number of strips and *I* is the second moment of area of each strip about the neutral axis.

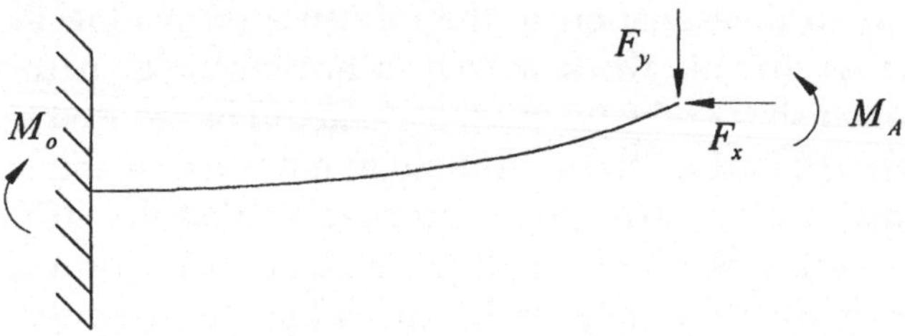

Figure 4.30: General mathematical model for a single flexure of the cross-strip pivot

Equation (4.114) implies that for small displacements the pivot point coincides with the initial intersection of the flexures irrespective of the angle of intersection. However, for small angles of intersection, the position of this intersection becomes both increasingly sensitive to deflection while the effect of manufacturing errors make it difficult to predict the exact position of the intersection upon assembly. Relatively large intersection angles, 2α, will be considered throughout.

The general applied loading for each beam is represented in figure 4.30. From this, taking bending moments to the left of an arbitrary position x, the bending equation is given by

$$EI\frac{d^2y}{dx^2} = M_0 - F_y x + F_x y \tag{4.116}$$

Again, defining

$$\begin{aligned} X &= \frac{x}{L} \\ 4\gamma^2 &= \frac{F_x L^2}{EI} \\ \varphi &= \frac{F_y L^2}{EI} \\ m_0 &= \frac{M_0 L}{EI} \end{aligned} \tag{4.117}$$

equation (4.116) can be rearranged in the form

$$\frac{d^2y}{dX^2} - 4\gamma^2 y = L(m_0 - \varphi x) \tag{4.118}$$

For boundary conditions of zero slope and deflection at the fixed end, the general solution is

$$y = \frac{Lm_0}{4\gamma^2}(\cosh(2\gamma X) - 1) - \frac{L\varphi}{8\gamma^3}(\sinh(2\gamma X) - 2\gamma X)$$

and

$$\begin{aligned} \frac{dy}{dx} &= \frac{m_0}{2\gamma}\sinh(2\gamma X) - \frac{\varphi}{4\gamma^2}(\cosh(2\gamma X) - 1) \\ \frac{d^2y}{dx^2} &= \frac{m_0}{L}\cosh(2\gamma X) - \frac{\varphi}{2\gamma L}\sinh(2\gamma X) \end{aligned} \tag{4.119}$$

These equations are similar to those of (4.46)-(4.52) and are identical with the parametric equations (4.69) and (4.71). For this isolated flexure the total bending moment M_l applied at the end of the beam is given by

$$M_l = M_0 - \frac{F_y L}{2} \tag{4.120}$$

which can be written in dimensionless form

$$m_l = \frac{M_l L}{EI} = m_0 - \frac{\varphi}{2} \tag{4.122}$$

Substitution of equations (4.114) and (4.120) into (4.69) leads to the relatively simple result

$$\frac{M_l}{\theta}\left(\frac{L}{EI}\right) = \frac{\gamma}{\tanh(\gamma)} + \gamma^2 \tag{4.123}$$

In series form this can be written as

$$\frac{M_l}{\theta}\left(\frac{L}{EI}\right) = 1 + \frac{4}{3}\gamma^2 - \frac{\gamma^4}{45}.. \tag{4.124}$$

which is different to the solution given in the paper of Haringx. For compressive loads through either a similar procedure or by the substitution

$$\gamma = \sqrt{\frac{F_x L^2}{EI}} = i\sqrt{\frac{-F_x L^2}{EI}} \tag{4.125}$$

Substituting (4.125) into (4.123), a solution is immediately given by

$$\frac{M_l}{\theta}\left(\frac{L}{EI}\right) = \frac{\gamma}{\tan(\gamma)} - \gamma^2 \tag{4.126}$$

or in series form

$$\frac{M_l}{\theta}\left(\frac{L}{EI}\right) = 1 - \frac{4}{3}\gamma^2 - \frac{\gamma^4}{45}... \tag{4.127}$$

Equations (4.123) and (4.126) have been plotted in figure 4.31. It is clear from this plot that for relatively low loads the simple equation for a cantilever stiffness (4.115) can be used with little error. For a high compressive load equation (4.126) will tend towards infinity (at $\gamma = \pi$) again resulting in the Euler buckling condition

$$F = \frac{\pi^2 EI}{4L^2} \tag{4.128}$$

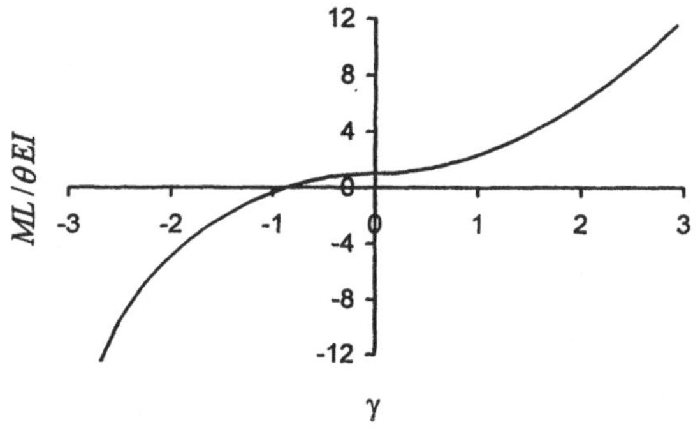

Figure 4.31: Plots of the moment equations (4.123) and (4.128) showing the effect of dimensionless axial load, γ, on the normalized bending stiffness of a single flexure of a cross strip hinge

Not surprisingly, this also corresponds to equation (4.68). From the series expansion of (4.127) or the graph of figure 4.31 it can be seen that the bending stiffness drops to zero at a value of

$$\gamma \approx \sqrt{\frac{3}{4}} \approx 0.866 \qquad (4.129)$$

It can be seen from figure 4.31 that for the relatively broad range of loads between $\gamma = \pm 1/4$ the bending stiffness will remain within 10 % of the nominal value with zero axial load.

It is normal to determine the stresses from simple bending theory. To account for the effect of the axial load, it is possible to determine the maximum stresses due to each individually and simply add these to obtain the resultant. These can be readily obtained from the formulae

$$\sigma_b = \frac{Et}{2L}\theta \qquad (4.130)$$

$$\sigma_a = \frac{F_x}{tb} \qquad (4.131)$$

from which the maximum total stress that occurs at the outside edge of the flexure is given by

$$\sigma = \frac{Et}{2L}\theta + \frac{F_x}{tb} \qquad (4.132)$$

At the limit of axial loading (i.e. $\gamma = 1/4$), the force is given by

$$F_x = \frac{EI}{4L^2} = \frac{Et^3 b}{48L^2} \qquad (4.133)$$

Assuming an upper limiting stress σ_{max} the maximum angular deflection θ_{max}, is

$$\theta_{\max} = \frac{2L}{Et}\sigma_{\max} - \frac{t}{24L} \tag{4.134}$$

In the above, the influence of stress concentrations at the junction of the flexure and rim have been ignored. Some issues are deferred to chapter 8, in which practical implementation and fabrication are discussed.

4.3.1.1 Center shift of the pivot

It is of interest to assess the stability of the pivot axis during deflection. For a pure bending moment, the flexures are distorted to arcs of a circle. For a total angular deflection, θ, the shift of the pivot, δ_p is always *towards* the moving platform and its magnitude for a simple cantilever is given by, Haringx, 1949

$$\frac{\delta_p}{L} = \frac{2\sin(\theta/2)}{\theta} - \cos(\theta/2) \tag{4.135}$$

For flexures crossing at an angle 2α, the deflection is modified to

$$\frac{\delta_p}{L} = \frac{1}{\cos\alpha}\left(\frac{2\sin(\theta/2)}{\theta} - \cos(\theta/2)\right) \tag{4.136}$$

It is common to use a cross strip pivot in which the flexure bisects at 90 degrees. In this case, expanding (4.136) into a series and, assuming the angular deflection to be small, this reduces to the approximate equation of Howells, 1996,

$$\frac{\delta_p}{L} = \frac{\sqrt{2}\theta^2}{12} \tag{4.137}$$

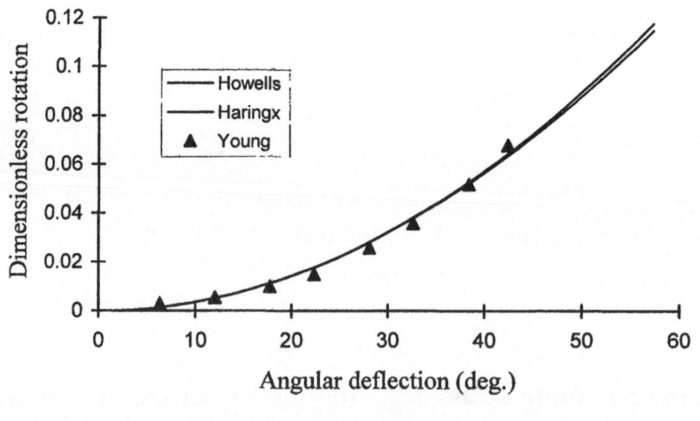

Figure 4.32: Center shift of the pivot point of a cross strip pivot ($\alpha = \pi/4$)

Equations (4.136) and (4.137) are plotted in figure 4.32 with the experimental data of Young, 1944, included for comparison. It can be seen that there is a reasonable correspondence with the approximate formula even for relatively large angular deflections.

4.3.2 The cartwheel hinge

The cartwheel hinge is similar to a cross strip pivot. Both consist of a pair of leaf type flexures, usually at right angles to each other, with the point of intersection being considered the pivot point. Cartwheel hinges, see figure 4.33, differ from cross strip pivots in that the flexures coincide at the hinge axis and are joined. Such a geometry is both symmetric and, unlike the cross strip pivot, amenable to manufacture using wire electro discharge machining. Because of this, the cartwheel hinge can be produced from a single monolithic piece of material. For some designs, the cartwheel hinge may be integrated within complete monolithic mechanisms and therefore reducing some of the problems associated with assembly. Following Howells *et al.*, 1996, because of the symmetry, it is possible to analyze a single *spoke* and apply superposition to assess the characteristics of the complete flexure. In the absence of radial loads, the deflections and force will be the same in each spoke. Consequently, the angular deflection of one spoke will be half of the deflection between the moving platform and base. Looking at the forces applied to a single spoke with no axial load, figure 4.34, it is possible to write the parametric equation

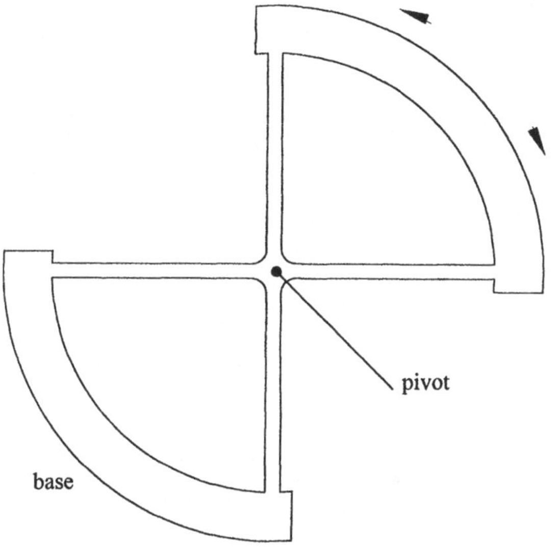

Figure 4.33: The cartwheel hinge

$$EI\frac{d^2y}{dx^2} = F_y(R - x) - M_A \qquad (4.138)$$

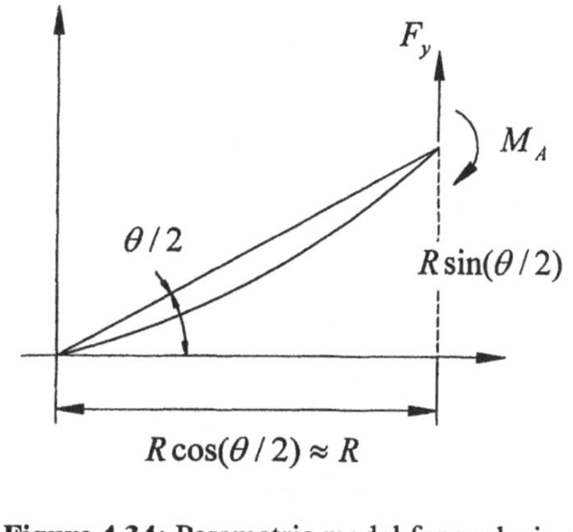

Figure 4.34: Parametric model for analysis of the cartwheel hinge. θ is the angular deflection and this is usually small.

where R is the half-length of a leaf spring, $=L/2$.

For zero slope and deflection at the hub, this can readily be integrated to yield

$$EI\frac{dy}{dx} = F_y Rx - F_y \frac{x^2}{2} - M_A x \tag{4.139}$$

$$EIy = F_y \frac{Rx^2}{2} - F_y \frac{x^3}{6} - M_A \frac{x^2}{2} \tag{4.140}$$

The boundary conditions at the rim (i.e. $x = L$) are

$$\begin{aligned} \left.\frac{dy}{dx}\right|_{x=R} &= \frac{\theta}{2} \\ y\big|_{x=R} &= R\sin\left(\theta/2\right) \end{aligned} \tag{4.141}$$

Again defining the dimensionless parameters

$$\begin{aligned} \varphi &= \frac{F_y R^2}{EI} \\ m &= \frac{M_A R}{EI} \end{aligned} \tag{4.142}$$

equations (4.139) to (4.142) can be combined to yield

$$\begin{aligned} \theta &= \varphi - 2m \\ \sin\left(\theta/2\right) &= \frac{\varphi}{3} - \frac{m}{2} \end{aligned} \tag{4.143}$$

Solving for m and φ finally produces

$$\varphi = 12\left[\sin\left(\theta/2\right) - \frac{\theta}{4}\right] \tag{4.144}$$

$$m = 6\left[\sin\left(\theta/2\right) - \frac{\theta}{3}\right] \tag{4.145}$$

Assuming small angular deflections these can be simplified to give

$$F_y \approx \frac{3EI}{R^2}\theta = F_y \approx \frac{12EI}{L^2}\theta \tag{4.146}$$

$$M_A \approx \frac{EI}{R}\theta = \frac{2EI}{L}\theta \tag{4.147}$$

where L is the length of the flexure spring from fixed to moving rim (i.e. diameter of the cartwheel). For small angular deflections, θ, the approximate equations (4.146) and (4.147) into (4.138) can be inserted into (4.140) to yield the complete set of equations for a single spoke of the cartwheel hinge

$$\frac{d^2y}{dx^2} = \frac{1}{R}\left[2 - \frac{3x}{R}\right]\theta \tag{4.148}$$

$$\frac{dy}{dx} = \frac{1}{R}\left[2x - \frac{3x^2}{2R}\right]\theta \tag{4.149}$$

$$y = \frac{1}{R}\left[x^2 - \frac{x^3}{2R}\right]\theta \tag{4.150}$$

From these it can be readily shown that the maximum bending moment, $M_{\max}$, occurs at the hub of the hinge and is given by

$$M_{\max} = \frac{2EI}{R}\theta \tag{4.151}$$

There is a point of inflection, corresponding to zero bending stress, at a position 2/3 of the distance from the hub to the rim.

4.3.2.1 Torsional stiffness of the cartwheel hinge

Because of the symmetry of the hinge, it is possible to determine the torque required to rotate one of the leaf springs and then double this for the complete cartwheel. The moment applied about the pivot for a single spoke must be capable of producing both the bending moment along the beam plus the applied force i.e.

$$M = FR + M_A \tag{4.152}$$

For the two combined spokes, the compliance's of each flexure will add resulting in one half of the torque. However, there are two sets of these springs and the *stiffness* of these will add. Consequently, the total torque is given by

$$M = \frac{4EI}{R}\theta \tag{4.153}$$

From this, the angular stiffness, $k_{M\theta}$, of the flexure is

$$k_{M\theta} = \frac{4EI}{R} \tag{4.154}$$

4.3.2.2 Center shift of the pivot

Following similar arguments for the derivation of the height deviation for a linear flexure, it shall be assumed that the length of the neutral axis along the spoke remains unaltered upon deflection. We can replace the original spoke with a new one of length R^* and integrate along the x-axis to its end point, which will be foreshortened to $R^* \cos(\theta/2)$. Consequently, for a single spoke we can calculate the length of the flexure, s, using the line integral

$$\begin{aligned} s &= \int_0^{R^*} \sqrt{1 + \left(\frac{dy}{dx}\right)^2}\, dx \\ &= \int_0^{R^*} 1 + \frac{1}{2}\left(\frac{dy}{dx}\right)^2 ..dx \end{aligned} \tag{4.155}$$

Substituting equation (4.150) into the above and integrating yields

$$\begin{aligned} s &= R^* \cos\left(\theta/2\right)\left(1 + \frac{17\theta^2}{120}..\right) \\ &\approx R^*\left(1 - \frac{\theta^2}{8}..\right)\left(1 + \frac{17\theta^2}{120}..\right) \approx R^*\left(1 + \frac{\theta^2}{60}..\right) \end{aligned} \tag{4.156}$$

Consequently, neglecting terms in θ of fourth order or above, the length of the spoke will remain unchanged if the new radius of the rim is shortened by an amount equal to the second term in parentheses in the last of equations (4.156). Taking into consideration the similar motion of the pivot towards the fixed rim and that the two springs intersect at an angle of $\pi/2$, the normalized displacement of the central pivot is given by

$$\begin{aligned} \frac{\delta_p}{R} &= \frac{\sqrt{2}\theta^2}{30} \\ &\text{or} \\ \frac{\delta_p}{L} &= \frac{\sqrt{2}\theta^2}{60} \end{aligned} \tag{4.157}$$

Comparison between the second of equations (4.157) with the pivot motion for a cross strip hinge given in (4.137) indicates that the cartwheel hinge

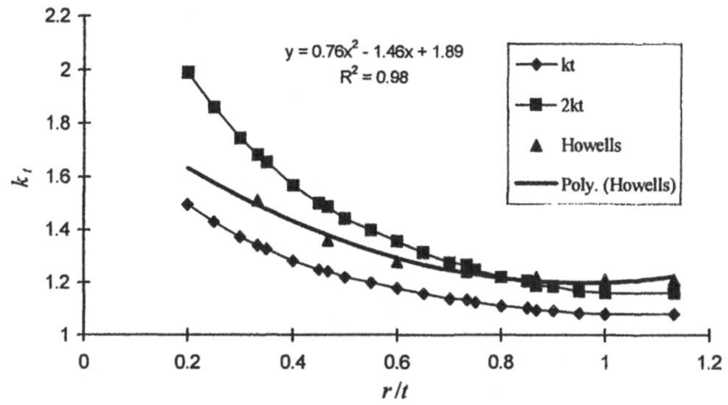

Figure 4.35: Stress concentration factors for a cartwheel hinge as a function of the normalized fillet radius *r/t*. Curves represent values for a simple cantilever with fillet from Frocht, 1951 and finite element results of Howells, 1996

represents a factor of 5 improvement in the stability of the center of rotation. It is also noted that the pivot point moves *towards* the stationary platform (or rim) which is in the opposite direction for that of the cross strip pivot.

4.3.2.3 Stresses in the hinge

From the above simple analysis the maximum bending moment in the springs can be used with the bending equation to provided an estimate of the maximum stress given by

$$\sigma_{\max} = \frac{M_{\max} t}{I2} = \frac{Et}{R}\theta \tag{4.158}$$

However, it is clear from the design that there will be a radius at the conjunction of hinges with an associated stress concentration. Assuming a radius ranging from somewhere between 1/3 and equal to the thickness of the hinge, it is reasonable to assume that this can be considered closely equivalent to a beam undergoing a pure bending moment with a fillet at the fixed end. Under these circumstances it is possible to use a graph of the stress concentration factor k_t shown in figure 4.35. For the cartwheel hinge, each fillet will be subject to stresses imposed from the moments of two spokes. Consequently, it is tentatively suggested that the resultant stress concentration will be twice that for a single beam and this is also plotted in figure 4.35. Finite element results derived from the paper of Howells *et al.* also indicate that the true value will be somewhere intermediate between the two values for small fillet radii and converges to within a few percent as the radius approaches the thickness of the hinges. From this graph, it can be seen that the stress concentration reaches a relatively constant value of 1.2 in the region at which the fillet radius is equal to the hinge thickness and this probably represents an optimal value. In fact, in the researches of Howells *et al.*, it was found the stress increases with fillet radius above this region. It is thought that this may be a consequence of the modeling. In this, cartwheels were used with spokes having a relatively low thickness to length ratio of 0.25/0.015. Consequently as the fillet radius becomes large, it will effectively shorten the spokes resulting in stiffening of the cartwheel. Analysis of stresses was performed by applying a fixed angular displacement and recording the maximum Von Mises stress. Clearly, the effective shortening of the spokes will increase the stress for a given distortion.

4.3.3 The cruciform hinge

It is occasionally desirable to provide a hinge with mounts displaced along a common rotation axis. One method of achieving such a mechanism is to construct the cruciform type hinge shown in figure 4.36. Such a hinge has relatively low torsion rigidity along its *z*-axis while maintaining the high resistance to bending about the two axes in the *xy* plane. In many civil engineering structures in which a high torsional rigidity is desired, such a design would, of course, be disastrous.

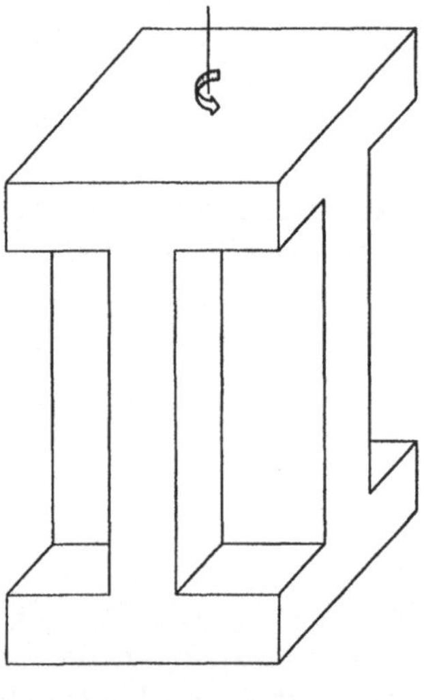

Figure 4.36: The cruciform hinge

Equations for predicting the behavior of such a mechanism do not appear to have been investigated. However, to achieve a rough *estimate* of the torsional stiffness of a prismatic beam with no constraints at the ends, it is possible to apply the principle of superposition and consider this to be equivalent to two beams at rights angles. For a single beam, we have already derived the relationship (2.104)

$$M = k_1 \frac{Gt^3 d}{3L}\theta \tag{4.159}$$

A reasonable estimate for the constant that will not depart by more than 4% (see figure 4.37) is given by the simple expression

$$k_1 = 1 - \frac{0.582t}{d} \tag{4.160}$$

Consequently, the torsional stiffness of the cruciform beam can be estimated from the equation

$$K_{\theta_z M_z} = 2\left(1 - \frac{0.582t}{d}\right)\frac{Gt^3 d}{3L} - 0.418\frac{Gt^4}{3L} \tag{4.161}$$

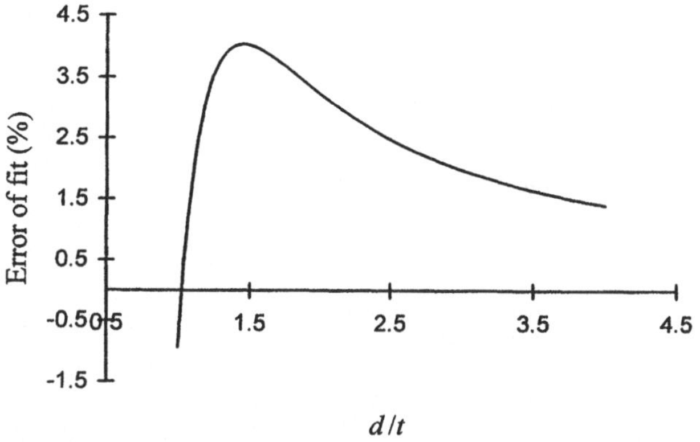

Figure 4.37: Percentage error between torsion constant and the simple curve fit of equation (4.166) as a function of depth to thickness ratio of a rectangular beam

This can be further rearranged to give

$$\begin{aligned} K_{\theta_z M_z} &= 2\left(\frac{d}{t} - 0.582\right)\frac{Gt^4}{3L} - 0.418\frac{Gt^4}{3L} \\ &= \left(\frac{d}{t} - 0.373\right)\frac{2Gt^4}{3L} \end{aligned} \qquad (4.162)$$

The bending stiffness about the x and y axes are identical for a 'square' cross and can be obtained from simple bending theory

$$\begin{aligned} K_{\theta_x M_x} &= \frac{EI_{yy}}{L} = \frac{E(dt^3 + td^3 - t^4)}{12L} \\ &= \frac{Gt^4(1+\upsilon)\left(\frac{d}{t} + \frac{d^3}{t^3} - 1\right)}{6L} \end{aligned} \qquad (4.163)$$

Temporarily defining the dimensionless parameter χ which is a function of the slenderness of the two strips in the cross section and always greater than unity

$$\chi = \frac{d}{t} \qquad (4.64)$$

The ratio of the bending stiffness about the x or y-axes to the torsional stiffness about the z-axis is given by

$$\frac{K_{\theta_x M_x}}{K_{\theta_z M_z}} = \frac{(1+\upsilon)\left(\chi + \chi^3 - 1\right)}{4(\chi - 0.373)} \qquad (4.165)$$

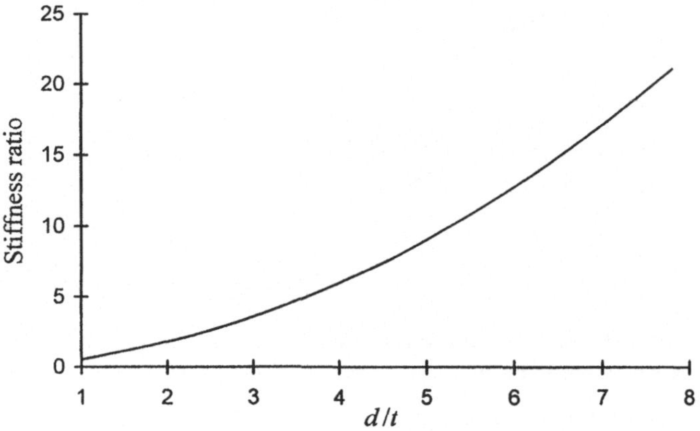

Figure 4.38: Ratio of bending to torsional stiffness for the cruciform hinge

Figure 4.38 shows that for values of d/t of 6 or more, the stiffness ratio is greater than an order of magnitude.

The objective of this simple analysis is to illustrate the essential feature of such a design. In reality, the beam is likely to be fixed at each end to rigid platforms and, as such, will be subject to constraints, which are **not** included in the above model. Rigid end constraints will not effect bending of the cruciform but are likely to increase the torsional stiffness and thus reduce the ratio predicted by equation (4.165). Additional work is required to derive a more complete understanding of this mechanism.

4.4 Two axis hinges

The following sections discuss elements to provide connections between rigid bodies with high compliance in two desired axes. As before, the axes providing compliance are considered to be the freedoms of the flexure element.

4.4.1 A simple two axis hinge (y, θ)

Attention has so far concentrated on the design of hinges to provide freedoms in one rotation coordinate with constraints in the other five. In a large number of applications it is desired that rigid bodies be connected by elements that provide combinations of linear and rotational freedoms in two, three or more axes. A relatively simple 'two axis' hinge, consisting of a slender beam fixed at each end, is shown in figure 4.39. Translation is provided by distortion perpendicular to the y-axis while rotation, θ_z, is achieved through the application of a pure bending moment at the midpoint. Constraint along the axis of the beam, the x-axis, is due to the axial compressive and tensile forces either side of such a load. Equations for the stiffness at the center of the beam in the plane of the drawing are

$$k_{\delta_y F_y} = \frac{192EI}{L^3} = \frac{16Ebt^3}{L^3}$$
$$k_{\theta_z M_z} = \frac{16EI}{L} = \frac{4Ebt^3}{3L} \qquad (4.166)$$
$$k_{\delta_x F_x} = \frac{Ebt}{L}$$

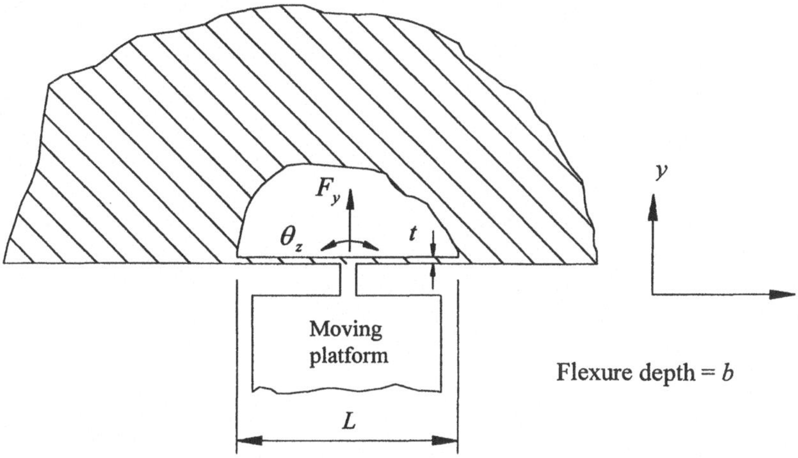

Figure 4.39: The simple, single beam, two-axis flexure

In the derivation of the second of equations (4.166), the singularity function method for solution of the bending equation is of utility. In practice, it is not possible to apply a bending moment at a point as this would induce an infinite shear stress resulting is discontinuous slopes and deflections at that point. Ignoring these difficulties, it is possible to solve the beam bending equation by matching slopes at the point of application of the moment. Unless the beam is slender, and great care is taken to avoid large stress concentrations at the point of application of the moment, it is unlikely that the predicted stiffness will be realized. Surprisingly, the maximum bending moment in the beam turns out to be one half of that applied, see also Young, 1989. It is possible to derive formulae for the stresses from simple bending theory. For a linear deflection in the y axis direction,

$$M_{y_{\max}} = \frac{F_y L}{8} \qquad (4.167)$$

In terms of the design stress σ_Y, the maximum deflection in the absence of other applied forces, moment or stress concentrations is limited to

$$\delta_{y_{\max}} = \frac{\sigma_Y L^2}{12Et} \quad (4.168)$$

Similarly, for the beam subject to a centrally applied moment, M_z, the maximum moment in the region at which it is applied is given by

$$M_{\max} = \frac{M_z}{2} = \frac{4Ebt^3}{L}\theta_z \quad (4.169)$$

In this case, the maximum angular deflection is limited to

$$\theta_{z_{\max}} = \frac{\sigma_Y L}{4Et} \quad (4.170)$$

For many applications, it is the ratio of relative stiffness values that provides a measure of the immunity to parasitic forces. For linear displacements, the ratios of stiffness in the desired y-axis to those in x and z are given by

$$\begin{aligned} \frac{k_{\delta_y F_y}}{k_{\delta_z F_z}} &= \left(\frac{t}{b}\right)^2 \\ \frac{k_{\delta_y F_y}}{k_{\delta_x F_x}} &= \left(\frac{4t}{L}\right)^2 \end{aligned} \quad (4.171)$$

Typically, ratios of thickness of the beam to the depth will be of the order of one tenth or greater resulting in immunity measured in relative displacements per unit force of 10 to 100:1.

4.4.2 The two beam, two degree of freedom flexure (y, θ)

Another common form of two-axis hinge designed to provide a single translation plus rotation is shown in figure 4.40. This mechanism consists of two cantilever beams connected to separate rigid bodies at one end and joined together at the 'free' end. Consequently, a y-axis displacement applied to the moving end will cause each of the beams to distort in the familiar 's' shape analyzed in section 4.1.5. Because the cantilevers are series connected between the rigid bodies, each will be subject to the same force and, therefore, displacement. The stiffness of the *complete* flexure will be half of that for each cantilever i.e.

$$k_{\delta_y F_y} = \frac{6EI}{L^3} \quad (4.172)$$

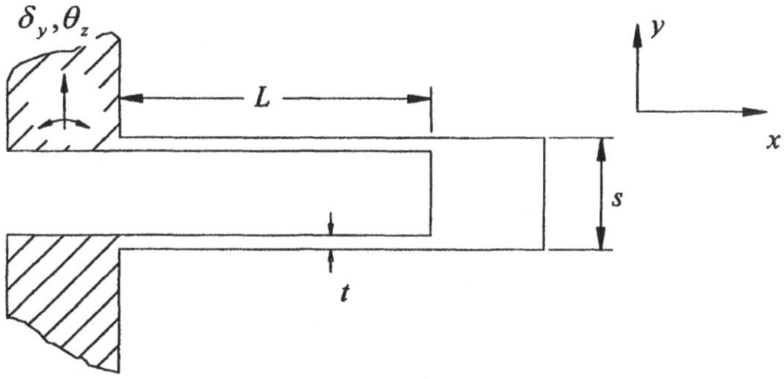

Figure 4.40: The two beam, two degree of freedom flexure

To prevent angular distortion about the z-axis while deflecting in the y-direction, a moment must be applied given by (remember that for a total displacement δ_y each beam will distort half of this)

$$M_z = \frac{3EI}{L^2}\delta_y \tag{4.173}$$

For angular deflections about the z-axis, each beam will bend as a simple cantilever with half total angle. Consequently, the bending moment required to induce an angular deflection θ_z is

$$\begin{aligned} M_z &= \frac{EI}{L}\frac{\theta_z}{2} \\ &\therefore \\ k_{\theta_z M_z} &= \frac{EI}{2L} = \frac{Ebt^3}{24L} \end{aligned} \tag{4.174}$$

Clearly, for a combination of displacement and rotation, the contributions from (4.173) and (4.174) must be appropriately summed.

Assuming that axial forces are small (i.e. $1-\cos(x/L) \approx \cosh(x/L)-1$), the stiffness of this element in the x-axis can be approximated by the angular twist at the 'free' ends of the cantilever given by

$$M_z = F_x s \tag{4.175}$$

From simple beam bending theory, the angular deflection at the end of the beam is given by

$$\theta\big|_{x=L} = \frac{F_x sL}{EI} \tag{4.176}$$

Upon such a deflection, there are two components of deformation that contribute to the displacement between the rigid bodies in the *x*-axis direction. The first is due to the combined equal extension and compression giving a displacement

$$\delta_x = \frac{2L}{Elbt}F_x = \frac{Lt^2}{6EI}F_x \tag{4.177}$$

The second contribution is due to the lever arm of length *s* and is given by

$$\delta_x = s\sin(\theta) \tag{4.178}$$

Again, assuming small deflections, equations (4.176) and (4.178) can be combined to yield the axial deflection

$$\delta_x = \frac{s^2 L}{EI}F_x \tag{4.179}$$

The total deflection is given by the sum of equations (4.177) and (4.179) from which the stiffness is

$$k_{\delta_x F_x} = \frac{6EI}{Lt^2 + 6s^2 L} \tag{4.180}$$

The separation between the cantilever beams, *s*, can not be less than the thickness *t* and the ratio s/t will typically be in the region of 1.5 - 2. It is therefore reasonable to ignore the term on the left in the denominator after which the stiffness is approximated as

$$k_{\delta_x F_x} = \frac{EI}{s^2 L} = \frac{Ebt}{12L}\left(\frac{t}{s}\right)^2 \tag{4.181}$$

In terms of a design stress σ_Y, the maximum deflections for such a design are

$$\theta_{z\max} = \frac{4\sigma_Y L}{Et} \tag{4.182}$$

$$\delta_{y\max} = \frac{2\sigma_Y L^2}{3Et} \tag{4.183}$$

Equations (4.182) and (4.183) have been derived with the assumption that the other displacements are absent. For combined linear translation and rotation, it is necessary to sum the stresses for both and reduce the range accordingly.

Table 4.4.2.1: A comparison of the stiffness formulae and maximum possible deflections for the simple one and two beam, two-degree of freedom hinge

Hinge parameter	Two beam hinge	Single beam hinge	Ratio $\frac{\text{Two beam hinge}}{\text{Single beam hinge}}$
$k_{\delta_y F_y}$	$\frac{6EI}{L^3}=\frac{Ebt^3}{2L^3}$	$\frac{192EI}{L^3}=\frac{16Ebt^3}{L^3}$	1/32
$k_{\theta_z M_z}$	$\frac{EI}{2L}=\frac{Ebt^3}{24L}$	$\frac{16EI}{L}=\frac{4Ebt^3}{3L}$	1/32
$k_{\delta_x F_x}$	$\frac{EI}{s^2L}=\frac{Ebt}{12L}\left(\frac{t}{s}\right)^2$	$\frac{Ebt}{L}$	$\frac{1}{12}\left(\frac{t}{s}\right)^2$
$\delta_{y\max}$	$\frac{2\sigma_Y L^2}{3Et}$	$\frac{\sigma_Y L^2}{12Et}$	8
$\theta_{z\max}$	$\frac{4\sigma_Y L}{Et}$	$\frac{\sigma_Y L}{4Et}$	16

Table 4.4.2.1 compares the stiffness and maximum deflection formulae for the one and two beam, two-axis hinge (reminder, two-axis implies that the hinge provides freedoms in two coordinates). In this table it is assumed that the geometric parameters are comparable, i.e. the lengths and thickness can be compared on a one is to one basis. From this simple analysis, it can be seen that the stiffness in the parasitic, x-axis direction is to the square of the ratio of t/s. Clearly, there is greater advantage in keeping s small thus favoring wire electro-discharge machining as a method for manufacture.

Torsion stiffness and shear stresses about the x-axis can be computed from the equations given in section 2.8.2 assuming the hinge to be a prismatic rectangular bar of length $2L$.

4.4.3 The two axis toroidal hinge (θ_x, θ_y)

A common technique for producing a two axis rotary hinge is to use a thin rod. In practice this can be achieved by simply drilling holes in both rigid bodies, inserting a rod and clamping at each end. To obtain two axes with equal compliance in both axes it is only necessary that the rod joining the rigid bodies is symmetric. Consequently, either a circular or square rod may be used. Because of this symmetry, each axis can be considered separately as a simple cantilever for which the previous analyses are applicable.

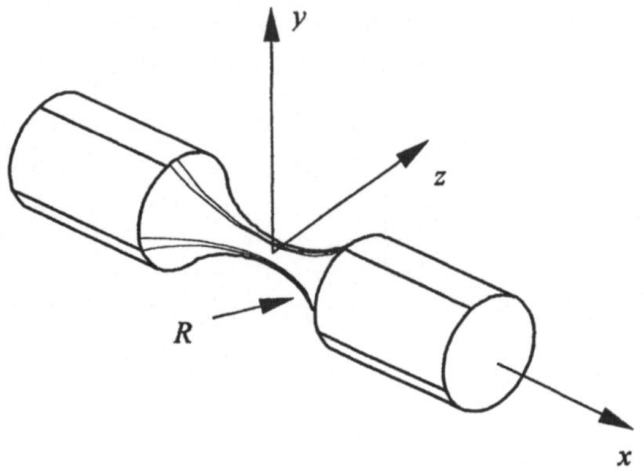

Figure 4.41: The two-axis circular, or toroidal, notch hinge, t = thickness at thinnest portion of the web

Analogous to the notch is the two-axis circular hinge of figure 4.41. Turning a circular notch into a solid rod can produce such a hinge. The equation for the angular stiffness in the freedom axes can be approximated from, Paros and Weisbord, 1965

$$\frac{M_y}{\theta_y} = \frac{M_z}{\theta_z} \approx \frac{ER^3(2\beta)^{7/2}}{20} = \frac{Et^{7/2}}{20R^{1/2}} \tag{4.184}$$

For a force applied at the end of the hinge, the corresponding angular displacement can be obtained from

$$\frac{F_z}{\theta_y} = \frac{F_y}{\theta_z} = \frac{Et^{7/2}}{20R^{3/2}\sqrt{1-\beta^2}} \approx \frac{Et^{7/2}}{20R^{3/2}} \tag{4.185}$$

The linear stiffness in the axial direction is

$$\frac{F_x}{\delta_x} \approx \frac{ER(2\beta)^{3/2}}{2} = \frac{Et^{3/2}}{2R^{1/2}} \tag{4.186}$$

Although shear compliance's are alluded to in the paper of Paros and Weisbord, 1965, it is not readily apparent how to derive such formulae. Additionally, there have been no systematic experiments to verify these equations. Being derived from the same base equations used for the single axis notch hinge, it would be prudent to budget for errors of around 20%.

Stresses can be computed from the stress concentration factors provided in the book of Pilkey, 1997. For bending of the hinge about either rotary axis θ_x or θ_y the maximum stress can be computed from

$$\sigma_{nom} = k_{tb}\frac{32M}{\pi t^3} \tag{4.187}$$

where the stress concentration factor for bending is given by

$$k_{tb} = 3.032 - 7.431\left(\frac{2\beta}{1+\beta}\right) + 10.390\left(\frac{2\beta}{1+\beta}\right)^2 - 5.009\left(\frac{2\beta}{1+\beta}\right)^3$$
$$\beta = \frac{t}{2R} \tag{4.188}$$

A torsional moment about the x-axis will produce a maximum shear stress given by

$$\tau_{nom} = k_{ts}\frac{16M_x}{\pi t^3} \tag{4.189}$$

where the stress concentration factor for a pure applied shear is given by

$$k_{ts} = 2.000 - 3.555\left(\frac{2\beta}{1+\beta}\right) + 4.898\left(\frac{2\beta}{1+\beta}\right)^2 - 2.365\left(\frac{2\beta}{1+\beta}\right)^3 \tag{4.190}$$

4.5 Case study 1: Force sensor for contact probe characterization

In contrast to the more common requirement for a mechanism to provide specified displacements, this case study represents an application in which the deflections of a single, two-axis flexure are used for the detection of applied forces, Periera *et al.* 1998. This work was motivated by a desire to determine contact forces between scanning probes as used in coordinate measuring machines and the surfaces of the component being measured, see, for example, Bosch, 1995. A coordinate measuring machine is an instrument that is used for the precise and accurate determination of the geometry of engineering components. Typically such a machine consists of three, mutually orthogonal linear motion slideways connected together in series. A touch sensitive probe is attached to the third slideway. In a scanning mode, the contact probe is maintained in contact with the surface at a nominally constant force while it

Figure 4.42: Digital image of three two-axis, dynamic force sensors shown with the mounting threads upwards. The central flexure consists of two simple notch hinges with orthogonal axes, the other two are toroidal two-axis flexures

moves around the component being measured. Subsequent displacements of the three axes are monitored and their paths used as a measure of the geometry of the component. However, at high scanning speeds sudden changes in geometry will result in large accelerations and the actual contact forces may vary considerably. To measure these forces, the monolithic flexure based sensors of figure 4.42 were constructed.

Three features of these monoliths are

- the mounting thread
- the flexure and
- a rigid cylinder

In normal operation the mounting thread screws into the base of the measurement frame with the flexure connecting between the base and an upper cylinder (note that the sensors are upside down in the figure). Two capacitance gages are attached to the rigid frame with the upper cylinders being used as the target electrodes. The capacitance gages are mounted in the frame with their measurement axes perpendicular to each other and both are radial to the axis of the flexure. To measure forces, the scanning probe is contacted on either the inside or outside surface of the cylinder at a prescribed height. Applied forces will cause a deflection of the cylinder in a direction parallel to that of the contact. Theoretically, the vector of this displacement can then be determined from the outputs of the two gages.

For this study, two flexure designs were chosen. One was the toroidal hinge discussed in section 4.4.3 while the other consisted of two, stacked, notch type hinges, see figure 4.43. Relevant design parameters for each will be discussed in the following two sections.

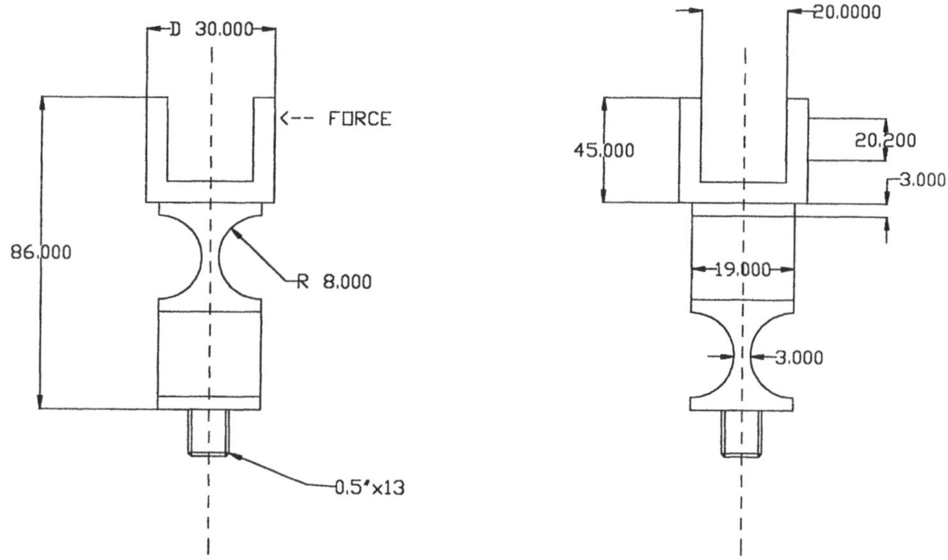

Figure 4.43: Dimensions of the two-axis force sensor

4.5.1 Toroidal notch type flexure

The first design was a circular notch type flexure, which is a commonly used two-axis flexure. An advantage of using such a flexure is that for an applied radial force the magnitude of deflection is independent of the radial direction. Because tool marks on the notch could be a source of stress concentration the notch was turned on a lathe from one end with a small feed.

In all analysis, the sensors are made from aluminum and this was assumed to have an elastic modulus, E of 68 GPa and a yield stress, σ_Y, of 255 MPa. For the design of this case study the notch radius R is 0.012 m and it has a thickness at the center of 0.005 m corresponding to a value $\beta = 0.208$. The axial distance, l, from the center of the hinge to the applied force is 0.048 m. During operation the maximum force from the probe is expected not to exceed 5 N. For the design to remain elastic, it is necessary to determine a maximum acceptable stress. Assuming that the elastic limit has a value of around 30% of the yield stress and including a safety factor of 2, the limiting stress becomes 39 MPa. Referring to Table 1 of chapter 2, it can be seen that this is very close to the lowest value of the endurance stress for aluminum alloys. Typically, calibration is an infrequent exercise and, at this design stress, fatigue is therefore considered unlikely. From equation (4.188) the stress concentration factor is 1.16. Consequently, if the bending stress is predicted from the bending equation, the limiting stress should be further reduced to 33.6 MPa.

Application of an applied radial force at the position on the axis as shown, will be modeled as a combined force plus a bending moment at the end of the hinge. Applying superposition, the resultant deflections will be considered to comprise a linear sum of the force and moments in isolation. From section 4.4.3, the angular and linear stiffness of the toroidal hinge is given by

$$\frac{M_z}{\theta_z} = E\frac{t^{7/2}}{20R^{1/2}} = 274 \qquad [\text{Nm rad}^{-1}]$$

$$\frac{F_y}{\theta_z} = E\frac{t^{7/2}}{20R^{3/2}\sqrt{1-\beta^2}} = 2.34\times10^4 \qquad [\text{N m}^{-1}]$$

Hence angular rotation, θ_M, due to the moment is

$$\begin{aligned}\theta_M &= \text{force}\times\text{force offset distance}\times\text{angular compliance}\\ &= 0.000875\end{aligned} \qquad [\text{rad}]$$

Correspondingly, the linear deflection due to the applied moment is

$$\begin{aligned}\delta_M &= \text{force offset distance}\times\theta_M\\ &= 42\times10^{-6}\end{aligned} \qquad [\text{m}]$$

Considering now the effect of the linear force component, the angular rotation, θ_F, is

$$\begin{aligned}\theta_F &= \text{force}\times\text{linear compliance}\\ &= 2.14\times10^{-4}\end{aligned} \qquad [\text{rad}]$$

From this, the linear deflection, δ_F, is

$$\begin{aligned}\delta_F &= \text{force offset distance}\times\text{angular rotation}\\ &= 10.27\times10^{-6}\end{aligned} \qquad [\mu\text{m}]$$

It is clear that the deflection due to the bending moment exceeds that due to the applied force by more than a factor of four. Applying superposition, the total deflection, δ, is simply the sum

$$\delta = \delta_M + \delta_F = 52.3 \qquad [\mu\text{m}]$$

In practice, the linear force will impose a shear stress in the hinge that is likely to be insignificant in comparison to the direct stress due to the bending moment. Consequently, this is considered to adequately represent the maximum stress in the hinge given by the bending equation (2.69). Noting that the second moment of area at the thinnest part of the hinge is $\pi t^4/64$, the bending stress, σ, is

$$\sigma = Mt/2I = 32Fl/\pi t^3 = 19.6 \qquad \text{[MPa]}$$

In view of the low estimate of the elastic limit and the safety factor, this flexure is expected to be linear for this application.

4.5.2 Two, stacked, notch-type flexures

In this design, the two-axis compliance is provided by two simple notch type hinges aligned with their axes perpendicular and in the plane of the overall flexure axis. A consequence of this stacked design is that the height from the applied force to the centers of the two respective notches is different. Consequently, if the two notches are of the same dimensions, the compliance at the point of the applied forces will depend on the direction of the applied force. Also there may be a problem with cross coupling between the two axes if the notches are not coincident both with each other and the cylinder.

The main advantage of this type of design is that it could be manufactured quickly by boring the two notches on a mill. For this design the same material and a scan force of 5 N are used as for the previous example. As can be seen from figure 4.43, the notch has a web thickness, t = 3 mm, a radius, R = 8 mm and a depth b = 19 mm. From the former of these two, the dimensionless factor β = 0.1875 and the linear and angular stiffness can be obtained from equation's (4.108) and (4.109)

$$\frac{F_y}{\theta_z} = k_{F\theta} = \frac{2Ebt^{5/2}}{9\pi R^{3/2}\sqrt{1-\beta^2}} = 6.41\times 10^4 \qquad \text{[N rad}^{-1}\text{]}$$

$$\frac{M_z}{\theta_z} = k_{M\theta} = \frac{2Ebt^{5/2}}{9\pi R^{1/2}} = 504 \qquad \text{[Nm rad}^{-1}\text{]}$$

To determine the compliance's for forces applied about the compliant axis of each of the notches, it is necessary to identify the upper (referred to as notch 1) and lower notch (notch 2). The distance, $l_{1,2}$, from the center of the notches to the applied force for each of the notches is

$$\begin{aligned} l_1 &= 0.051 \\ l_2 &= 0.070 \end{aligned} \qquad \text{[m]}$$

Proceeding as the previous example, the maximum bending stresses in the two notches are 1.8 MPa for notch 1 and 7.5 MPa for notch 2. Clearly, both are well within the design limits for this material. The stiffness, *k*, at the point of application of the force in the direction perpendicular to the axis of notch 1 is

$$k = \frac{1}{\dfrac{l_1}{k_{F\theta}} + \dfrac{l_1^2}{k_{M\theta}}} = 167900 \qquad [\mathrm{N\ m^{-1}}]$$

For notch 2

$$k = \frac{1}{\dfrac{l_2}{k_{F\theta}} + \dfrac{l_2^2}{k_{M\theta}}} = 92470 \qquad [\mathrm{N\ m^{-1}}]$$

For an applied force of 5 N, the maximum deflections in the two perpendicular directions are 30 and 54 μm for deflections of notches 1 and 2 respectively.

References

Bosch J.A. (Ed.), 1995, *Coordinate Measuring Machines and Systems*, Vol. **42** of Manufacturing Engineering and Materials Processing, Marcel Dekker.

Eastman F.S., 1937, The design of flexure pivots, *J. Aero. Sci.*, **5**(1), 16-21

Frocht M.M., 1951, *Strength of Materials*, The Ronald Press, NY, 237-238

Frocht M.M., 1935, Factors of stress concentration photoelastically determined, *Trans. ASME, J. Appl. Mech.*, **57**, A67

- Frocht M. M., Guernsey R. and Landsberg D., 1952, A photoelastic reexamination of notched tension bars, *ibid..*, **74**, 124.

Goodier J.N., 1941, An extension of the photoelastic method of stress measurement to plates in transverse bending, *J. Appl. Mech..*, **63**, A27-A29 (1941)

Haringx J. A., 1949, The cross strip pivot as a constructional element, *Appl. Sci. Res.; series A*, **A1**, 313-332

Howells M. R., Duarte R. and McGill R., 1996, Properties of the cartwheel type flexural hinge, personal communication

Jones R. V., 1951, Parallel and rectilinear spring movements, *J. Sci. Instrum.*, **28**, 38-41

Jones R.V., 1970, The pursuit of measurement, *Proc. IEE*, **117**(6), 1185-1191

Ling C. B., 1952, On the stresses in a notched strip, *Trans. ASME: Appl. Mech.*, **74**, 141-146

Ling C. B., 1968, On stress concentration factor in a notched strip, *Trans. ASME: Appl. Mech.*, **90**, 833-835

Paros J.M. and Weisbord L., 1965, How to design flexure hinges, *Machine Design*, 151-156 (Nov.)

Pereira P.H., Muralidhar A., Hocken R.J., Miller J.A. and Smith S.T., A two-axis, static and dynamic force characterization device, patent application in progress, April 1998

Pilkey W.D., 1997, *Peterson's Stress Concentration Factors*, John Wiley and Sons, New York (1974), pages 104, 122, 128

Plainevaux J. E., 1956, Etude des déformations d'une lame de suspension élastique, *Nuovo Cimento*, **4**, 922-928 (In French).

Smith S.T. and Chetwynd D.G., 1992, *Foundations of ultraprecision mechanism design*, Gordon and Breach, London.

Smith S.T., Badami V. G., Dale J. S. and Xu Y., 1997, Elliptical flexure hinges, *Rev. Sci. Instrum*, **68**(3), 1474-1483

Speake C.C., 1987, Fundamental limits to mass comparison by means of a beam balance, *Proc. Roy. Soc. Lond.*, **A414**, 333-358

- Keyser P.T. and Jefferts S.R., 1989, Magnetic susceptibility of some materials used for apparatus construction (at 295K), *Rev. Sci. Instrum.*, **60**(8), 2711-2714
- Flanders P.J., 1990, A vertical force alternating-gradient magnetometer, *Rev. Sci. Instrum.*, **61**(2), 839-847
- Quinn T.J., 1992, The beam balance as an instrument for very precise weighing, *Meas. Sci. Technol.*, **3**, 141-159

Thorpe, II, A.G., 1953, Flexure pivots – Design formulas and charts, *Product Engineering*, Feb., 192 - 200

Wittrick W.H., 1948, The theory of symmetrical crossed flexure pivots, *Aust. J. Sci. Res.*, A**1**(2), 121-134

Wittrick W.H., 1948, The properties of crossed flexure pivots and the influence of the point at which the strips cross, *The Aeronautical Quarterly*, **2**, 272-292

Young W.C., 1989, *Roark's Formulas for Stress and Strain*, 6th ed., McGraw-Hill Book Company, NY.

Young W. E., 1944, An investigation of the cross strip pivot, *J. Appl. Mech.*, **11**, A113-A120

5 Flexure systems

5.0 Overview

Previous chapters have concentrated on individual flexure elements and techniques for the analysis of stress, strain or, equivalently, forces and subsequent deflections. In this chapter, flexure elements combined to produce complete mechanisms are assessed. Static and dynamic performance of these systems is evaluated using methods developed in chapters 2 and 3. A particularly important mechanism in flexure design is the so-called four-bar link and, as such, this occupies the opening discussions. Not only is this commonly used in its simplest form, many of the more complex mechanisms can be constructed through appropriately combining successive four bar linkages. Throughout this book, flexures are commonly modeled as a series of infinitely rigid links connected by massless flexure elements. For applications requiring precise rectilinear motion, it is necessary to consider the effects of the finite stiffness of the links. In particular, the moments imposed by the flexure elements often result in significant angular distortion of the links connecting them, sometimes resulting in unacceptable tilt errors. To illustrate this, an analysis of platform distortions in a four bar link is presented plus some discussion on the optimal position of the location of the drive to reduce parasitic tilts at specific locations around the mechanism. Following this, mobility analysis of general planar mechanisms is discussed, in the process returning to the ideal link/joint models. Using examples of some of the more common mechanisms, Lagrangian analysis for the development of closed form analytic models is demonstrated. Analysis of static and dynamic properties of these mechanisms provides a background to the following section that discusses generalized approaches for dynamic analysis. D'Alembert's principle is used to derive equations of motion for flexure mechanisms of arbitrary geometry and, for small motions, these are expressed in matrix form. Analysis of planar mechanisms using matrix techniques is then introduced and illustrated with a case study examination of the four bar link connected by flexure elements for which compliances in all freedoms are included. Correspondence between the generalized and simplified models of the preceding sections is demonstrated. Finally, the transformation matrices necessary to extend this analysis to three-dimensional dynamic models are provided.

5.1 The four bar link

Already demonstrated in chapter 4, for small displacements, the four bar link with equal length supports provides a simple mechanism for achieving an approximate rectilinear motion. Probably the two most commonly encountered forms of this are the notch and leaf-type mechanisms shown in figure 5.1. As stated in section 4.1.5, if the upper platform of the parallel spring mechanism is perfectly rigid, the optimal drive should be positioned mid-way between the moving and stationary platforms. When the moving platform has finite rigidity additional considerations may be necessary, some of which are outlined in section 5.2. Issues associated with the manufacture and assembly of these types of flexure are discussed in chapter 8. To introduce the reader to some of the analytic methods for assessment of linear spring performance, the following two sections consider the leaf and notch type mechanisms respectively. In many cases, similar analytic models can be developed for mechanisms comprised of either flexure type. Because of this, the reader can assume that where analysis is not included for one type, it is safe to follow the procedures for the other.

5.1.1 The simple leaf type rectilinear spring

For the simple leaf type flexure mechanism shown in figure 5.1, the total stiffness in the direction of motion, x, is given by

$$k_{\delta_x F_x} = \frac{24EI_{zz}}{L^3} = 2Eb\left(\frac{t}{L}\right)^3 \tag{5.1}$$

where I refers to the second moment of area of a single leaf spring about the neutral axis.

Equation (3.254) can be used to estimate the lowest mode, free natural frequency corresponding to linear motion in the x-axis direction given by

$$\omega_{n_x} = \sqrt{\frac{k_{\delta_x F_x}}{M + \frac{26}{35}m}} = \sqrt{\frac{24EI_{zz}}{L^3\left(M + \frac{26}{35}m\right)}} \quad \text{(rad s}^{-1}\text{)} \tag{5.2}$$

where M is the mass of the moving platform and m corresponds to the mass of each flexure.

There will, of course, be other frequencies that might be significant. In general, most of these are likely to be higher, see section 5.5. Possible exceptions correspond to the torsional mode natural frequency about the y-axis plus cantilever bending about the x-axis. For the simple spring as shown, assuming that the upper surface of the moving platform is normal to the gravity field (i.e. horizontal), an observer on this surface would denote these yaw, θ_y, and roll, θ_x, respectively. Rotations about the z-axis correspond to pitch (θ_z, in the present

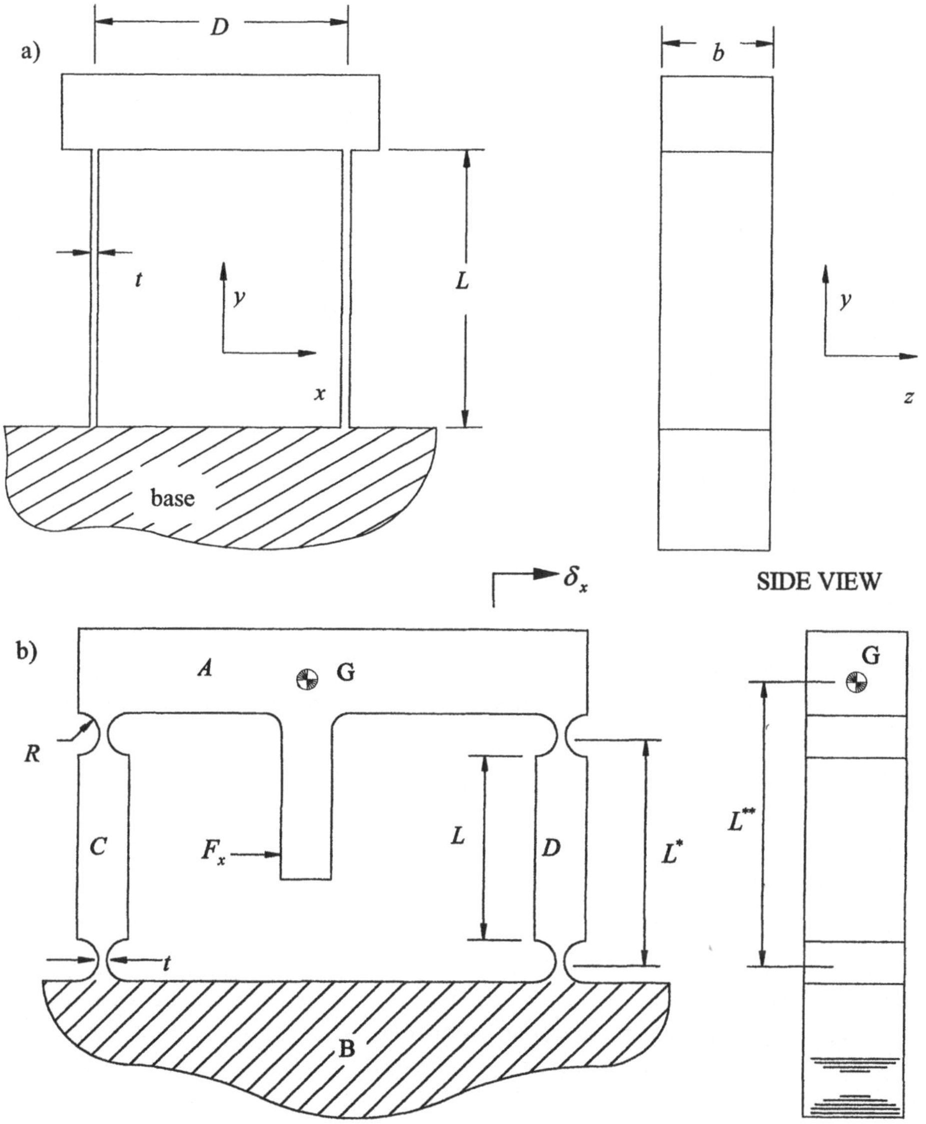

Figure 5.1: Two common linear spring mechanisms, a) leaf spring based flexure, b) notch hinge flexure

example). An estimate of the natural frequency for bending about the x-axis can be obtained from the beam bending equation for a cantilever and Rayleigh's method for the determination of the natural frequency of a cantilever with a mass attached at the free end, equation (3.259)

$$\omega_{n_{\theta_x}} = \sqrt{\frac{k_{\delta_z F_z}}{M + 2\frac{33}{140}m}} = \sqrt{\frac{6EI_{xx}}{L^3\left(M + \frac{66}{140}m\right)}} \tag{5.3}$$

(Note that the total mass, $2m$, and stiffness, $2k$, for both flexures is included in this calculation while the second moment of area is computed for a single leaf)

Clearly for small values of m relative to the platform mass, the ratio of the two natural frequencies can be approximated from the equation

$$\begin{aligned}\frac{\omega_{n_{\theta_x}}}{\omega_{n_x}} &= \sqrt{\frac{6EI_{xx}\left(M + \frac{26}{35}m\right)}{24EI_{zz}\left(M + \frac{66}{140}m\right)}} \\ &\approx \sqrt{\frac{b^2}{4t^2}}\end{aligned} \tag{5.4}$$

In most applications, b is much greater than t and, in such as case, the assumption that equation (5.2) corresponds to the lowest mode natural frequency is valid.

The stiffness in the z-axis direction in response to a force applied at the upper platform can be computed from

$$k_{\delta_z F_z} = \frac{6EI_{xx}}{L^3} = \frac{Eb^3t}{2L^3} \tag{5.5}$$

Considering the linear elastic properties of flexures, and for small distortions in comparison with the beam dimensions, it is assumed that the principle of superposition applies. Therefore, the subsequent distortion as a consequence of a combination of forces can be computed separately and, thereafter, added vectorially. Another assumption is that of isotropic elastic properties. In most (but certainly not all) applications, the spring is likely to be constructed from readily machinable materials such as steel or aluminum which are invariably considered isotropic.

The torsional stiffness about the z-axis, is found by applying equal tensile and compressive axial forces to the flexures which are spaced D apart, therefore

$$k_{\theta_z M_z} = \frac{EA}{2L}D^2 \tag{5.6}$$

Torsional stiffness about the y-axis is more difficult to compute since several effects combine. Rotation of the platform about this axis will cause equal and opposite deflections in x and z and each spring will twist about its y-axis. All these deflections contribute to the resisting torque. The torsional stiffness of each spring about its longitudinal axis is, Timoshenko and Goodier, 1970,

$$k^{*}_{\theta_y M_y} = k_1 \frac{Gbt^3}{3L} \tag{5.7}$$

where G is the modulus of rigidity and k_1 is a shape factor that can be obtained from equation (2.104), section 2.8.2.

For small rotations the deflections at the end of each beam are given by

$$\varepsilon_z = \frac{D}{2}\sin\theta_y \approx \frac{D\theta_y}{2} \tag{5.8}$$

$$\varepsilon_x = \frac{D}{2}(1-\cos\theta_y) \approx \frac{D\theta_y{}^2}{4} \tag{5.9}$$

The end forces to cause these deflections can be reinterpreted as couples on the spacing of D and $D\theta$ respectively. For the x direction, equation (5.1) gives exactly twice the linear stiffness since we are only considering one beam at this point. However, in the z-direction, the platform rotation constrains the springs into the s-shaped configuration, and so, the related stiffness is twice that given by equation (5.5). The torsional resistance as a result of these deflections in these two axes is

$$k^{**}_{\theta_y M_y} = \frac{Eb^3tD^2}{2L^3} + \frac{Ebt^3D^2\theta_y^2}{4L} \tag{5.10}$$

Since θ_y is small and $t<b$, the contributions from the second term in equation (5.10) can, in most cases, be neglected. Therefore, the overall torsional stiffness for the two-spring flexure may be taken as

$$k_{\theta_y M_y} = k^{*}_{\theta_y M_y} + 2k^{**}_{\theta_y M_y} \approx \frac{bt}{L}\left[\frac{Ed^2D^2}{2L^2} + \frac{2k_1Gt^2}{3}\right] \tag{5.11}$$

Noting that $G=E/2(1+\nu)\approx E/2.6$ for typical values of Poisson's Ratio, ν, that $b>t$ and that, provided $b\approx 3t$ or more, k_1 will be of the order of 0.9, it can be seen that the first term in equation (5.11) will dominate. Substituting the above values and rounding to simple numbers gives

$$k_{\theta_y M_v} = \frac{Eb^3t}{L}\left(\frac{D^2}{2L^2} + \frac{t^2}{5d^2}\right) \tag{5.12}$$

Commonly, D and d are roughly equal and b is almost always greater than $2.5t$, so it seems unlikely that there will be a contribution of as much as 10% extra stiffness due to twisting of the flexures. Potentially more significant is the effect of shear deflection in the z-direction. This will be proportionally larger than is the case for simple cantilever bending. Taking the simplest model of shear using a shape factor of 1.5 corresponding to a rectangular cross section, the deflection of the beam in shear is

$$\delta_s = \frac{1.5FL}{btG} \tag{5.13}$$

where F is the shear force, which is equal to the deflecting force in this application. This deflection is additional to that predicted for bending so the effective stiffness is reduced. The s-shaped deflection stiffness in the y-direction for each spring will become

$$k_{\theta_y M_y} = \frac{Eb^3t}{L^3}\left(1 + \frac{1.5Eb^2}{GL^2}\right)^{-1} \tag{5.14}$$

If the ratio of length to breadth becomes significantly low, then this expression should be used as the first term in parentheses in equation (5.12). Even if this ratio falls as low as 4, the correction still only contributes about 20% and will be compensated a little by the torsional stiffness of the springs. Consequently, for basic design modeling, it is generally reasonable to use the simpler form shown in equation (5.12).

Note that, using this approximation,

$$\frac{k_{\delta_x F_x}}{k_{\theta_y M_y}} = \left(\frac{2t}{bD}\right)^2 \tag{5.15}$$

There is considerable scope, through choice of the geometric parameters, for varying the relative values of linear and torsional stiffness to match a desired stiffness ratio. A case study in which this torsional compliance is exploited to create a two-axis actuator can be found in Smith *et al.*, 1994.

5.1.2 The simple notch type linear spring

A dynamic analysis of the simple notch type mechanism is presented and this provides an introduction to the application of Lagrange's equation to flexure mechanisms. The objective of the simple spring is to provide an approximate straight-line motion in the x-axis direction. For perfect hinges rotating about their mid points and, assuming small displacements, the upper moving platform, A, will approximate rectilinear motion. It is also apparent that the support legs, C & D, will rotate so that the rectangular centers of the four hinges transform to a parallelogram, see Figure 5.1(b). Considered to be a single degree of freedom system, a single coordinate is all that is necessary to completely describe the motion of all points in the flexure. If x represents the linear displacement of the platform A, the total kinetic energy of the spring is the sum of linear motion of this platform plus the sum of the kinetic energy due to rotation of the two identical support legs, C and D, about the centers of the lower, stationary hinges. As a first step in the determination of the Lagrangian, it is necessary to derive an expression for the total kinetic energy of the flexure. In this case, being convenient, two coordinates representing the displacement and rotation are temporarily used

$$T = \frac{M_A}{2}\dot{x}^2 + 2\left(\frac{I_C\dot{\theta}_z^2}{2}\right) \tag{5.16}$$

However, these coordinates are not independent. For small deflections, the angular rotation of the flexure can be approximated from

$$\theta_z = \sin\left(\frac{x}{L^*}\right) \approx \frac{x}{L^*} \tag{5.17}$$

Therefore (5.16) can be expressed in terms of the single coordinate x by

$$\begin{aligned} T &= \frac{M_A}{2}\dot{x}^2 + \frac{I_C\dot{x}^2}{L^{*2}} \\ &= \left(\frac{M_A}{2} + \frac{I_C}{L^{*2}}\right)\dot{x}^2 \end{aligned} \tag{5.17}$$

Changes in potential due to this displacement are considered to occur primarily in the necked portion of the hinges. In reality, there will be some distortion of the support platforms. Assuming a high compliance of the hinge in comparison with the supports, in most designs, it is reasonable to neglect these distortions. Because of the symmetry of this mechanism, all hinges will undergo identical angular deflections. Ignoring shear compliance in the x-axis, the component of strain energy due to pure bending is simply

$$V = 4\left(\frac{K_{\theta_z M_z}\theta_z^2}{2}\right) \approx \frac{2K_{\theta_z M_z}x^2}{L^{*2}} \tag{5.18}$$

For this single degree of freedom system, Lagrange's equation can immediately be written in the form

$$\frac{d}{dt}\left(\frac{\partial T}{\partial \dot{x}}\right)-\frac{\partial T}{\partial x}+\frac{\partial V}{\partial x}=F_x \tag{5.19}$$

Substituting (5.17) and (5.18) into (5.19), the equation governing motion of this flexure is given by

$$\left(M_A+\frac{2I_C}{L^{*2}}\right)\ddot{x}+\frac{4k_{\theta_z M_z}}{L^{*2}}x=F_x$$

or (5.20)

$$\ddot{x}+\frac{4k_{\theta_z M_z}}{L^{*2}\left(M_A+\frac{2I_C}{L^{*2}}\right)}x=\frac{F_x}{\left(M_A+\frac{2I_C}{L^{*2}}\right)}$$

The left-hand side of the latter of equations (5.20) is immediately recognizable as a simple, homogeneous, second order differential equation of the form

$$\ddot{x}+\omega_n^2 x=0 \tag{5.21}$$

In its simplest form the natural frequency for the notch type linear spring is given by

$$\omega_n^2=\frac{4k_{\theta_z M_z}}{L^{*2}\left(M_A+\frac{2I_C}{L^{*2}}\right)} \tag{5.22}$$

For the notch type hinge, the angular stiffness can be approximated from equation (4.91) *i.e.*

$$K_{\theta_z M_z}=\frac{2Ebt^{5/2}}{9\pi R^{1/2}} \tag{5.23}$$

Determination of the second moment of mass, or moment of inertia, is not so straightforward. However, if it is assumed that the support legs can be adequately modeled as long, uniform, thin rods and the mass of material in the notches is ignored, the second moment of mass about the centroid of the support legs is

$$I' = \frac{M_C L^2}{12} \tag{5.24}$$

To determine the second moment of mass about the pivots of the lower hinges, it is convenient to use the parallel axis theorem from which

$$\begin{aligned} I &= \frac{M_C L^2}{12} + \frac{M_C L^{*2}}{4} \\ &= \frac{M_C L^2}{12} + \frac{M_C (L+2R)^2}{4} \\ &= \frac{M_C L^2}{3} + M_C\left(LR + 4R^2\right) \end{aligned} \tag{5.25}$$

Substituting the two above expressions into (5.23) yields

$$\omega_n^2 = \sqrt{\frac{\dfrac{8Ebt^{5/2}}{9\pi R^{1/2}}}{M_A L^{*2} + 2M_C L^2\left(\dfrac{1}{3} + \left(\dfrac{R}{L} + 4\dfrac{R^2}{L^2}\right)\right)}} \tag{5.26}$$

Equation (5.26) can be simplified further if it is assumed that $R << L$ after which it corresponds to equation (4.36) in the book of Smith and Chetwynd, 1992.

The total 'static' linear stiffness of the flexure can be obtained by setting the acceleration to zero, and is

$$\frac{F_x}{x} = \frac{4k_{\theta_z M_z}}{L^{*2}} = \frac{8Ebt^{5/2}}{9\pi R^{1/2} L^{*2}} \tag{5.27}$$

Although, theoretically, for a single degree of freedom system there can be only one mode of vibration, in practice there will exist an infinite number of frequencies and mode shapes. It is important to identify some of the frequencies that may be of comparable value to that in the most compliant direction i.e. the x-axis. For most designs, the second lowest natural frequency corresponds to the flexure oscillating as a pendulum with the pivot being at the mid-point of the lower notches and rotating about an x-axis at the center. This is probably more easily visualized by looking at the end view of the flexure in figure 5.1(b). For this particular mode of vibration, the top platform will be moving about an arc in the z-axis direction, the aforementioned pivot point being at the center of the lower notches.

Yet again, to determine an approximate equation for the natural frequency it is necessary to make a number of assumptions. Firstly, it will be assumed that the second moment of mass (moment of inertia) of the upper platform about the upper notch hinges is small in comparison to its second moment of mass from the x-axis pivot point in the lower notches. Consequently, it will be expected that the inertial forces on the upper notches about the x-axis and also linear forces in the z-direction will induce negligible distortion in comparison to the lower notches. Based on this assumption, it is possible to model the complete linear spring as a rigid body attached to the base by the two notches.

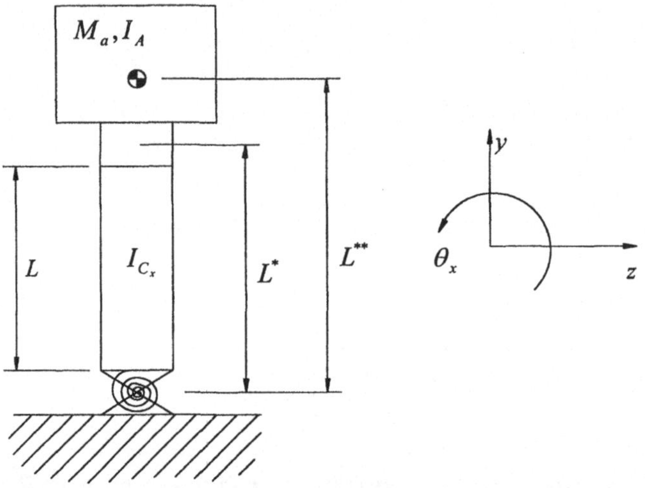

Figure 5.2: Lumped model of the simple notch type linear spring for computation of the fundamental mode frequency of oscillation about an axis passing through the centroids of the two lower notches

When oscillating in the x-axis direction, subsequent motions in the y-z plane do not effect the x position of any part of the flexure. Consequently, it may be assumed that the natural frequency of vibration computed previously, which involved predominantly x-axis motion of elements in the flexure, will have little influence on the mode considered here and vice versa[1]. Based on these assumptions, the simple spring can be modeled as comprising two long thin rods with a mass attached at the free end, see figure 5.2. The kinetic and potential energy corresponding to this mode of vibration are given by

$$T = \left(2\frac{I_{C_x}}{2} + \frac{M_A L^{**2}}{2} + \frac{1}{2} I_{A_x}\right)\dot{\theta}_x^2$$
$$V = 2\frac{k_{\theta_x M_x}}{2}\theta_x^2 \tag{5.28}$$

[1] In reality, both modes of vibration induce motion of components of the spring in the y direction. As a consequence, there will be some coupling of these two modes and this will have an effect on the natural frequency. In all likelihood this effect will oft times be small.

From equation (4.111), the bending stiffness of each notch is given by

$$k_{\theta_x M_x} = \frac{Eb^3}{12}\left[\pi\left(\frac{1}{2\beta}\right)^{1/2} - \frac{\sqrt{2\pi}}{\sqrt{3}}\right] \tag{5.29}$$

Substitution of equation (5.28) into Lagrange's equation readily provides a solution for the natural frequency

$$\omega_n^2 = \frac{2k_{\theta_x M_x}}{\left(2I_{C_x} + M_A L^{**2} + I_{A_x}\right)} = \frac{\dfrac{Eb^3}{6}\left[\pi\left(\dfrac{1}{2\beta}\right)^{1/2} - \dfrac{\sqrt{2\pi}}{\sqrt{3}}\right]}{\left(2I_{C_x} + M_A L^{**2} + I_{A_x}\right)} \tag{5.30}$$

The numerator of equation (5.30) is the effective bending stiffness of the flexure about the x-axis in response to a moment applied to the platform.

5.1.3 The virtual center

So far, it has been assumed that all four-bar links are comprised from a rectangular distribution of hinges, therefore resulting in a parallelogram motion between the upper platform and base link. A generalized mechanism modeled as

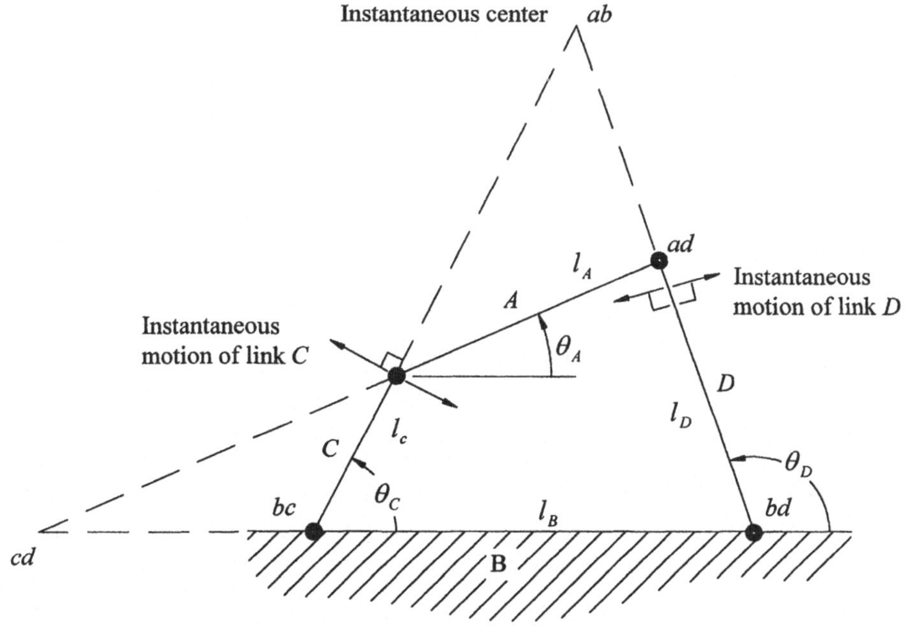

Figure 5.3: The general four-bar link

four rigid links connected by single freedom pivots is shown in Figure 5.3. Of interest to this section is analysis of the motion of the moving platform relative to the base and the dynamic characteristics of symmetric forms of this mechanism.

Before discussing the generalized four-bar mechanism, study of a single link will reveal some important principles of rigid body motion. Consider a rigid link free to move in space, upon this are marked two, fixed points, *A* & *B*, figure 5.4. Also indicated by arrows at points *A* and *B* on this link are the instantaneous displacements. Clearly, if an observer were to be positioned at *A* but with a fixed orientation relative to a stationary link then, although the observer would be moving with *A*, the point *B* must always remain a constant distance away. If the link has any rotation then point *B* can move relative to *A*. However, the constraint of constant length imposes only tangential relative motion. Therefore

> *Theorem I; the relative motion between any two points in a rigid body is always perpendicular to a straight line connecting them.*

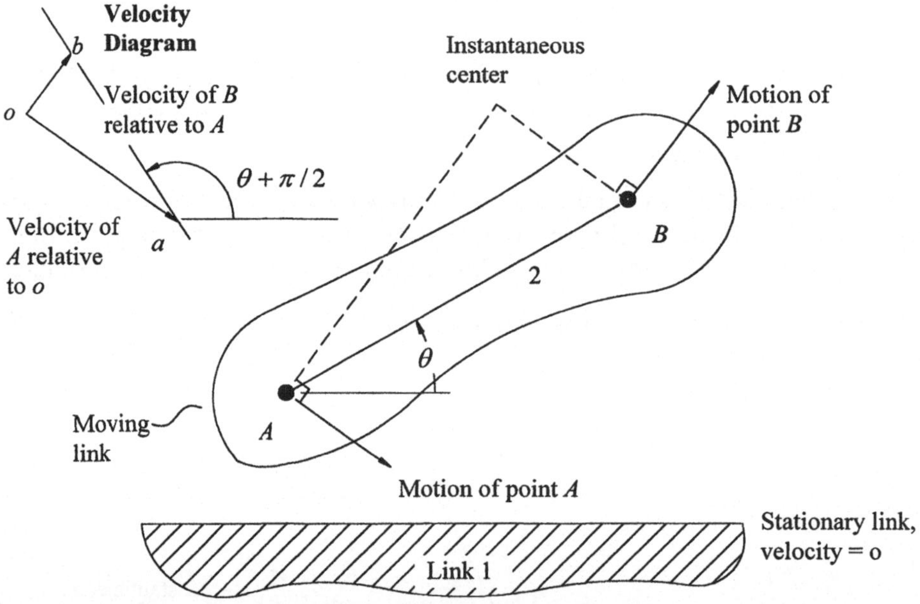

Figure 5.4: Geometry of generalized, planar, rigid body motion

In many ways, the above statement might be an obvious definition of a rigid body. However, the above theorem can be extended to include a statement about the nature of the motion relative to a point fixed in the fixed (often referred to as the 'stationary') frame. If it so happened that the rigid body is rotating about the fixed point, *o*, then it is obvious that the lengths between *oA* and *oB* must be constant and also that their motion will be perpendicular to the line connecting these points. Consequently, a second theorem can be postulated

> *Theorem II; if a rigid body moves about a stationary point, o, in such a way that any two arbitrary points, A & B, on that body each describe an*

instantaneous motion perpendicular to the lines oA and oB, the point o lies on the instantaneous axis of pure rotation of the rigid body.

In the case of planar mechanisms, the point *o* is the instantaneous pivot of the rigid body. Referring back to the four-bar link of figure 5.3, it is obvious that the pivots connecting each end of link *A* to links *C* & *D* will be moving perpendicular to the axes of links *C* & *D*. Extending the axes of the links *C* & *D* they eventually cross at the point marked *ab*. Clearly, at this point the two ends of the link *A* will also be moving perpendicular to the line connecting them and therefore by theorem II above, the link *A* is instantaneously rotating about *ab*. The notation *ab* for the instantaneous center indicates the rotation of link *A* relative to the stationary link *B*. Consequently, for small motions, it is possible to use a four bar link to create a virtual center of rotation. This can be of great practical use for designs in which a pure rotation about a virtual pivot is required in the instance where the pivot point is to be vacant to enable the placement of specimens etc. Also, by theorems I and II the instantaneous center of motion for the link of figure 5.4 is shown.

The concept of the instantaneous center is not new and the theory of mechanisms has been developed for over one hundred years (for introductory texts see Keown, 1921, or Mabie and Reinholtz, 1987). In general, for a *planar* system of *n* links connected by single freedom joints (either slides or pivots although only the latter are discussed in this book) there will be $n - 1$ instantaneous centers relative to each link. However, at each location there are two centers. Consequently, the total number of locations, *N*, of instantaneous centers is given by

$$N = \frac{n(n-1)}{2} \tag{5.31}$$

For the four bar link of figure 5.3, there are 6 instantaneous centers (sometimes called *centros* in older texts on kinematics of mechanisms). Four of these, *bc ca ad bd*, occur at the pivots. As already shown the other two can be found at the intersection of lines co-linear with the axes of the two connecting links that provide the motion constraints. The construction for the instantaneous centers *ab* & *cd* is shown in figure 5.3. The astute observer will realize that any centro is located on a straight line with two others which complies with Kennedy's more general theorem

Theorem III; for three independent bodies in general plane motion the three virtual centers lie on a common straight line

Grubler's equation (2.108) for a planar mechanism immediately gives a mobility of 1 for the four-bar link connected by single freedom hinges. Consequently, in view of the fact that complete motion is defined by a single coördinate, one may erroneously be lead to the belief that geometric analysis would be correspondingly simple. Unfortunately, given the lengths of the links between hinges and one of the angles between two links, the remaining two angles can only be determined from transcendental equations. To see this, the vector equation that must be satisfied if the mechanism is to connect at the pivots can be expressed in complex form by the requirement that

$$l_A e^{i\theta_A} = l_B + l_D e^{i\theta_D} - l_C e^{i\theta_C} \tag{5.32}$$

An equally valid form is given by the conjugate functions (these are merely a multiplication of –1 in the imaginary direction)

$$l_A e^{-i\theta_A} = l_B + l_D e^{-i\theta_D} - l_C e^{-i\theta_C} \tag{5.33}$$

Multiplying (5.32) by (5.33) eliminates the angular term in θ_A to give

$$\begin{aligned} l_A^2 = l_B^2 + l_C^2 + l_D^2 + l_B l_D\left(e^{i\theta_D} + e^{-i\theta_D}\right) - l_B l_C\left(e^{i\theta_C} + e^{-i\theta_C}\right) \\ - l_C l_D\left(e^{i\theta_C} e^{-i\theta_D} + e^{-i\theta_C} e^{i\theta_D}\right) \end{aligned} \tag{5.34}$$

Expanding the complex exponents by De Moivre's theorem produces

$$\begin{aligned} l_A^2 = l_B^2 + l_C^2 + l_D^2 + 2 l_B l_D \cos\theta_D - 2 l_B l_C \sin\theta_C - \\ 2 l_C l_D\left(\cos\theta_C \cos\theta_D + \sin\theta_C \sin\theta_D\right) \end{aligned} \tag{5.35}$$

Consequently, if the lengths of the links and one of the angles are known, the other can be determined from (5.35). Assuming angle θ_C is known and using the standard t substitution

$$\begin{aligned} t &= \tan\left(\frac{\theta_D}{2}\right) \\ \cos\theta_D &= \frac{1-t^2}{1+t^2} \\ \sin\theta_D &= \frac{2t}{1+t^2} \end{aligned} \tag{5.36}$$

the angle θ_D can be determined from the quadratic

$$at^2 + bt + c = 0 \tag{5.37}$$

where

$$\begin{aligned} a &= l_A^2 - l_B^2 - l_C^2 - l_D^2 + 2l_B l_C \cos\theta_C + 2l_B l_D - 2l_C l_D \cos\theta_C \\ b &= 4l_C l_D \sin\theta_C \\ c &= l_A^2 - l_B^2 - l_C^2 - l_D^2 + 2l_B l_C \cos\theta_C + 2l_B l_D + 2l_C l_D \cos\theta_C \end{aligned} \tag{5.38}$$

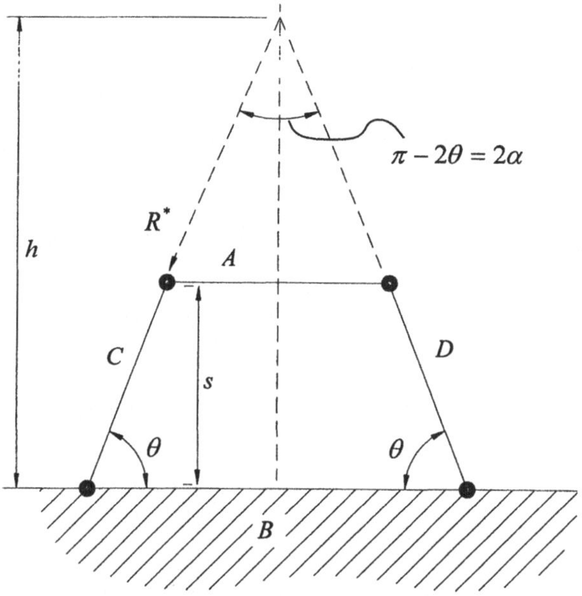

Figure 5.5: The trapezoidal four-bar link

Fortunately, most flexures are symmetric with links A and B parallel so that the joints form a trapezoid. In this case the added symmetry of the trapezoidal four-bar link shown in figure 5.5 greatly simplifies analysis. In this case, the height, h, of the centro from the base is simply

$$h = \frac{l_B s}{l_B - l_A} \tag{5.39}$$

The radius of rotation, R^*, of the link A is given by

$$R^* = \frac{l_A s}{(l_B - l_A)\cos\left(\frac{\pi}{2} - \theta\right)} = \frac{l_A l_C}{(l_B - l_A)} \tag{5.40}$$

Because of the common motion at the moving pivots (in this case study, the pivots are assumed to be at the centers of rotation of the flexure hinges), the relationship between the rotation of R^* and the support links is simply

$$\delta\alpha = \frac{l_C \delta\theta}{R^*} \tag{5.41}$$

The second moment of mass about the instantaneous center, I, can be obtained using the parallel axis theorem

$$I = I_{A0} + M_A\left(R^* \cos(\pi/2 - \theta)\right)^2 = I_{A0} + M_A\left(\frac{l_A s}{l_B - l_A}\right)^2 \tag{5.42}$$

I_{A0} is the second moment of mass of platform A about the point mid-way between the two moving pivots.

Using (5.41) and (5.42), the total kinetic energy for a virtual velocity, $\delta\dot{\theta}$, is

$$\begin{aligned} T &= \left(\frac{I}{2}\left(\frac{l_C}{R^*}\right)^2 + I_C\right)\delta\dot{\theta}^2 \\ &= \left(\frac{I}{2}\left(\frac{l_B - l_A}{l_A}\right)^2 + I_C\right)\delta\dot{\theta}^2 \end{aligned} \tag{5.42}$$

Correspondingly, the potential energy is simply

$$\begin{aligned} V &= 2\left(\frac{k}{2}\right)(\delta\theta + \delta\alpha)^2 + 2\left(\frac{k}{2}\right)(\delta\theta)^2 = k\left[\left(1 + \frac{l_C}{R^*}\right)^2 + 1\right](\delta\theta)^2 \\ &= k\left[\left(\frac{l_B}{l_A}\right)^2 + 1\right](\delta\theta)^2 \end{aligned} \tag{5.43}$$

In the absence of applied forces, Lagranges equation immediately gives

$$\left(I\left(\frac{l_B - l_A}{l_A}\right)^2 + 2I_C\right)\delta\ddot{\theta} + 2k\left[\left(\frac{l_B}{l_A}\right)^2 + 1\right]\delta\theta = 0 \tag{5.44}$$

From this, the undamped natural frequency, ω_n, is

$$\omega_n^2 = \frac{2k\left[\left(\frac{l_B}{l_A}\right)^2 + 1\right]}{\left(I\left(\frac{l_B - l_A}{l_A}\right)^2 + 2I_C\right)} \tag{5.45}$$

Often, it might be reasonable to assume that the links can be modeled as thin rods for which the second moment of mass about the center is simply $ml^2/3$. For a mechanism with a rectangular distribution of hinges, equation (5.45) reduces to

$$\omega_n^2 = \frac{4k}{\left(M_A s^2 + 2I_C\right)} \tag{5.46}$$

Replacing s with L^*, this is identical to equation (5.22)

5.1.4 Effect of the drive on the natural frequency

So far, it has been assumed that the drive applied to the flexure asserts no influence on the natural frequency. In reality, drives often add significant stiffness to the mechanism. Consider, for example a piezoelectric or micrometer drive. This must be contacted with the mechanism via a coupling (maybe including a lever mechanism, see chapter 7) of finite stiffness between the output displacement and actuator, k_d, see chapter 8. If electromagnetic drives are employed, this additional stiffness can, in principle be reduced to zero, Smith and Chetwynd, 1990. For closed loop systems the added compliance can range from large negative to large positive values over limited ranges of frequency.

For a single degree of freedom system undergoing small displacements, the natural frequency can always be expressed in the form

$$\omega_n^2 = \frac{k_e}{m_e} \tag{5.47}$$

Assuming that the drive adds little mass, (5.47) is simply modified to include the drive stiffness, k_d, so that

$$\omega_n^2 = \frac{k_e + k_d}{m_e} \tag{5.48}$$

5.2 Optimal geometry for the rectilinear motion of components on a simple linear spring

In many applications, it is desired to produce small, rotation-free motions. Examples include displacement of an optical wave front by translating a plane mirror, small motion of a mechanical measuring scale etc. Attaching a mirror to the platform of a flexure is an obvious solution for the former application. When it is necessary to achieve precise rectilinear motions through use of a flexure stage, the distortion of the moving platform and other 'rigid' links may be significant. As an illustration, this section discusses the optimal design of a simple linear spring mechanism and the effects of drive location on the rectilinearity of motion of the upper platform when induced elastic deformations cannot be neglected.

5.2.1 Simple linear spring mechanisms

Lateral distortion of a prismatic beam due to applied bending moments can be obtained using Macauley's method, see section 2.7. A mathematical representation of the flexure analyzed in this section is shown in figure 5.6. From this, the moment equation along the beam can be written as

$$-M = M_1 + R_1 x - M_3[x-a]^0 \tag{5.49}$$

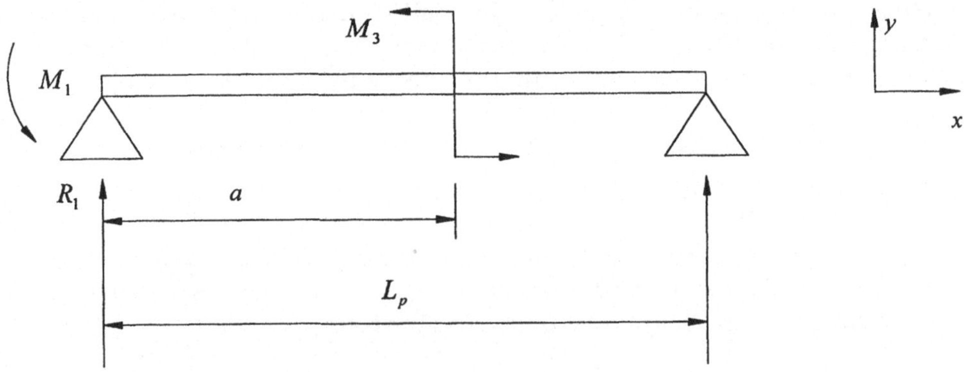

Figure 5.6: Bending moments applied to the moving platform of a simple linear spring

Integrating the bending equation, the displacement of the beam is given by

$$-EI_p y = M_1\frac{x^2}{2} + R_1\frac{x^3}{6} - \frac{M_3}{2}[x-a]^2 + C_1 x + C_2 \tag{5.50}$$

The expression in square parentheses is the singularity function and defined such that it only exists for positive values, see section 2.7.4. For a simply supported beam, the boundary conditions correspond to zero deflection at each end. Consequently, for $x = 0$ it is simple to show that $C_2 = 0$. At the other end, $x = L_p$, the constant C_1 is given by

$$C_1 = \frac{M_3}{2L_p}(L_p - a)^2 - M_1\frac{L_p}{2} - R_1\frac{L_p^2}{6} \tag{5.51}$$

In this particular beam configuration, the *applied* bending moments originate from the distortion of the leaf spring type flexures plus the moment due to the applied force from the drive. Assuming that displacements are small, the slope of the platform at the connections to the applied loads is ignored. Under these conditions, it is possible to calculate the magnitude of the applied bending moments from the reactions of the simple flexure spring, for which the relevant parameters are shown in figure 5.7. The bending moments, M_1 at each end of the platform are given by the equation

$$M_1 = \frac{6EI_f}{L_f^2}\delta \tag{5.52}$$

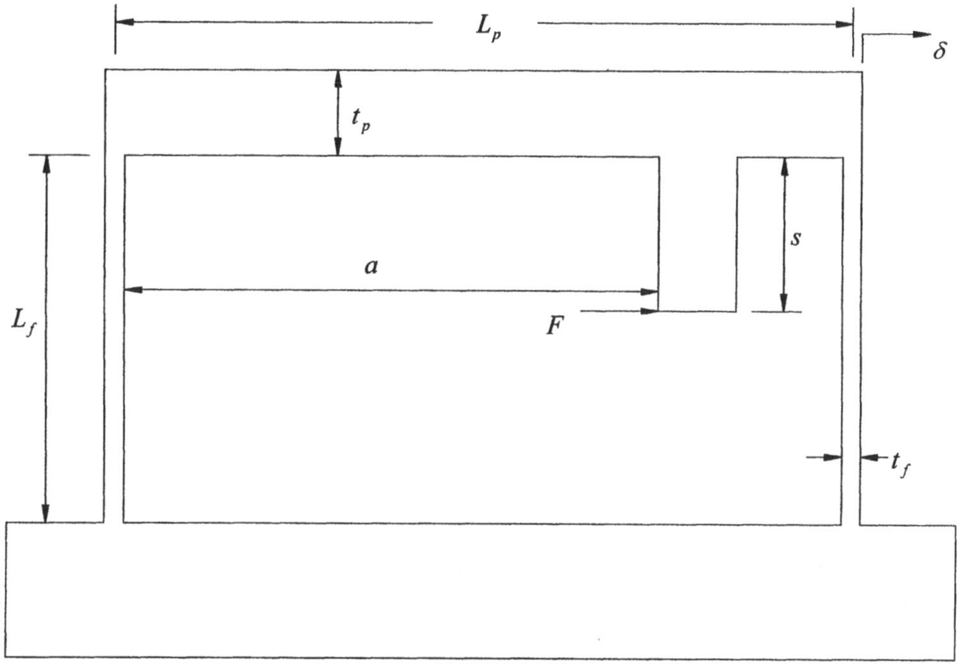

Figure 5.7: Parametric model of a simple linear spring mechanism

These moments are introduced from the horizontal force externally applied to the platform. The moment arm of length *s* combined with the drive force itself introduces a bending moment, M_3 at a distance *a* from one end. This moment can be expressed in terms of the displacement of the platform

$$M_3 = \frac{24EI_f s}{L_f^3}\delta \tag{5.53}$$

Clearly, these two applied moments given by equations (5.52) and (5.53) are related by the simple equation

$$M_3 = \frac{4s}{L_f}M_1 = kM_1 \tag{5.54}$$

where $k = 4s / L_f$

The reaction at the support, R_1 can be obtained by taking moments about the opposite end of the beam and equating it to zero to yield

$$R_1 = \frac{M_3 - 2M_1}{L_p} = \frac{M_1}{L_p}(k-2) \tag{5.55}$$

Equations (5.54) and (5.55) can be substituted into (5.50) and (5.51) to yield

$$-EI_p y = M_1\frac{x^2}{2} + \frac{M_1}{L_p}(k-2)\frac{x^3}{6} - \frac{kM_1}{2}[x-a]^2 + C_1 x \tag{5.56}$$

$$\begin{aligned} C_1 &= M_1\left[\frac{k}{2L_p}\left(L_p - a\right)^2 - \frac{L_p}{2} - (k-2)\frac{L_p}{6}\right] \\ &= M_1\left[\frac{kL_p}{3} + \frac{ka^2}{2L_p} - ka - \frac{L_p}{6}\right] \end{aligned} \tag{5.57}$$

In many instances, it may be desirable that the slope at either end of the beam is zero. The slope at an arbitrary point x along the beam can be computed from the equation

$$\begin{aligned} -EI_p\frac{dy}{dx} &= M_1 x + \frac{M_1}{L_p}(k-2)\frac{x^2}{2} - kM_1[x-a] + C_1 \\ &= M_1\left[x + \frac{(k-2)}{L_p}\frac{x^2}{2} - k[x-a] + \frac{kL_p}{3} + \frac{ka^2}{2L_p} - \frac{L_p}{6}\right] \end{aligned} \tag{5.58}$$

At the left-hand support this is given by

$$\begin{aligned} -EI_p\frac{dy}{dx} &= M_1\left[\frac{kL_p}{3} + \frac{ka^2}{2L_p} - ka - \frac{L_p}{6}\right] \\ &= M_1 L_p\left[\frac{4}{3}\left(\frac{s}{L_f}\right) + 2\left(\frac{s}{L_f}\right)\left(\frac{a}{L_p}\right)^2 - 4\left(\frac{s}{L_f}\right)\left(\frac{a}{L_p}\right) - \frac{1}{6}\right] \end{aligned} \tag{5.59}$$

Defining the ratio's

$$\alpha = \frac{s}{L_f}$$
$$\beta = \frac{a}{L_p} \tag{5.60}$$

The condition for zero slope at the left support is

$$\alpha = \frac{1}{8 - 24\beta + 12\beta^2} \tag{5.61}$$

For example, choosing the value $a = L_p/2$, $\beta = 0.5$, from which we find $\alpha = -1$, corresponding to $s = -L_f$ or, in words, the actuating force is at a position above the flexures at a distance equal to their length. A plot of the normal distortion along the platform for such a position of the actuator is shown in figure 5.8.

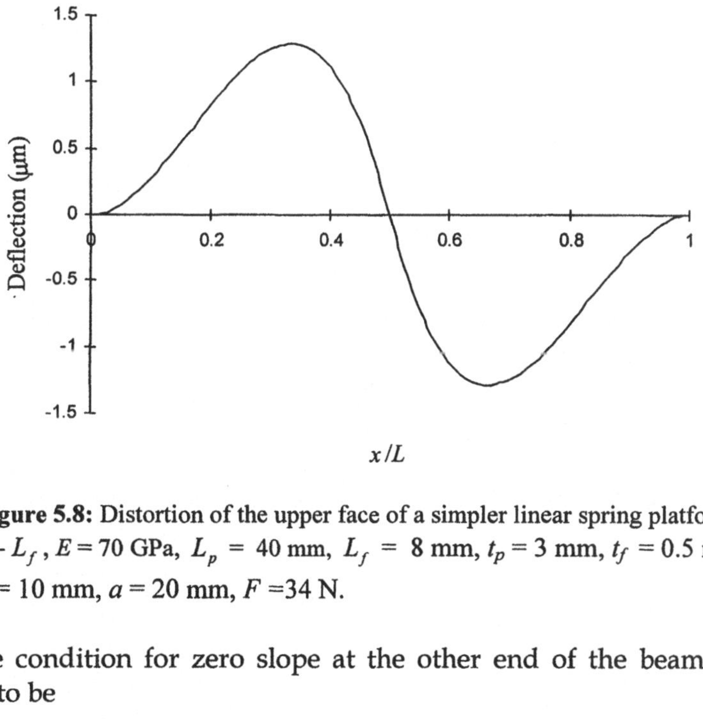

Figure 5.8: Distortion of the upper face of a simpler linear spring platform, $s = -L_f$, $E = 70$ GPa, $L_p = 40$ mm, $L_f = 8$ mm, $t_p = 3$ mm, $t_f = 0.5$ mm, $d = 10$ mm, $a = 20$ mm, $F = 34$ N.

The condition for zero slope at the other end of the beam can be readily shown to be

$$\alpha = \frac{1}{12\beta^2 - 4} \tag{5.62}$$

As a further example, choosing a value of $a = 3L_p/4$, $\beta = 0.75$, from which equation (5.62) gives a value for α of 4/11 (contrast with -4/13 for zero slope at the other end). Figure 5.9 shows the subsequent deflection given by equation (5.56) using these values for zero slope. Because, in this case, the point of application of the drive is inside the flexure and produces a smaller bending moment than the previous case, such a design is likely to be more desirable.

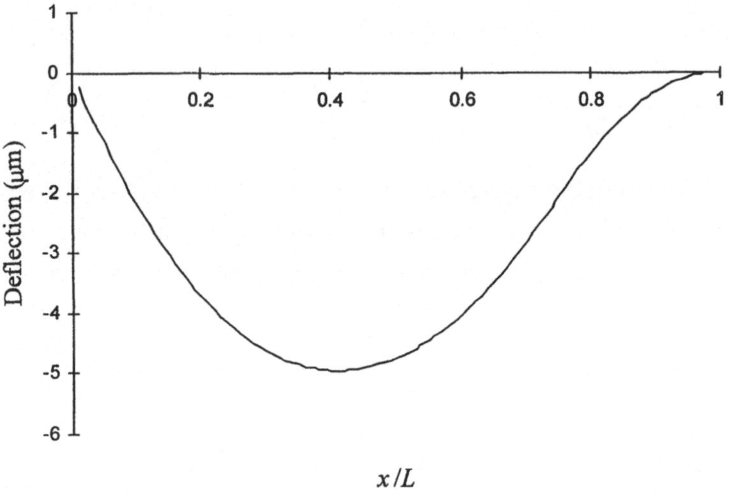

Figure 5.9: Deflection of the upper platform for the values $s = 2.91$ mm, $\alpha = 4/11$ and $\beta = 3/4$

5.2.2 Effect of axial strains in the flexure supports

In the previous analysis, the equal and opposite reaction forces, R_1 & R_2, along the axis of the flexures due to the moment produced by the drive have been ignored. This will result in a net tilt of the complete platform given by the difference between the axial extensions of each flexure divided by the length, L_p, of the platform separating them. Again, assuming small deflections are, the angular tilt (anti-clockwise positive) is

$$\theta = (R_1 - R_2)\frac{L_f}{L_p E A_f} = 2R_1 \frac{L_f}{L_p E A_f}$$
$$= \frac{2M_1(k-2)}{L_p E A_f} = \frac{t_f^2}{L_p^2 L_f}(k-2)\delta \tag{5.63}$$

This reduces to zero when the drive is positioned midway between the platform and base, a conclusion in agreement with the more complete analytical results of Plainevaux, 1956. It should also be noted that, in this equation, the slope is independent of the cross sectional area of the platform while its length reduces value in inverse proportion. For the parameters used to generate figure 5.9, the slope predicted by equation (5.63) for a displacement of 100 μm is -17 μrads. Current design goals for precision flexures are in the region of 1 μrad of angular error and this is often considered a significant effect.

5.2.3 Combined effects

For small deflections, the principle of superposition can be applied to refine the analysis of slope deflection of the platform support. It should be noted that this ignores secondary effects, such as, the influence of the platform slope on the values of bending moments from equation (5.52) which relies on a zero slope boundary condition in its derivation. Summing equation (5.58) and (5.63), for small deflections, the total slope of the platform is given by

$$\frac{dy}{dx} = \left[\frac{t_f^2}{L_p}(k-2) - \frac{6I_f}{I_p}\left(x + \frac{(k-2)}{L_p}\frac{x^2}{2} - k[x-a] + \frac{kL_p}{3} + \frac{ka^2}{2L_p} - ka - \frac{L_p}{6}\right)\right]\frac{\delta}{L_f^2}$$

$$= \left[\frac{(k-2)}{L_p} - \frac{6t_f}{t_p^3}\left(x + \frac{(k-2)}{L_p}\frac{x^2}{2} - k[x-a] + \frac{kL_p}{3} + \frac{ka^2}{2L_p} - ka - \frac{L_p}{6}\right)\right]\frac{t_f^2\delta}{L_f^2} \qquad (5.64)$$

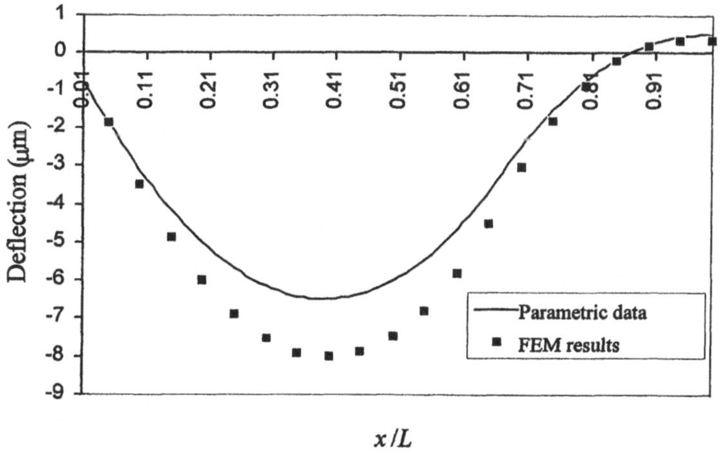

Figure 5.10: Theoretical prediction for the full equations and a comparison with results from finite element analysis. Parameters are given in the caption of figure 5.8 except for the modified geometric values discussed in the text

From this rather lengthy expression, determination of the geometric conditions for zero slope at either end of the platform support are not as straightforward. However, a number of important features are still readily apparent. The slope is independent of the material from which the monolithic it is comprised, is proportional to the flexure displacement, and inversely proportional to the square of the flexure length. All of these parameters do not influence the conditions for zero slope.

Figure 5.10 shows the theoretical prediction of the platform distortion with results from finite element analysis for comparison. Although there is a relatively large discrepancy between displacements, the theoretical model appears to be a reasonable predictor of the end slopes of the platform.

5.3 Planar mechanisms

A general approach to the design of planar mechanisms is presented here.

For a planar mechanism, it has already been shown, section 2.9, that the mobility is given by

$$\begin{aligned} M_3 &= 3(n-1)-\sum_{i=1}^{j} c_i = 3(n-1)-\sum_{i=1}^{j}(3-f_i) \\ &= 3(n-j-1)+\sum_{i=1}^{j} f_i \end{aligned} \tag{5.65}$$

where n is the number of links, j is the number of joints and f is the number of freedoms of the ith joint and c represents the corresponding constraints $(3-f)$.

For a mechanism made up entirely of single degree of freedom joints such as the notch hinge, this reduces to Grubler's equation

$$M = 3(n-1)-2j \tag{5.66}$$

Clearly, based on this analysis, the four bar link (j = 4, n = 4) represents a single degree of freedom mechanism. It is important to maintain some caution when applying this formula to real systems. The planar assumption reduces the problem to motion in a two dimensional space. For a more complete analysis, the mobility should be computed from the three dimensional formula.

$$M_6 = 6(n-j-1)+\sum_{i=1}^{j} f_i \tag{5.67}$$

In three dimensions, analysis of the four bar link yields M = -2 indicating overconstraint. As a consequence, it is to be expected that the performance of such a mechanism will rely upon precision of manufacture. For most applications, over-constraint will be accommodated by the off-axis compliance of the flexures and such errors are often insignificant. This surprising result illuminates some of the pitfalls that, in some cases, might introduce significant errors in a design. Consequently, the following section discusses the reasons for the differences between planar and three dimensional mobility analysis.

To create a mechanism, it is necessary to connect a number of rigid links using hinges. In the design process, there are a number of parameters that can be selected. While mobility analysis is a direct measure of the feasibility of a mechanism in terms of the freedoms it may possess, it is also necessary to consider other parameters that can be varied to produce an optimal design. Before discussing planar mobility analysis in more detail, some other important factors that effect the function of a mechanism are

1. The depth of the mechanism. Increasing the depth results in an increased stiffness both in and out the plane of motion. Consequently, this will increase the load capacity of the flexure and reduce the influence of masses being added to moving platforms. However, the linear increase in stiffness with depth is matched by an equally proportionate increase in mass therefore nulling any increase in natural frequency. An increased stiffness also requires more work from the drive. As a rule, it is assumed that deeper flexures are better. Some manufacturing issues are addressed in chapter 8.
2. The loads to be supported and the length of translation will dominate the size of the mechanism. Consequently, planar flexures can range in size from micro-mechanical systems to large structural supports weighing many tons, see for example Thorpe, 1953.
3. Manufacture. Using computer-controlled machine tools, wire electro-discharge machines and micro-electronic manufacturing facilities, planar shapes of almost arbitrary complexity can now be produced over an enormous range of dimensions.
4. As shown in the previous section, forces will distribute throughout the mechanism resulting in deformation of *all* component parts. An alternative view of a flexure mechanism is that of a series of connected flexible elements of varying compliance. Such a continuum approach to flexure design can be modeled using the method of finite elements. However, except where stated, subsequent discussions will assume that links are infinitely rigid and freedoms of connecting hinges provide combinations of pure rotation and translation.

5.3.1 Discussion of the simplifying assumptions for reduction of mobility analysis to planar mechanisms

Reduction of mobility analysis to planar mechanisms should be approached with caution. Consider again the simple four-bar link. Implicit to this two-dimensional mechanism is the assumption that the compliant axes of all joints are parallel. Assuming such manufacturing perfection is possible, it is no longer necessary to consider the effects of out-of-plane constraints. Alternatively, it may be assumed that the perfection of manufacture infers additional freedoms of the joints in view of the fact that some theoretical constraints will not be 'tested' in practice. With this caveat, consider first the conceptualized planar mechanism shown in figure 5.11. This indicates the essential features of a model for assessing mechanism mobility in which individual joints are capable of providing one to three freedoms.

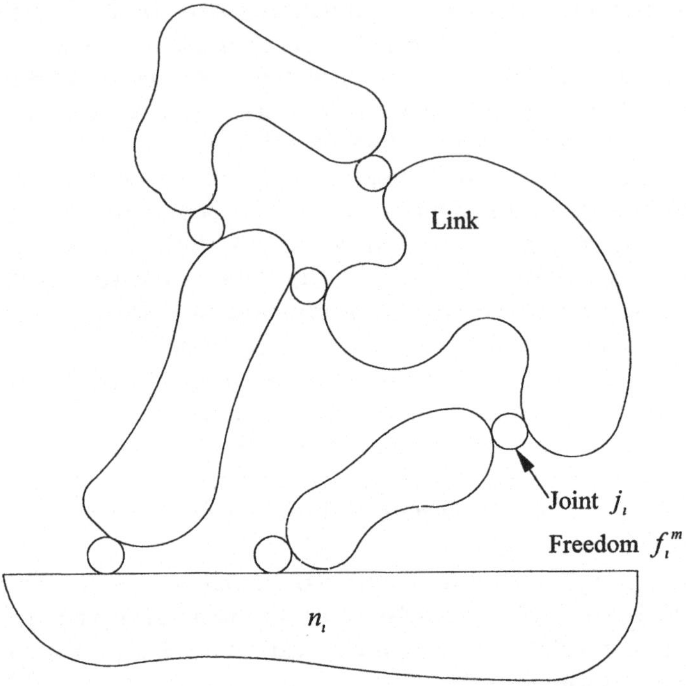

Figure 5.11: A conceptualized, general planar mechanism consisting of 5 links and 6 joints

Considering the case of a mechanism in which all connections utilize single freedom joints. Available joints providing a single freedom are flexure hinges, slideways (including screws) or a more complex mechanism that, of itself, forms a single freedom joint when used to connect two links. Using equation (5.66) it can be shown that the minimum number of links to form a mechanism from single degree of freedom joints is 2 with 1 joint (it can also be seen from (5.67) that this is a single freedom mechanism in three dimensions). An example might be a rigid pendulum hanging from a hinge. The next available lowest number of links is 4 with 4 joints which is, of course, the four bar link. In general, it is relatively simple to show that the relationship between the number of links and joints for mobility 1 is given by

$$j = 3n/2 - 2 \qquad n = 2, 4 :. \infty \qquad (5.68)$$

Figure 5.12 shows a possible mechanism that comprises 6 links and 7 joints. Increasing the number of links so that $n = 8$ and $j = 10$ results in the somewhat modified form of the Peaucellier mechanism shown in figure 5.13. Detailed kinematic analysis reveals that, by selection of suitable length of link and distribution of hinge, such a mechanism can be produce perfect rectilinear

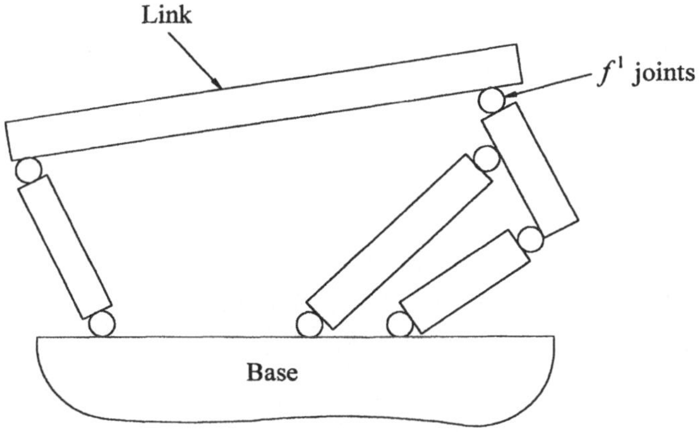

Figure 5.12: A mechanism of mobility $M = 1$ comprising 6 links and 7 joints

motion of the point P in the figure, see for example Bevans, 1948. Another mechanism of interest is that comprising 12 links and 16 joints which is commonly referred to as the double compound rectilinear spring, see figure 5.14. This has the particular advantage of a high degree of symmetry about its center and also along the axis of motion.

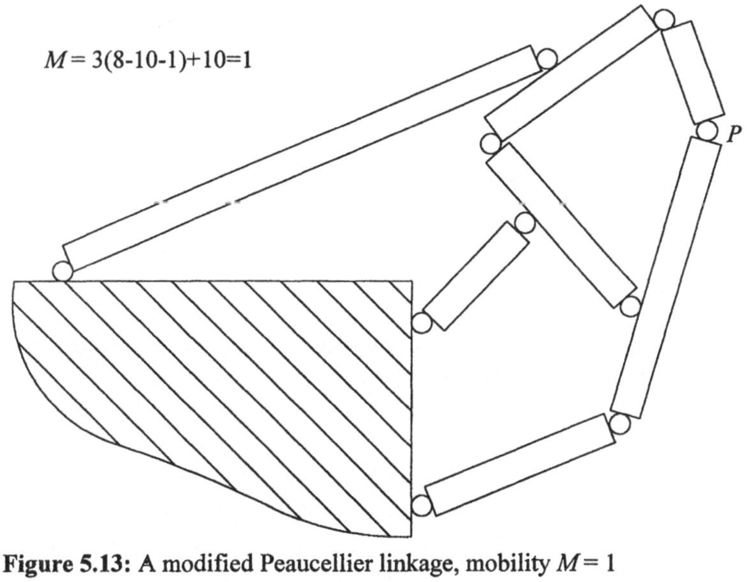

Figure 5.13: A modified Peaucellier linkage, mobility $M = 1$

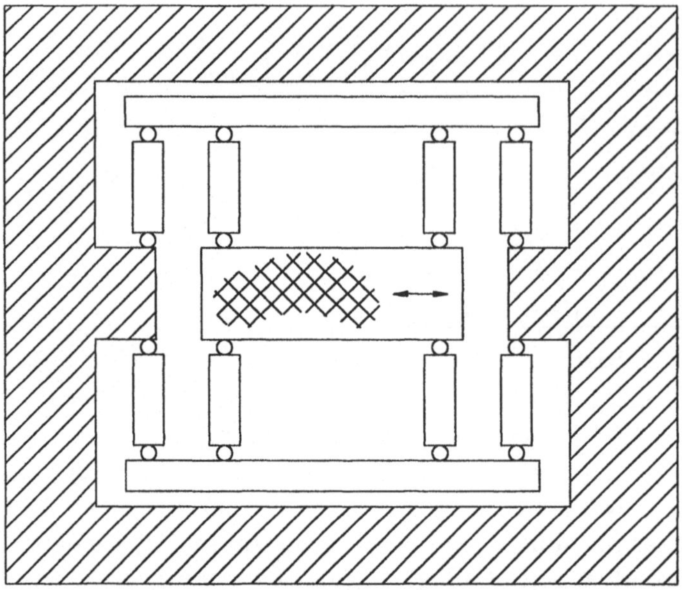

Figure 5.14: Schematic representation of the double compound rectilinear spring, $M = 3(12\text{-}16\text{-}1)+16 = 1$.

Table 5.3.1.1: Number links and single freedom joints to construct a planar mechanism of mobility 1

Links	Joints
2	1
4	4
6	7
8	10
10	13

For mobility 2, a minimum of 3 links and 2 joints are required. An example of this is the chaotic double pendulum. Another combination might be 5 links and 5 joints or the more interesting 7 links and 7 joints. An example of the latter is the compound linear spring shown in figure 5.15.

Using the full equation (5.65) for planar mechanisms, the above analysis can be extended to include 2 and 3 freedom hinges such as those outlined in chapter 4. Considering only planar mechanisms, a two axis joint must comprise either two orthogonal translations (not necessarily rectilinear) or a rotation plus one translation. Obviously a three-axis hinge contains all three and is considered free in the plane (and therefore non-existent!). The utility of the three-freedom hinge is in its ability to withstand loads normal to the plane of the mechanism. To reduce the mobility of a mechanism, the objective of most flexures, it is apparent that the fewer links the better. For single mobility, a possible combination is 2

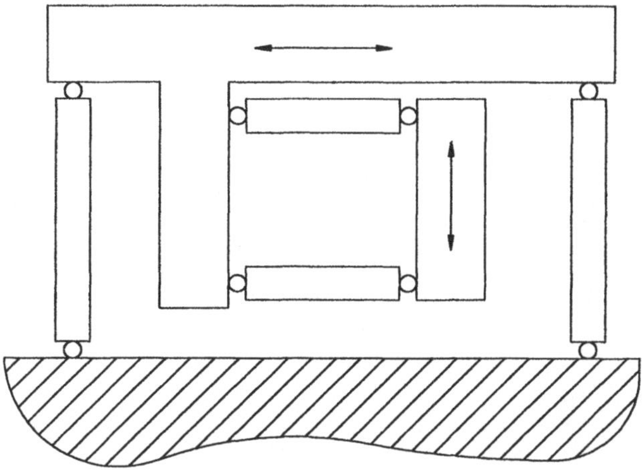

Figure 5.15: A compound linear spring of mobility 2 providing two orthogonal linear translations i.e. an *x-y* stage.

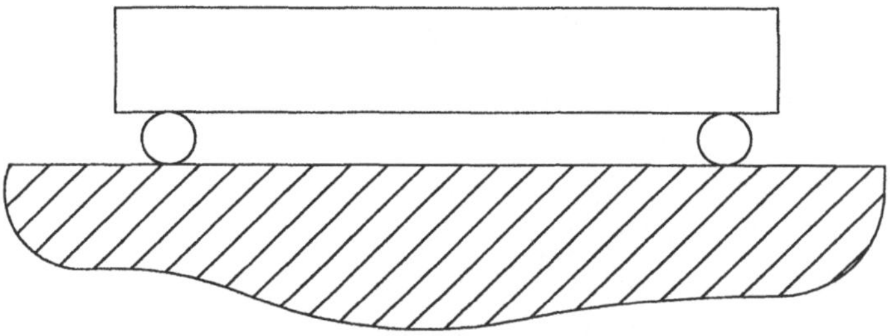

Figure 5.16: A planar mechanism of mobility 1 comprising 2 links and 2 two-freedom joints ($M = 3(2\text{-}2\text{-}1)\text{+}4$)

links and 2 two-freedom joints, figure 5.16. A similar single degree of freedom mechanism can be produced using a single degree of freedom hinge connecting 2 joints. However, there is an essential difference between these two. That is, using a single degree of freedom rotational hinge, the only remaining freedom is a rotation whereas two 2-freedom joints provide the possibility of either a rotation *or* translation. Of course adding together combinations of single freedom rotational hinges and links can produce linear mechanisms but this is at the expense of added materials, complexity, and production cost. In an attempt to produce a more symmetric mechanism, figure 5.17 shows a mechanism with 2 links and four joints. Using 2 freedom joints this is slightly over constrained while three freedom joints result in a mobility of 3 and therefore the links can adopt arbitrary relative positions within the plane. Accepting that this requires precise manufacture, a possible pseudo single degree of freedom mechanism is shown in figure 5.18. This utilizes the two-beam, two-axis hinges discussed in section 4.4.2.

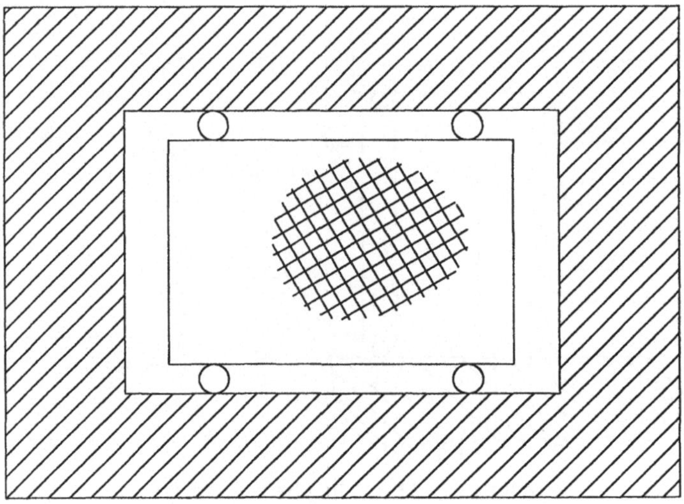

$$M = 3(2-4-1)+\sum f$$
$$= -1\ (f^2 \text{ joints})$$
$$= 3\ (f^3 \text{ joints})$$

Figure 5.17: A mechanism having 2 links and 4 joints

Some three axis hinges, or joints, are shown in figure 5.19. Literature on three axis flexures is scarce and detailed analysis awaits further research.

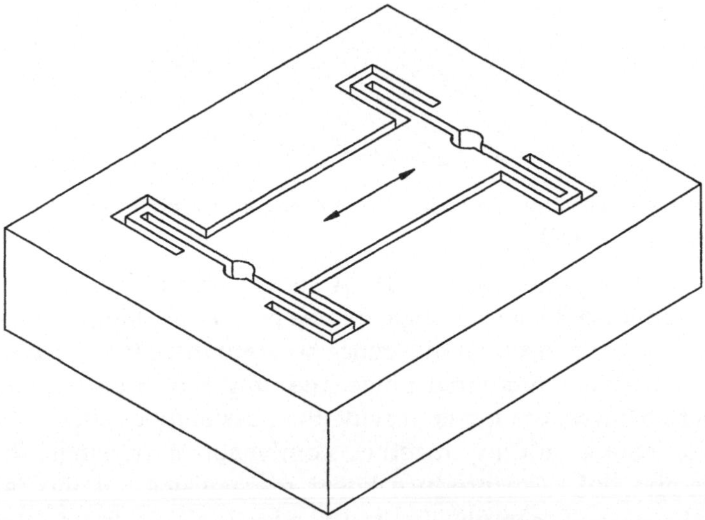

Figure 5.18: A translation mechanism using two-beam, two-axis hinges

As evidenced by the few mechanisms shown in previous figures, there are an endless array of mechanisms that can be produced using combinations of links and joints. The interested reader will find some stimulating designs presented in papers listed in the bibliography of interesting flexures at the end of this chapter.

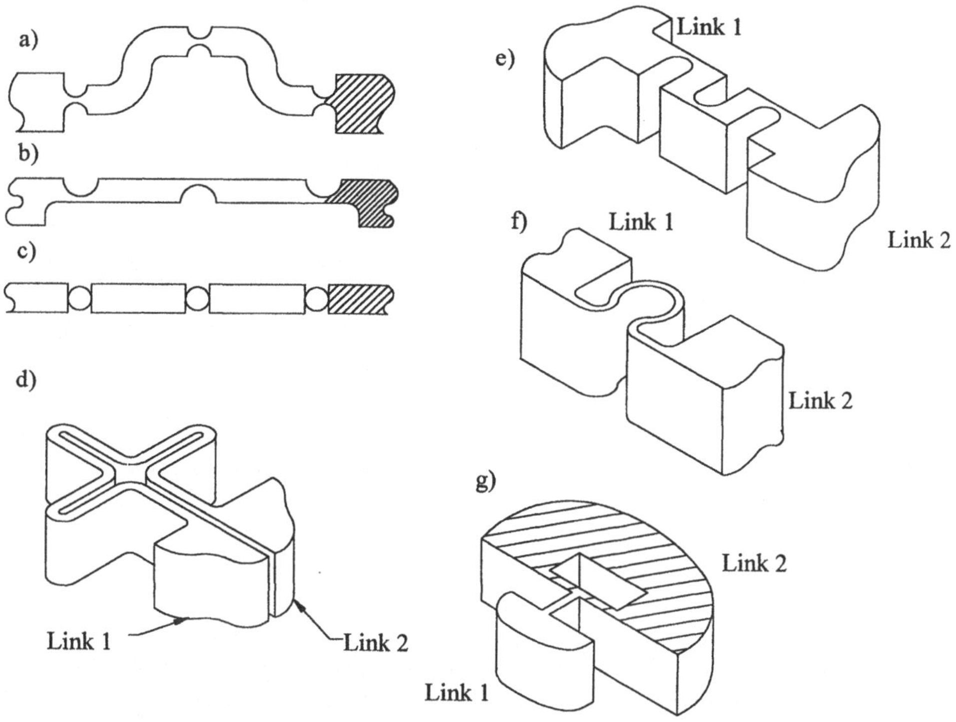

Figure 5.19: Some three degree of freedom, planar joints, a) series notch type, b) series half notches, c) mobility model with three links and three single freedom joints i.e. $M = 3(3-1)+3 = 3$, d) symmetric array of two-beam, two-axis hinges (note that the legs connecting to the links are not necessary for this joint to provide the requisite three freedoms), e) isometric diagram of three half notch joint produced by slotting, f) hinge formed from a cylindrical band, g) combined single beam, two axis hinge plus linear flexure

5.4 Dynamics of ideal planar flexures (some common mechanisms)

In the following sections some general methods for assessing the performance of planar and three-dimensional flexures will be introduced. Such generality necessarily results in more complex and abstract mathematical formulisms. As a consequence it is likely that the underlying principles may be lost. Fortunately, in many applications, flexures contain symmetries that considerably simplify analysis. This section presents a dynamic model for some of the more common geometries used for planar mechanisms. The simple four-bar link for parallelogram motion of the upper platform assessed in section 5.1 will not be further assessed in this section. However, in practice, flexure hinges will have finite stiffness values in the so-called constraint directions. Therefore, a complete planar analysis of even the simple four-bar link should consider these compliances. Necessarily, each joint must then be modeled as a provider of three degrees of freedom leading to a mechanism of high mobility. In this case generalized dynamic models become necessary and these are discussed in section 5.5 where the apparently simple four-bar link is reassessed.

5.4.1 The compound rectilinear spring

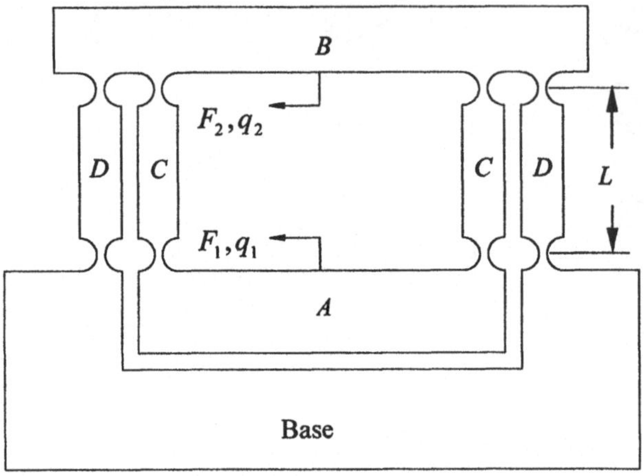

Figure 5.20: Mathematical model for the compound rectilinear spring

One of the major concerns with the simple linear spring is the arcuate motion of the upper platform relative to the base. One configuration that may be used to overcome this limitation is achieved by attaching two such hinges together to produce the geometry shown in figure 5.20. Assuming all hinges are of equal compliance and all support legs are of the same length with parallel axes, it is relatively easy to see that, if platform A is the only element of the mechanism subject to a static, externally applied force it will undergo a purely rectilinear motion. Platform B can also be shown to displace exactly half as far as A when both displacements are measured relative to the base. However, when considering free vibrations of such a structure, the rectilinear motions of platforms A and B must be considered to be independent and are given the coordinates q_1 and q_2 respectively (i.e. this is a mechanism of mobility = 3(7-1)+2(8) = 2). To derive expressions for kinetic and potential energy in the flexure, although not necessary, the following assumptions have been made for compactness of the solution

1. All motions are small
2. The support legs (C and D) are prismatic with the central axis passing through the centers of the hinges at each end.
3. As a consequence of 2, the center of mass for each support leg lies on the axis of the leg at a height midway between the two notch hinges.
4. Second moments of mass have been measured about center of mass
5. For the time being, it will be assumed that the rotational stiffness (Nm rad^{-1}) for the notches are equal on support legs C and D, although the stiffness of the notches on C need not necessarily be equal to those on support legs D.

Based on the above considerations, it is relatively straightforward to derive expressions for the kinetic, T, and potential, V, energy of the flexure

$$T = \frac{1}{2}M'_A\dot{q}_1^2 + \frac{1}{2}M_B\dot{q}_2^2 + 2\frac{1}{2}I_D\left(\frac{\dot{q}_2}{L}\right)^2 + 2\frac{1}{2}I_C\left(\frac{\dot{q}_2 - \dot{q}_1}{L}\right)^2 + 2\frac{1}{2}M_D\left(\frac{\dot{q}_2}{2}\right)^2 + 2\frac{1}{2}M_C\left(\frac{\dot{q}_2 + \dot{q}_1}{2}\right)^2 \tag{5.69}$$

$$V = 4\frac{1}{2}k_D\left(\frac{q_2}{L}\right)^2 + 4\frac{1}{2}k_C\left(\frac{q_2 - q_1}{L}\right)^2 \tag{5.70}$$

Considering, for example, the coordinate corresponding to displacement of the platform A, the various elements of Lagrange's equation are

$$\frac{d}{dt}\left(\frac{\partial T}{\partial \dot{q}_1}\right) = \left(M_A + \frac{M_c}{2} + \frac{2I_c}{L^2}\right)\ddot{q}_1 + \left(\frac{M_c}{2} - \frac{2I_c}{L^2}\right)\ddot{q}_2$$

$$\frac{\partial T}{\partial q_1} = 0$$

$$\frac{\partial V}{\partial q_1} = \frac{4k_c}{L^2}(q_1 - q_2)$$

Proceeding in like manner for the second coordinate, from Lagrange's equation, the equations governing motion of this flexure are given by

$$\begin{aligned} &\left(M_A + \frac{M_c}{2} + \frac{2I_c}{L^2}\right)\ddot{q}_1 + \left(\frac{M_c}{2} - \frac{2I_c}{L^2}\right)\ddot{q}_2 + \frac{4k_c}{L^2}q_1 - \frac{4k_c}{L^2}q_2 = F_1 \\ &\left(\frac{M_c}{2} - \frac{2I_c}{L^2}\right)\ddot{q}_1 + \left(M_B + \frac{(M_c + M_D)}{2} + \frac{2(I_c + I_D)}{L^2}\right)\ddot{q}_2 - \frac{4k_c}{L^2}q_1 \\ &\qquad \ldots + \frac{4(k_c + k_D)}{L^2}q_2 = F_2 \end{aligned} \tag{5.71}$$

This rather cumbersome equation can be expressed in the simpler form

$$\begin{aligned} a_{11}\ddot{q}_1 + a_{12}\ddot{q}_2 + c_{11}q_1 + c_{12}q_2 &= F_1 \\ a_{21}\ddot{q}_1 + a_{22}\ddot{q}_2 + c_{21}q_1 + c_{22}q_2 &= F_2 \end{aligned} \tag{5.72}$$

where the a's and c's are constant and it is noted that $a_{12} = a_{21}$ and $c_{12} = c_{21}$. Equation (5.72) can be further simplified by writing it in the matrix form

$$\begin{bmatrix} a_{11} & a_{12} \\ a_{21} & a_{22} \end{bmatrix} \begin{Bmatrix} \ddot{q}_1 \\ \ddot{q}_2 \end{Bmatrix} + \begin{bmatrix} c_{11} & c_{12} \\ c_{21} & c_{22} \end{bmatrix} \begin{Bmatrix} q_1 \\ q_2 \end{Bmatrix} = \begin{Bmatrix} F_1 \\ F_2 \end{Bmatrix}$$

or (5.73)

$$[\mathbf{M}]\{\ddot{\mathbf{q}}\} + [\mathbf{K}]\{\mathbf{q}\} = \{\mathbf{F}\}$$

For the homogeneous (or steady state) solution, equation (5.73) can be rearranged and expressed in the form of an eigen equation

$$\begin{aligned} &[\mathbf{M}]\{\ddot{\mathbf{q}}\} + [\mathbf{K}]\{\mathbf{q}\} = \{\mathbf{0}\} \\ &[\mathbf{I}]\{\ddot{\mathbf{q}}\} + [\mathbf{M}]^{-1}[\mathbf{K}]\{\mathbf{q}\} = \{\mathbf{0}\} \end{aligned} \tag{5.74}$$

Following the techniques outlined in chapter 3, equations (5.73) and (5.74) can be used to derive the two natural frequencies (eigenvalues) and mode shapes (eigenvectors) and the four steady-state frequency response functions. Extension of this analysis for systems containing damping involves little additional difficulty although the aforementioned parameters must be expressed in complex form. To demonstrate this, a solution is readily obtained assuming displacements of each coordinate of the form

$$\{\mathbf{q}\} = \mathrm{Re}\left(\{\Phi\}e^{i\omega t}\right) \tag{5.75}$$

To produce the eigen-equation of the system, it is necessary to substitute equation (5.75) into (5.74) after which

$$\begin{aligned} &([\mathbf{A}] - \omega^2[\mathbf{I}])\{\Phi\} = \\ &([\mathbf{A}] - \lambda[\mathbf{I}])\{\Phi\} = \{\mathbf{0}\} \end{aligned} \tag{5.76}$$

Solutions to equation (5.76) for the two eigenvalues ($\lambda_{1,2} = \omega_{1,2}^2$) and the four eigen vectors (Φ) can be readily obtained using matrix solution algorithms. Alternatively, equation (5.76) can be written in expanded form

$$\begin{aligned} &\left(c_{11} - a_{11}\omega^2\right)\Phi_1 + \left(c_{12} - a_{12}\omega^2\right)\Phi_2 = 0 \\ &\left(c_{21} - a_{21}\omega^2\right)\Phi_1 + \left(c_{22} - a_{22}\omega^2\right)\Phi_2 = 0 \end{aligned} \tag{5.77}$$

From which the ratio of the two eigenvectors is readily obtained

$$\frac{\Phi_1}{\Phi_2} = \frac{a_{12}\omega_{1,2}^2 - c_{12}}{c_{11} - a_{11}\omega_{1,2}^2} \tag{5.78}$$

However, it is necessary to know the values of the natural frequencies before this can be computed. These can be readily solved by substituting (5.78) into the second of (5.77) to yield

$$\begin{aligned} &\left(c_{21}-a_{21}\omega^2\right)\left(\frac{a_{12}\omega^2-c_{12}}{c_{11}-a_{11}\omega^2}\right)\Phi_2+\left(c_{22}-a_{22}\omega^2\right)\Phi_2 \\ &=\left(c_{21}-a_{21}\omega^2\right)\left(a_{12}\omega^2-c_{12}\right)+\left(c_{11}-a_{11}\omega^2\right)\left(c_{22}-a_{22}\omega^2\right) \\ &=\omega^4\left(a_{11}a_{22}-a_{12}^2\right)-\omega^2\left(a_{11}c_{22}+a_{22}c_{11}\right)+c_{11}c_{22}-c_{12}^2=0 \end{aligned} \tag{5.79}$$

The last of equation (5.79) is a quadratic in $\omega^2(=\lambda)$ for which the two solutions can be readily computed.

The steady-state response for an arbitrary, harmonically varying input force is similarly obtained. In this case it is only necessary to assume that the response is linearly related to the input. For example assuming an applied force at coordinate *1*, of the form

$$F_1=\mathrm{Re}\left(F_1 e^{i\omega t}\right) \tag{5.80}$$

the response can be readily written as

$$\begin{aligned} q_1&=\mathrm{Re}\left(H_{11}(i\omega)F_1 e^{i\omega t}\right) \\ q_2&=\mathrm{Re}\left(H_{21}(i\omega)F_1 e^{i\omega t}\right) \end{aligned} \tag{5.81}$$

From (5.73), it is relatively simple to obtain an equation of the form

$$\begin{aligned} &\left(c_{11}-a_{11}\omega^2\right)H_{11}(i\omega)+\left(c_{12}-a_{12}\omega^2\right)H_{21}(i\omega)=1 \\ &\left(c_{21}-a_{21}\omega^2\right)H_{11}(i\omega)+\left(c_{22}-a_{22}\omega^2\right)H_{21}(i\omega)=0 \end{aligned} \tag{5.82}$$

This can be expressed in abbreviated matrix form

$$\begin{bmatrix} e_{11}(\omega) & e_{21}(\omega) \\ e_{21}(\omega) & e_{22}(\omega) \end{bmatrix}\begin{Bmatrix} H_{11}(\omega) \\ H_{21}(\omega) \end{Bmatrix}=\begin{Bmatrix} 1 \\ 0 \end{Bmatrix} \tag{5.83}$$

Note also that $e_{12}=e_{21}$. Two of the solutions for the values of the $H(\omega)$'s at any frequency can be readily obtained using Cramer's rule for solving these simple simultaneous equations. Similar methods can be used to determine the remaining two frequency responses for a force applied at the second coordinate. Analytic expressions for the eigenvalues can be obtained by setting the determinant of the left-hand matrix in (5.83) to zero. Chapter 3 contains a more complete discussion of these methods.

5.4.2 The double compound linear spring (including the lever driven spring)

An advantage of the previous design is that, provided the support legs are all of equal length and the notches of equal stiffness, then it is possible to achieve perfect rectilinear motion. This is due to the fact that the arcuate motion of the first simple spring is matched by an equal arcuate motion of the other in a way that the mutual approaches of each moving platform to its respective base both cancel with respect to the stationary frame. However, a penalty for this is the introduction of a second independent coordinate. Figure 5.21 shows a more symmetric design consisting of two compound springs attached at the moving platform. Such a mechanism is called a double compound linear spring. As a consequence of the two compound springs being joined together, it can be seen that a deviation from rectilinear motion is no longer possible in this design. This is confirmed by a mobility analysis for which

$$M = 3(n-1) - 2j = 3(12-1) - 2(16) = 1 \quad (5.84)$$

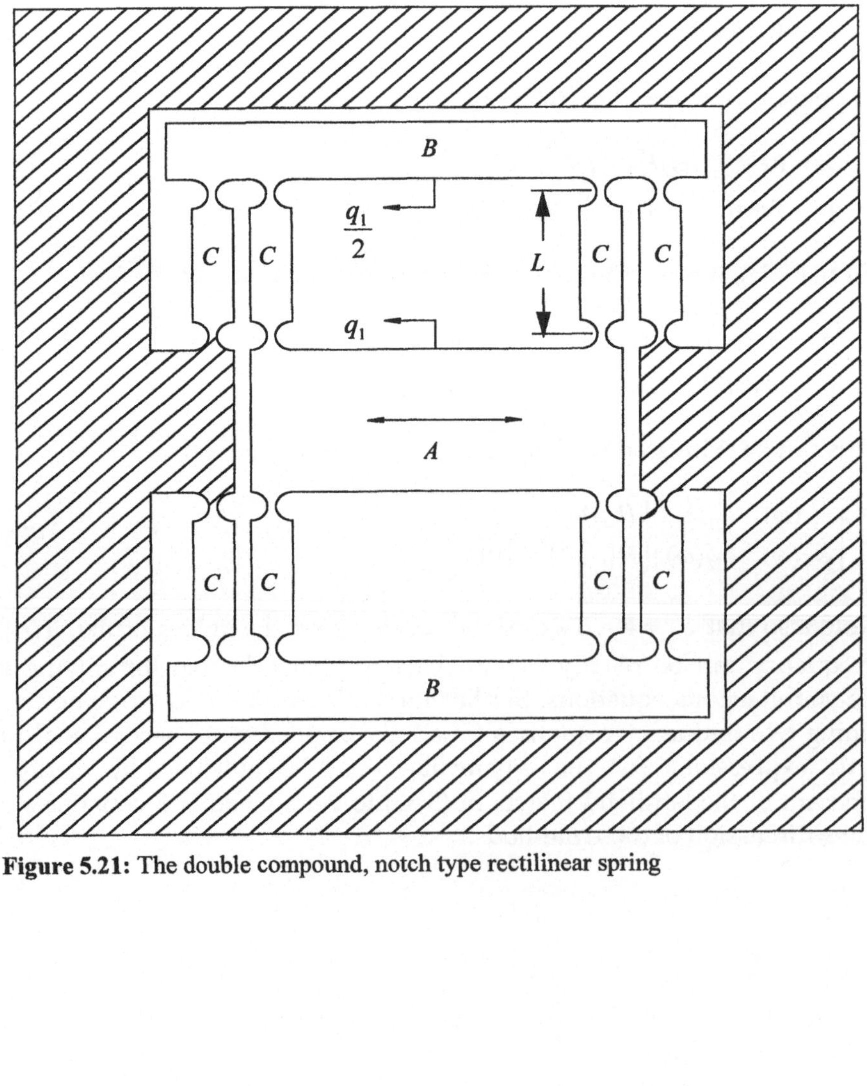

Figure 5.21: The double compound, notch type rectilinear spring

Consequently, one coordinate is sufficient to determine the kinetic and potential energy of all elements of the mechanism which, in the absence of a driver are given by

$$T = \frac{1}{2}M_A\dot{q}_1^2 + 2\frac{1}{2}M_B\left(\frac{\dot{q}_1}{2}\right)^2 + 4\frac{1}{2}M_C\left(\frac{\dot{q}_1 + \frac{\dot{q}_1}{2}}{2}\right)^2$$

$$..+8\frac{1}{2}I_C\left(\frac{\dot{q}_1}{2L}\right)^2 + 4\frac{1}{2}M_C\left(\frac{\dot{q}_1}{2.2}\right)^2 \quad (5.85)$$

$$= \frac{1}{2}\left(M_A + \frac{M_B}{2} + \frac{5M_c}{2} + \frac{2I_c}{L^2}\right)\dot{q}_1^2$$

$$V = 16\frac{1}{2}k\left(\frac{q_1}{2L}\right)^2$$

The derivation of the kinetic energy is easily seen by considering each of the terms in the first of equation's (5.85). Consecutively, these terms correspond to; linear motion of the primary platform, approximate linear motion of the two secondary platforms (assuming all lengths of the support legs are equal the secondary platforms will experience half of the velocity of the primary), linear motion of the centriod of the four central support legs, the common angular velocity of all the support legs and the linear motion of the center of mass of the four outer support legs. Substitution of equation (5.85) into Lagrange's equation gives the equation governing motion

$$\left(M_A + \frac{M_B}{2} + \frac{5M_c}{2} + \frac{2I_c}{L^2}\right)\ddot{q}_1 + \frac{4}{L^2}kq_1 = F_1 \quad (5.86)$$

The natural frequency of this flexure immediately follows

$$\omega_n^2 = \frac{4k}{L^2\left(M_A + \frac{M_B}{2} + \frac{5M_c}{2} + \frac{2I_c}{L^2}\right)} \quad (5.87)$$

Modeling the support legs as thin rods, the second moment of mass about the center of mass (i.e. at the center of the leg) is

$$I_C = \frac{M_C L^2}{12} \quad (5.88)$$

After which equation (5.87) becomes

$$\omega_n^2 = \frac{4k}{L^2\left(M_A + \frac{M_B}{2} + \frac{8M_c}{3}\right)} \tag{5.89}$$

This is the same as the result given in Smith and Chetwynd, 1991.

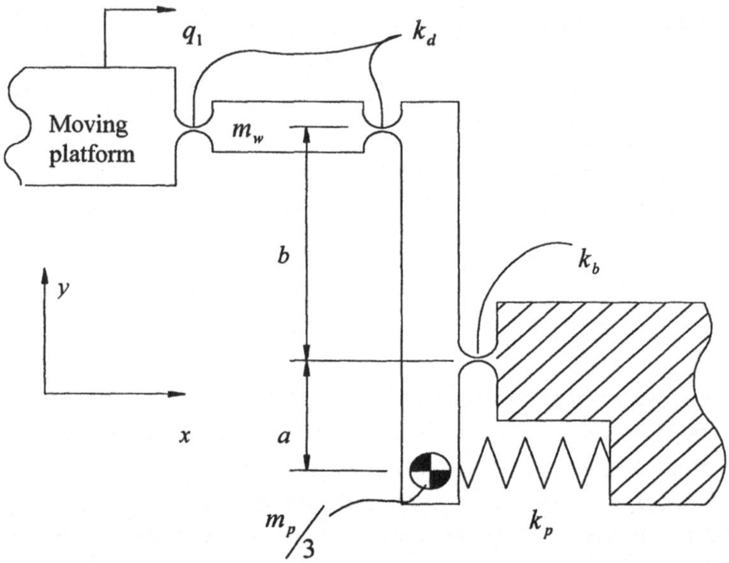

Figure 5.22: A simple lever drive applied to the moving platform of the double compound rectilinear spring

The addition of a lever drive merely adds to both the potential and kinetic energy terms of the Lagrangian. Considering the simple lever of figure 5.22, there will be three additional compliance's contributing to the potential energy plus the additional inertia of the lever, driver and 'wobble pin'. The lever pivot and primary platform drive flexures have angular stiffness values denoted by k_b & k_d respectively. Assuming that the lever is driven by a piezoelectric element which is rigidly coupled to the lever (unlikely in practice, see chapter 7) this particular drive can be considered as a linear spring having stiffness k_p. From Rayleigh's method (section 3.11.2.3), it would seem reasonable to lump one third of the mass of the piezoelectric element, m_p, to the lever at the point of connection as shown in the figure. A wobble-pin is necessary to provide freedom for the y-axis component of the motion of the lever. Assuming that the notch flexures behave as perfect notches, the contribution of the platform notch to the potential energy may reasonably be neglected for small motions. Based on these assumptions, the *additional* kinetic, T_a, and potential, V_a, energy terms are

$$\begin{aligned} T_a &= \frac{1}{2} I_b \left(\frac{\dot{q}_1}{b} \right)^2 + \frac{1}{2} \frac{m_p}{3} \left(\frac{\dot{q}_1}{n} \right)^2 + \frac{1}{2} m_w \dot{q}_1^2 \\ &= \frac{1}{2} \left(\frac{I_b}{b^2} + \frac{m_p}{3n^2} + m_w \right) \dot{q}_1^2 \\ V_a &= \frac{1}{2} (k_b + k_d) \left(\frac{q_1}{b} \right)^2 + \frac{1}{2} k_p \left(\frac{q_1}{n} \right)^2 \\ &= \frac{1}{2} \left(\frac{k_b + k_d}{b^2} + \frac{1}{2} \frac{k_p}{n^2} \right) q_1^2 \end{aligned} \tag{5.90}$$

where n is the lever ratio b/a.

Consequently, the fundamental mode natural frequency of the lever driven, double compound linear spring becomes

$$\omega_n^2 = \frac{\dfrac{4k}{L^2} + \dfrac{k_b + k_d}{b^2} + \dfrac{1}{2} \dfrac{k_p}{n^2}}{M_A + \dfrac{M_B}{2} + \dfrac{8M_c}{3} + \dfrac{I_b}{b^2} + \dfrac{m_p}{3n^2} + m_w} \tag{5.91}$$

Clearly, whether the fundamental frequency increases or reduces will depend on the relative values of the lever ratio, stiffness and mass components of the lever mechanism.

5.4.3 A coupled two-axis flexure

Consider the flexure mechanism of figure 5.23 overleaf that possesses 5 links and 5 joints. Grubler's equation immediately gives

$$M = 3(5-1) - 2(5) = 2 \tag{5.92}$$

Having established that this is a two-degree of freedom system, the next step is to decide on a suitable set of generalized coordinates. Recognizing that the two independent freedoms of the central mass are a translation in the vertical direction and a rotation about its mid-point which, in this case, coincides with its center of mass, suggests a suitable origin for the generalized coordinates. As shown in figure 5.23(b), these coordinates are designated x_o for the translation and θ_o for the rotation. Input drives are located at the two lever points and, for small motions, the input translations are given the temporary coordinates x_1 & x_2. To derive the Lagrangian of this system in terms of the generalized coordinates, it is necessary to determine the functional relationships between these coordinates. For small motions the transformations are given by the equations

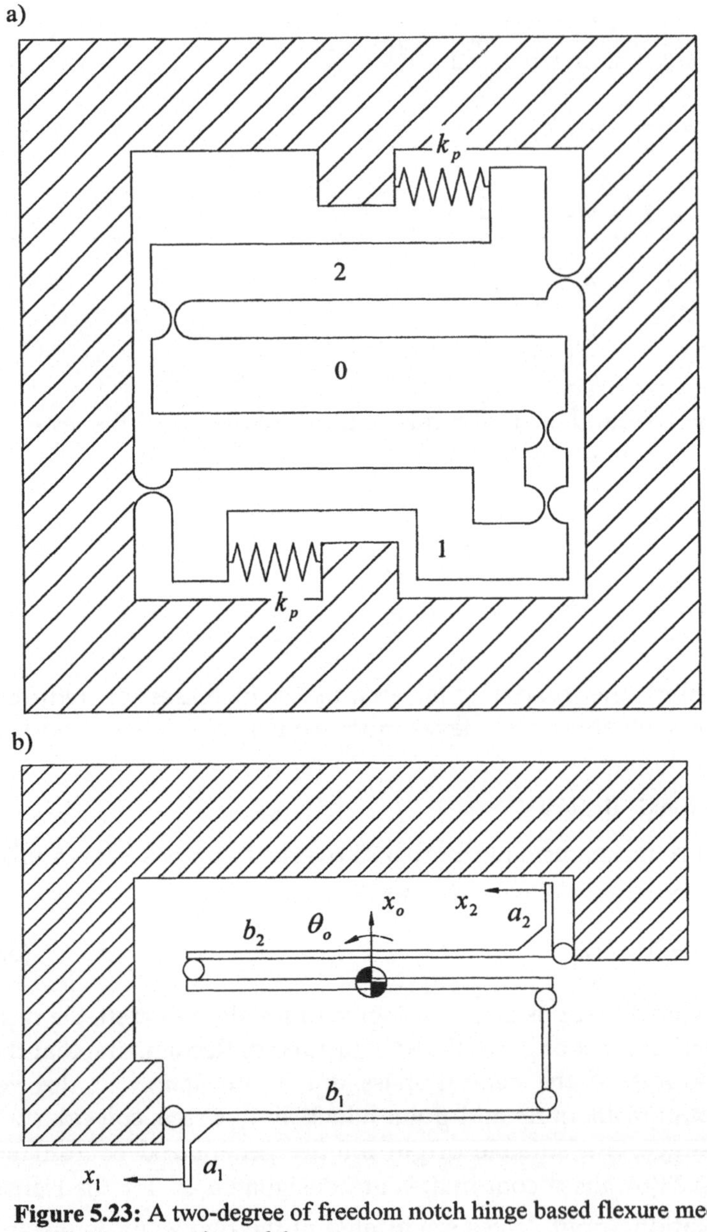

Figure 5.23: A two-degree of freedom notch hinge based flexure mechanism, a) solid model, b) parametric model

$$
\begin{aligned}
x_o &= \frac{n_1 x_1 + n_2 x_2}{2} \\
\theta_o &= \frac{n_2 x_2 - n_1 x_1}{l}
\end{aligned}
\tag{5.93}
$$

The inverse transformation is

$$
\begin{aligned}
x_1 &= \frac{2x_o - l\theta_o}{2n_1} \\
x_2 &= \frac{l\theta_o + 2x_o}{2n_2}
\end{aligned}
\tag{5.94}
$$

Defining the displacement ratios and angular deflections (two further temporary coordinates) of the levers arms

$$
\begin{aligned}
n_1 &= \frac{b_1}{a_1} \\
n_2 &= \frac{b_2}{a_2} \\
\theta_1 &= -\frac{x_1}{a_1} \\
\theta_2 &= \frac{x_2}{a_2}
\end{aligned}
\tag{5.95}
$$

the kinetic energy of this flexure is given by

$$
\begin{aligned}
T &= \frac{1}{2}\left[I_1\dot{\theta}_1^2 + I_2\dot{\theta}_2^2 + m_{1'}(n_1\dot{x}_1)^2 + m_o\dot{x}_o^2 + I_o\dot{\theta}_o^2\right] \\
&= \frac{1}{2}\left[\begin{array}{l} I_1\left(\dfrac{2\dot{x}_o - l\dot{\theta}_o}{2n_1a_1}\right)^2 + I_2\left(\dfrac{l\dot{\theta}_o + 2\dot{x}_o}{2n_2a_2}\right)^2 + \\ m_{1'}\left(\dfrac{2\dot{x}_o - l\dot{\theta}_o}{2}\right)^2 + m_o\dot{x}_o^2 + I_o\dot{\theta}_o^2 \end{array}\right]
\end{aligned}
\tag{5.96}
$$

Assuming the lever and platform have similar geometry (i.e. $a_1 = a_2 = a$, $n_1 = n_2 = n = l/a$, $I_1 = I_2 = I$, $b_1 = b_2 = l$), equation (5.96) reduces to

$$T=\frac{1}{2}\left[\frac{I}{2n^2a^2}\left(4\dot{x}_o^2+l^2\dot{\theta}_o^2\right)+m_{1'}\left(\frac{2\dot{x}_o-l\dot{\theta}_o}{2}\right)^2+m_o\dot{x}_o^2+I_o\dot{\theta}_o^2\right]$$

$$=\frac{1}{2}\left[\frac{I}{2n^2a^2}\left(4\dot{x}_o^2+l^2\dot{\theta}_o^2\right)+\frac{m_{1'}}{4}\left(4\dot{x}_o^2-4\dot{x}_o l\dot{\theta}_o+l^2\dot{\theta}_o^2\right)+m_o\dot{x}_o^2+I_o\dot{\theta}_o^2\right] \quad (5.97)$$

$$=\frac{1}{2}\left[\left(\frac{2I}{l^2}+m_{1'}+m_o\right)\dot{x}_o^2+\left(\frac{I}{2}+\frac{m_{1'}l^2}{4}+I_o\right)\dot{\theta}_o^2-\left(m_{1'}l\right)\dot{x}_o\dot{\theta}_o\right]$$

Similarly, it can be shown that the potential energy is

$$V=\frac{1}{2}\left[k_p\left(x_1^2+x_2^2\right)+k_n\left(2\theta_1^2+\theta_2^2\right)+k_n\left(\theta_o-\theta_2\right)^2+k_n\theta_o^2\right]$$

$$=\frac{1}{2}\left[k_p\left(x_1^2+x_2^2\right)+2k_n\left(\theta_1^2+\theta_2^2+\theta_o^2-\theta_o\theta_2\right)\right] \quad (5.98)$$

Again, assuming similar geometry of the two levers, (5.98) can be rearranged to give

$$V=\frac{1}{2}\left[k_p\left(x_1^2+x_2^2\right)+\frac{2k_n}{a^2}\left(x_1^2+x_2^2+\theta_o^2-a\theta_o x_2\right)\right] \quad (5.99)$$

Substituting (5.94) into (5.99) gives the potential energy in terms of the generalized coordinates

$$V=\frac{1}{2}\left[x_o^2\left(\frac{2k_p}{n^2}+\frac{4k_n}{l^2}\right)+\theta_o^2\left(\frac{k_pl^2}{2n^2}+k_n\right)-x_o\theta_o\frac{2k_n}{l}\right] \quad (5.100)$$

Equation's (5.97) and (5.100) can be written in the generalized forms

$$T=\frac{1}{2}a_{xx}\dot{x}_o^2+\frac{1}{2}a_{\theta\theta}\dot{\theta}_o^2+a_{x\theta}\dot{x}_o\dot{\theta}_o$$

$$V=\frac{1}{2}c_{xx}x_o^2+\frac{1}{2}c_{\theta\theta}\theta_o^2+c_{x\theta}x_o\theta_o \quad (5.101)$$

As usual the coefficients show reciprocity i.e. $a_{x\theta}=a_{\theta x}$ & $c_{x\theta}=c_{\theta x}$. Substituting equations (5.101) into Lagrange's equation, the equations governing free motion of this system are given by

$$a_{xx}\ddot{x}_o+a_{x\theta}\ddot{\theta}_o+c_{xx}x_o+c_{x\theta}\theta=0$$

$$a_{x\theta}\ddot{x}_o+a_{\theta\theta}\ddot{\theta}_o+c_{\theta\theta}\theta_o+c_{x\theta}x_o=0 \quad (5.102)$$

This can be more conveniently expressed in the familiar matrix form

$$\begin{bmatrix} a_{xx} & a_{x\theta} \\ a_{x\theta} & a_{\theta\theta} \end{bmatrix} \begin{Bmatrix} \ddot{x}_o \\ \ddot{\theta}_o \end{Bmatrix} + \begin{bmatrix} c_{xx} & c_{x\theta} \\ c_{x\theta} & c_{\theta\theta} \end{bmatrix} \begin{Bmatrix} x_o \\ \theta_o \end{Bmatrix} = \{\mathbf{0}\} \tag{5.103}$$

In this simplified form, this can be rearranged in the form of an eigen equation from which two natural frequencies can be determined. To determine the frequency response due to forces applied at the drives, it is necessary to transform these to equivalent forces in the generalized coordinates.

5.5 General model for dynamics of planar flexures

While many flexures can be adequately modeled as the ideal mechanisms considered above, in some cases the effects of flexure compliance in off-axis directions may be of some concern. In such a case, the flexure must be modeled as an element that provides an additional freedom in each compliant axis. Since most mechanisms consist of a number of rigid bodies connected by flexure 'hinges', the addition of degrees of freedom to each hinge results in a considerable increase in the overall mobility. As a consequence, even the simplest flexure mechanism must necessarily be considered a multi-degree-of-freedom system.

Either as a consequence of planar manufacture or inherent geometric symmetries, it is often reasonable to consider the important characteristics of a flexure system to take place in a plane. As a consequence, it is possible to derive much useful information from an analysis of the dynamics of such a planar system. If the correct plane is considered (as is most likely), results of this analysis will include the lowest mode frequencies therefore representing limiting dynamic performance. While still relatively simple, other designs may require a considerably more complex three-dimensional analysis. It so happens that the techniques necessary for planar analysis are merely simplifications of the more complete three-dimensional case. Because the steps involved in the former can be relatively easily followed and visualized, the analysis of this section can be considered as an introduction to the more abstract, generalized dynamics of compliantly coupled, rigid bodies presented in the following section.

In view of the large number of degrees of freedom for the general analysis of planar mechanisms, it is important to introduce a rigorous definition of the various coordinates of the flexure. To illustrate the general principles involved in the determination of the static and dynamic equations governing motion of a planar mechanism, consider the four-bar link modeled in figure 5.24. Each of the four rigid links is connected by compliant elements. To ease analysis, a number of assumptions have been made. For more advanced applications, the book of Goldstein, 1980, provides a more complete continuum analysis of rigid bodies as well as some of the transformations relevant to this and the following section.

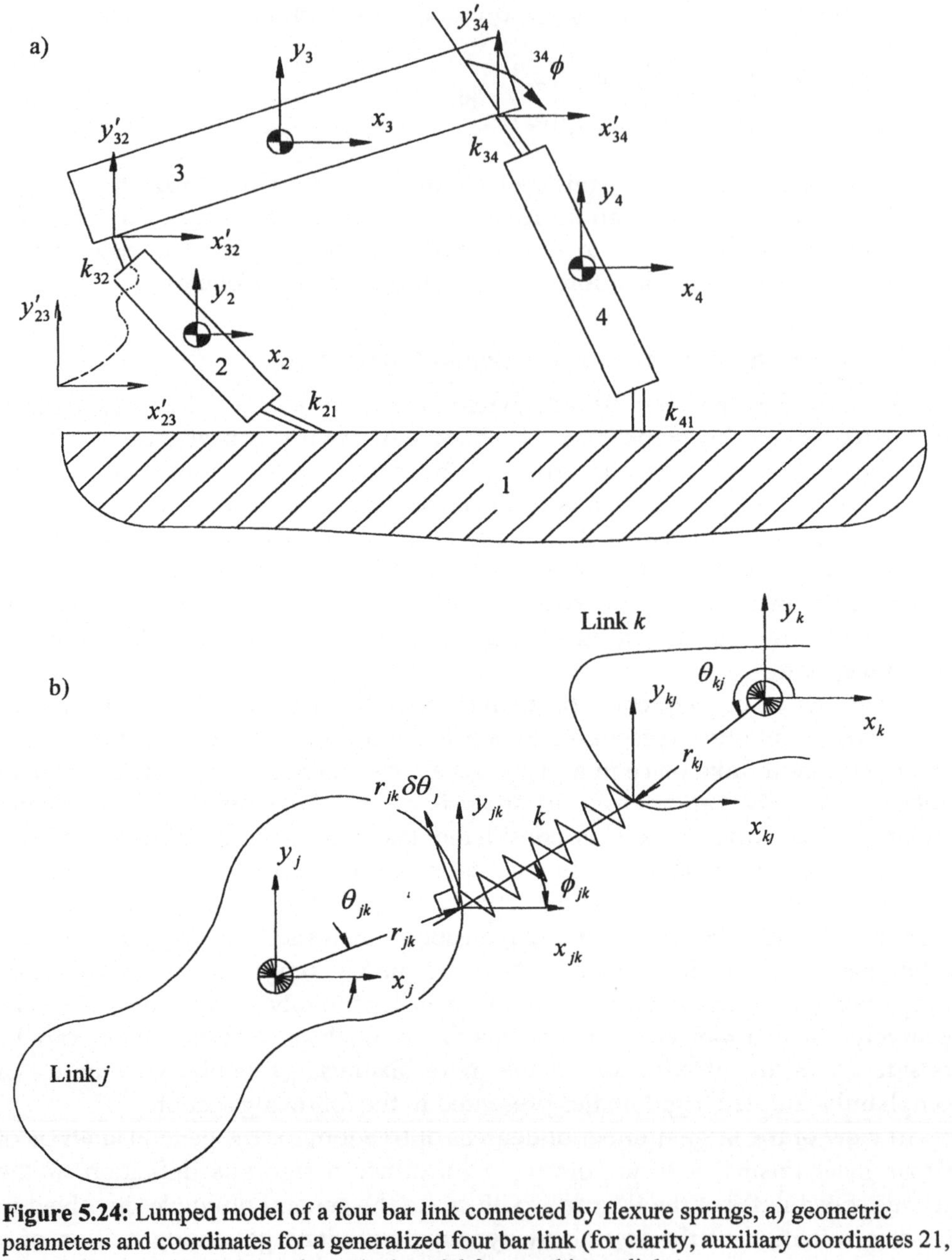

Figure 5.24: Lumped model of a four bar link connected by flexure springs, a) geometric parameters and coordinates for a generalized four bar link (for clarity, auxiliary coordinates 21, 43 and 41 are omitted), b) mathematical model for an arbitrary link j.

For each rigid body of the planar mechanism, three coordinates (two translations plus one rotation) are necessary to completely specify its position. To determine the kinetic energy in the moving body, it is also necessary to know both the geometry (more correctly, the mass distribution) of the rigid body and the position of the origin of the local coordinate system. Before proceeding, it is necessary that the meaning of the local coordinate system be clearly understood. By the local coordinate system it is implied that a fixed origin is attached to a link in its initial state. At this time, the position of each element within the link is known. The origin of this coordinate system is placed at an appropriate point (to be discussed) and thereafter remains stationary with respect to the global coordinate system. Motion of any element on the link is then measured as the difference between the start point and its instantaneous position as monitored by the local coordinate system. Being separate entities, each link can be identified by a unique single valued number. In the example shown in figure 5.24, links are numbered consecutively from 1 to 4, the first link representing the rigid body to which all other links are referred.

Compliant elements connecting the links of the mechanism can also be ascribed a unique number for identification purposes. However, in this case it is necessary to identify which two links are being connected by the flexure element. One further complication arises. This is due to the fact that hinge compliances may be more easily determined in a direction that is not coincident with its adjacent local coordinates. Before considering how to deal with this, it is probably timely to define the coordinate systems and notation used to describe the position of the elements within the flexure mechanism.

5.5.1 Coordinate systems

In this and all subsequent analysis it will be necessary to discriminate three different coordinate systems. These are

1. The global coordinate system
2. The principal local coordinates of a link
3. Auxiliary local coordinates

The global coordinate system refers to the inertial frame to which the 'stationary' link is affixed. Invariably this will correspond to the earth or some vehicle, travelling at constant velocity, to which the flexure is attached. Throughout this text, the stationary link, containing the global coordinates, is ascribed the number 1. All motions of flexure elements will be referred to this coordinate set.

For the purpose of deriving the total kinetic and potential energy, it is convenient to define a local coordinate system for each of the rigid links within the flexure. As already mentioned, choice of the origin of this coordinate system is arbitrary. However, once made, it is important to use inertial properties measured about this point. For planar motion, the inertia corresponding to rotation about the *z*-axis perpendicular to the plane of operation of the mechanism is simply computed from

$$I_{zz} = \int_V \rho(\mathbf{r}) r^2 dV \tag{5.104}$$

where the radius vector **r** is measured from the *z*-axis of the local coordinate system in this case.

For a homogeneous material of constant mass density, equation (5.104) can be expressed in the alternative form

$$I_{zz} = \int_V \rho\left(x^2 + y^2\right) dV \tag{5.105}$$

In much of the following, the inertia properties are known and it is not necessary to compute the integral of equation (5.105). For the three dimensional analysis there will be nine components of inertia relative to the local coordinate system. Denoting the general coordinates by the symbol q, the elements of the inertia matrix can be computed from the equation

$$I_{ij} = \int_V \rho(\mathbf{r})\left(r^2 \delta_{ij} - q_i q_j\right) dV \tag{5.106}$$

where δ_{ij} represents the Kronecker delta of matrix and vector analysis.

Motion of the rigid link in terms of the local coordinates may be visualized by initially marking an imaginary dot on the rigid body at the origin. Motion of the body will produce a displacement of the dot relative to the local coordinate system the origin of which remains stationary with respect to the global. Translation and rotation of a point at the origin will subsequently define the position of all points in the rigid body, thereby effecting a complete vector description of the flexure system. Coordinate transformations represent the linear motion of a this dot in each coordinate system. By definition, rotation measured at any point on the rigid link will be the same.

To compute the total forces in the flexure system, it is necessary to determine the position and relative motion of the individual flexure springs at the two points of attachment. For this purpose, an auxiliary coordinate set is defined at these points. In practice, the position of this point is fully defined by the principal local coordinates and it is only necessary to determine the relevant transformations to express all forces in terms of the principal coordinates.

5.5.2 Notation

Clearly, in view of the large number of variables involved in this analysis, it is necessary to carefully choose a unique notation that enables a full geometric description of a general planar mechanism.

For each link of the mechanism it is necessary to define its position in terms of one rotation and two translation coordinates. For link j of a flexure, the unique principal local coordinates are given by

$$\{\mathbf{x}\}_j = \begin{Bmatrix} x_j \\ y_j \\ \theta_j \end{Bmatrix} \tag{5.107}$$

To conveniently separate the components of rotational and translational kinetic energy, the origin of this coordinate system has been chosen to be at the center of mass of the link. In practice this axis will often coincide with the principal axes of the inertia ellipsoid. In such a case, the mass matrix (to be defined) is diagonal which greatly simplifies analysis.

To compute natural frequencies, it is necessary to describe the geometry of the flexure mechanism in its initial state. For this purpose, two parameters are required, one providing the angle of the link relative to the points of connection of each flexure element, the other the orientation of the flexure axis. The initial angle (anti-clockwise positive) between the x-axis of the link, j, and a straight line between the origins of the link coordinate system and the auxilliary coordinate system of the flexure connecting to an adjacent link k is, see figure 5.27,

$$\theta_{0\,jk} \tag{5.108}$$

For each flexure, there will be three stiffness values. In practice, these are likely to be known in the direction of the flexure axis, which is unlikely to be coincident with that of the local coordinates. The three-valued stiffness vector for the flexure connecting links j to k in the direction of *its axis* is denoted

$$[\mathbf{k}]^{jk} \tag{5.109}$$

To compute the potential energy stored in the spring it is more convenient to express the stiffness in terms of the components in the direction of the local coordinate system. In this case, the transformed stiffness is denoted

$$[\mathbf{k}]_{jk} \tag{5.110}$$

Using this notation, the angle between the flexure element connecting links j and k and the x-axis of the local coordinate system is denoted

$${}^{jk}\phi \tag{5.111}$$

Finally, the auxiliary local coordinates of link j at the point of attachment to a flexure connecting link k are marked primes. Hence, an auxiliary coordinate system is given by

$$\{\mathbf{x}'\}_{jk} = \begin{Bmatrix} x'_{jk} \\ y'_{jk} \\ \theta'_j \end{Bmatrix} = \begin{Bmatrix} x'_{jk} \\ y'_{jk} \\ \theta_j \end{Bmatrix} \tag{5.112}$$

In principle, equations (5.104) to (5.112) contain the complete set of geometric parameters necessary to describe the geometry of the flexure system and subsequently determine the total potential and kinetic energy for substitution into Lagrange's equation.

Before it is possible to derive the equations of motion, it is necessary to determine the forces in the system as a function of arbitrary displacements in all degrees of freedom. Because each rigid body, except for the stationary link, possesses three degrees of freedom the potential energy must reduce to a matrix equation containing independent coordinates numbering

$$3(n-1) \tag{5.113}$$

where n is the number of links in the mechanism. To reduce the number of variables to this value, the following transformations are necessary.

5.5.3 Transformations

I. Stiffness transformations

Consider the simple leaf type flexure of figure 5.25 for which the stiffness equations relating isolated forces and subsequent displacements are

$$\begin{aligned} \frac{F_y}{y} &= k_y = \frac{12EI}{L^3} \\ \frac{F_x}{x} &= k_x = \frac{EA}{L} \\ \frac{M_\theta}{\theta} &= k_\theta = \frac{EI}{L} \end{aligned} \tag{5.114}$$

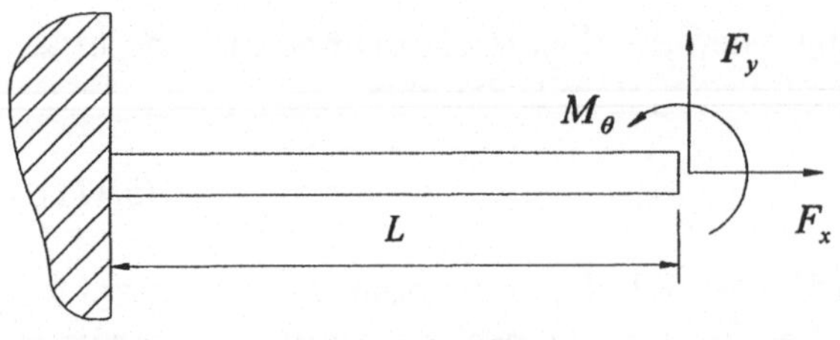

Figure 5.25: A simple leaf type flexure

As a word of caution, for combined loads, more precise stiffness equations should be computed from the trigonometric functions presented in chapter 3. However, for relatively low loads it may tentatively be assumed that the principle of superposition applies and the flexure can be modeled in terms of ideal springs as shown in figure 5.26. Also shown in this figure are the two coordinates indicating the flexure (denoted x', y' and θ) and local coordinate axes (denoted x, y and θ). Considering the measurement of point P in this figure, it can be readily shown that the relation between displacements and rotations in the two coordinate sets are given by

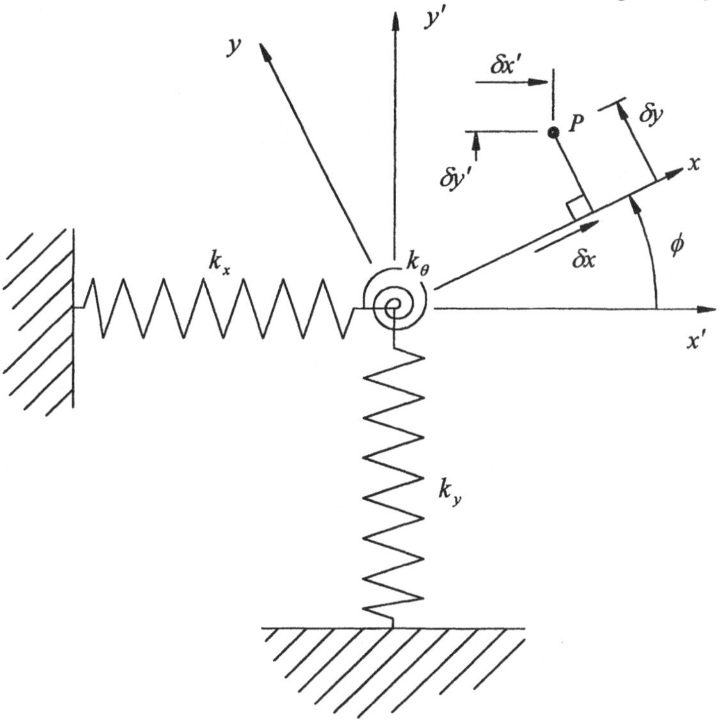

Figure 5.26: Idealized model of the simple leaf-type flexure

$$\begin{aligned} \delta x' &= \delta x\cos(\phi) - \delta y\sin(\phi) \\ \delta y' &= \delta y\cos(\phi) + \delta x\sin(\phi) \\ \delta\theta' &= \delta\theta \end{aligned} \qquad (5.115)$$

In matrix form

$$\begin{Bmatrix} \delta x' \\ \delta y' \\ \delta\theta' \end{Bmatrix} = \begin{bmatrix} \cos(\phi) & -\sin(\phi) & 0 \\ \sin(\phi) & \cos(\phi) & 0 \\ 0 & 0 & 1 \end{bmatrix} \begin{Bmatrix} \delta x \\ \delta y \\ \delta\theta \end{Bmatrix} \qquad (5.116)$$
$$\{\delta \mathbf{x}'\} = [\mathbf{A}_1]^{\mathrm{T}}\{\delta \mathbf{x}\}$$

where

$$[\mathbf{A}_1]=\begin{bmatrix}\cos(\phi) & \sin(\phi) & 0\\ -\sin(\phi) & \cos(\phi) & 0\\ 0 & 0 & 1\end{bmatrix} \tag{5.117}$$

To determine the generalized component of force in the direction of the local coordinates, it is necessary to derive an expression for the force in the generalized coordinate directions as a function of the corresponding displacements. Because ${}^{jk}\phi={}^{kj}\phi$ it is relatively simple to show that the generalized forces are given by

$$\begin{Bmatrix}{}^{jk}F_x\\ {}^{jk}F_y\\ {}^{jk}M_\theta\end{Bmatrix} = \begin{bmatrix}\cos({}^{jk}\phi) & \sin({}^{jk}\phi) & 0\\ -\sin({}^{jk}\phi) & \cos({}^{jk}\phi) & 0\\ 0 & 0 & 1\end{bmatrix}\begin{bmatrix}k_x^{jk} & 0 & 0\\ 0 & k_y^{jk} & 0\\ 0 & 0 & k_\theta^{jk}\end{bmatrix}\begin{bmatrix}\cos({}^{jk}\phi) & \sin({}^{jk}\phi) & 0\\ -\sin({}^{jk}\phi) & \cos({}^{jk}\phi) & 0\\ 0 & 0 & 1\end{bmatrix}\begin{Bmatrix}\delta x'_{jk}-\delta x'_{kj}\\ \delta y'_{jk}-\delta y'_{kj}\\ \delta\theta_j-\delta\theta_k\end{Bmatrix} \tag{5.118}$$

This rather cumbersome equation can be written more succinctly in the matrix form

$$\begin{aligned}{}^{jk}\{\mathbf{F}\} &= -[\mathbf{A}_1]_{jk}[\mathbf{k}]^{jk}[\mathbf{A}_1]^{\mathrm{T}}{}_{jk}\{\{\delta\mathbf{x}'_{\mathbf{jk}}\}-\{\delta\mathbf{x}'_{\mathbf{kj}}\}\}\\ &= -[\mathbf{K}_\mathbf{e}]_{jk}\{\{\delta\mathbf{x}'_{\mathbf{jk}}\}-\{\delta\mathbf{x}'_{\mathbf{kj}}\}\}\end{aligned} \tag{5.119}$$

All that remains is to express the forces as they apply to the center of mass of the link (i.e. the forces applied to the origin of the principal local coordinates). With reference to figure 5.27, it can be seen that the forces applied to the center of mass of the link j due to forces applied by the flexure connecting to link k are given by

$$\begin{aligned}\begin{Bmatrix}F_{x_{jk}}\\ F_{y_{jk}}\\ M_{\theta_j}\end{Bmatrix} &= \begin{bmatrix}1 & 0 & 0\\ 0 & 1 & 0\\ -r_{jk}\sin\theta_{0jk} & r_{jk}\cos\theta_{0jk} & 1\end{bmatrix}\begin{Bmatrix}F_{x'_{jk}}\\ F_{y'_{jk}}\\ M_{\theta_j}\end{Bmatrix}\\ &= [\mathbf{A}_3]\{\mathbf{F}'_{\mathbf{jk}}\}\end{aligned} \tag{5.120}$$

Combining equations (5.119) and (5.120), the externally applied forces on link j due to the flexure jk can be expressed in terms of the principal local coordinates by

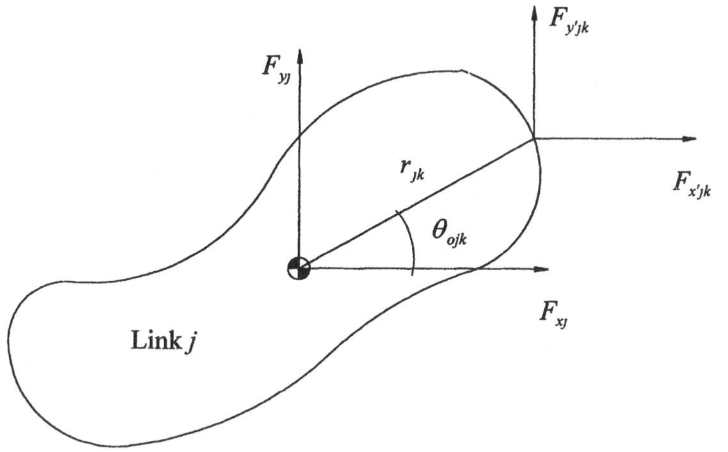

Figure 5.27: Planar forces applied to a rigid link

$$\{\mathbf{F}_{jk}\} = -[\mathbf{A}_3]_{jk}[\mathbf{A}_1]_{jk}[\mathbf{k}]^{jk}[\mathbf{A}_1]_{jk}^{\mathbf{T}}\{\{\delta\mathbf{x}'\}_{jk} - \{\delta\mathbf{x}'\}_{kj}\} \tag{5.121}$$

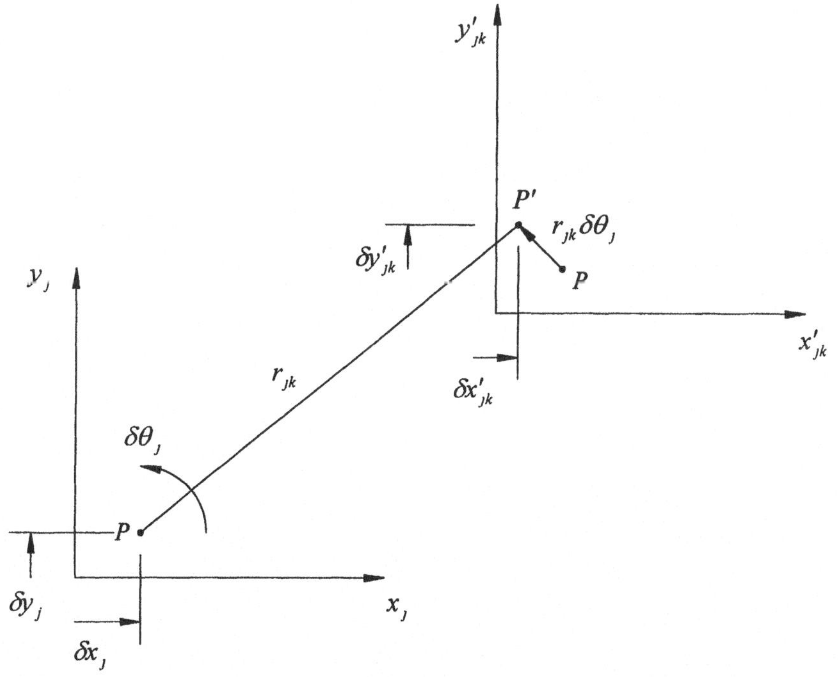

Figure 5.28: Illustration of the geometric relationship between the auxiliary and principal local coordinates

Equation (5.120) still contains the auxiliary local coordinates. Consequently, with reference to figure 5.28 it is necessary to translate these to the principle coordinates using the transformations

$$\begin{Bmatrix} \delta x'_{jk} \\ \delta y'_{jk} \\ \delta\theta_j \end{Bmatrix} = \begin{bmatrix} 1 & 0 & -r_{jk}\sin\theta_{0jk} \\ 0 & 1 & r_{jk}\cos\theta_{0jk} \\ 0 & 0 & 1 \end{bmatrix} \begin{Bmatrix} \delta x_j \\ \delta y_j \\ \delta\theta_j \end{Bmatrix} = [\mathbf{A}_3]_{jk}{}^{\mathrm{T}} \{\delta\mathbf{x}_j\} \tag{5.122}$$

Consequently, (5.121) can be expressed in terms of the principal local coordinates for the forces at link j due to a flexure connected to an adjacent link k given by

$$\begin{aligned} \{\mathbf{F}_{jk}\} &= -[\mathbf{A}_3]_{jk}[\mathbf{K}_e]_{jk}[\mathbf{A}_3]^{\mathrm{T}}_{jk}\{\delta\mathbf{x}\}_j \\ &+ [\mathbf{A}_3]_{jk}[\mathbf{K}_e]_{jk}[\mathbf{A}_3]^{\mathrm{T}}{}_{kj}\{\delta\mathbf{x}\}_k \end{aligned} \tag{5.123}$$

Finally, for each link, the total applied force will be the sum of forces from all flexures connecting adjacent links. If link j connects to l_j adjacent links, k, the total applied force is given by

$$\begin{aligned} \{\mathbf{F}_j\} &= \sum_{k=1}^{l} \{\mathbf{F}_{jk}\} \\ &= +\sum_{k=1}^{l} \begin{Bmatrix} -[\mathbf{A}_3]_{jk}[\mathbf{K}_e]_{jk}[\mathbf{A}_3]^{\mathrm{T}}{}_{jk}\{\delta\mathbf{x}\}_j \\ +[\mathbf{A}_3]_{jk}[\mathbf{K}_e]_{jk}[\mathbf{A}_3]^{\mathrm{T}}{}_{kj}\{\delta\mathbf{x}\}_k \end{Bmatrix} \end{aligned} \tag{5.124}$$

Clearly, a link cannot connect to itself so that $j \neq k$. The inertial reaction to the sum of all applied forces is

$$\begin{bmatrix} m_j & 0 & 0 \\ 0 & m_j & 0 \\ 0 & 0 & I_j \end{bmatrix} \begin{Bmatrix} \ddot{x}_j \\ \ddot{y}_j \\ \ddot{\theta}_j \end{Bmatrix} = [\mathbf{M}]_j\{\ddot{\mathbf{x}}\}_j \tag{5.125}$$

From D'Alembert's principle, and assuming the incremental displacements are both small and occur about the initial state, the requirement for dynamic equilibrium of the n links gives

$$\sum_{j=1}^{n-1} \left([\mathbf{M}]_j\{\ddot{\mathbf{x}}\}_j + \sum_{k=1}^{l_j} \begin{Bmatrix} [\mathbf{A}_3]_{jk}[\mathbf{K}_e]_{jk}[\mathbf{A}_3]^{\mathrm{T}}{}_{jk}\{\mathbf{x}\}_j \\ -[\mathbf{A}_3]_{jk}[\mathbf{K}_e]_{jk}[\mathbf{A}_3]^{\mathrm{T}}_{kj}\{\mathbf{x}\}_k \end{Bmatrix} \right) \tag{5.126}$$

Again, some of the terms can be collected to form a more compact matrix equation of the form

$$\sum_{j=1}^{n-1}\left([\mathbf{M}]_j\{\ddot{\mathbf{x}}\}_j+\sum_{k=1}^{l_j}\begin{Bmatrix}[\mathbf{A}_4]_{jk}\{\mathbf{x}\}_j\\-[\mathbf{A}_5]_{jk}\{\mathbf{x}\}_k\end{Bmatrix}\right) \tag{5.127}$$

Note that the difference in the two **A** matrices is to be found in the order of the subscripts and both are distinctly different matrices, the first involving terms that contain geometric variables for the link j only while the second includes the geometry of the adjacent links.

Equation (5.127) indicates the elements of a larger matrix equation of the form

$$\begin{bmatrix}\mathbf{M}_1 & 0 & \cdots & 0\\ 0 & \ddots & & \\ & & \ddots & \\ 0 & & 0 & \mathbf{M}_{n-1}\end{bmatrix}\begin{Bmatrix}\ddot{\mathbf{x}}_1\\ \vdots\\ \vdots\\ \ddot{\mathbf{x}}_{n-1}\end{Bmatrix}+\begin{bmatrix}[\mathbf{A}_4]_{11} & -[\mathbf{A}_5]_{12} & \cdots & -[\mathbf{A}_5]_{1,n-1}\\ -[\mathbf{A}_5]_{21} & \ddots & & \vdots\\ \vdots & & \ddots & \vdots\\ -[\mathbf{A}_5]_{n-1,1} & \cdots & \cdots & [\mathbf{A}_4]_{n-1,n-1}\end{bmatrix}\begin{Bmatrix}\mathbf{x}_1\\ \vdots\\ \vdots\\ \mathbf{x}_{n-1}\end{Bmatrix}=\{\mathbf{0}\} \tag{5.128}$$

Again, this can be written in condensed form by the familiar second order, linear, differential equation

$$[\mathbf{M}]\{\ddot{\mathbf{x}}\}+[\mathbf{K}]\{\mathbf{x}\}=\{\mathbf{0}\} \tag{5.129}$$

In this equation, the inertia matrix is simple diagonal while the stiffness matrix is symmetric.

Damping elements at the individual hinges can be included using the same transformations as those for the flexures, after which the general equation of motion is of the form

$$[\mathbf{M}]\{\ddot{\mathbf{x}}\}+[\mathbf{B}]\{\dot{\mathbf{x}}\}+[\mathbf{K}]\{\mathbf{x}\}=\{\mathbf{0}\} \tag{5.130}$$

Having derived this linear form, the eigenvalues and subsequent frequency response can be obtained using the methods outlined in section 3.9. In particular, equations (5.129) and (5.130) can readily be reduced to an eigen equation of the form

$$[\mathbf{A}]\{\mathbf{Z}\}=\lambda\{\mathbf{Z}\} \tag{5.131}$$

where

$$[\mathbf{A}]=\begin{bmatrix}\mathbf{0} & \mathbf{I}\\ -[\mathbf{M}]^{-1}[\mathbf{K}] & -[\mathbf{M}]^{-1}[\mathbf{B}]\end{bmatrix} \tag{5.132}$$

In view of the complexity of the above algebra, this method is probably best illustrated by selecting the simplest of examples

5.5.4 Case study I: The simple linear spring flexure

As an example of the implementation of the above matrix procedure, the simple parallel spring utilizing leaf type hinges is revisited. Figure 5.29 shows the geometry of the flexure with relevant dimensions included in the caption. Comprising a simple four-bar link, even this simple mechanism possesses 9 degrees of freedom. Fortunately, because all of the flexure hinges are of identical geometry (and orientation, a factor that will be made use of shortly), the three orthogonal stiffness values for each of the hinges are identical. Consequently the mechanical characteristics of this flexure can be given by the values in table 5.5.4.1

Table 5.5.4.1: Mechanical properties of the simple flexure

Parameter	Value
$m_2 = m_3$ (kg)	0.012
m_3 (kg)	0.032
$I_2 = I_4$ (kg m^2)	9.0x10^{-7}
I_3 (kg m^2)	6.667x10^{-6}
k_x (N m^{-1})	2.1x10^8
k_y (N m^{-1})	2.1x10^6
k_θ (Nm rad^{-1})	4.375

Again, the symmetry of this mechanism considerably reduces the complexity of the transformation matrices. In particular, because all hinges are of the same orientation, the angle of the axis of the flexures relative to the local coordinates is the same for each hinge and of value $-\pi/2$. Consequently, the rotation matrices are all equal i.e.

$$[\mathbf{A}_1] = \begin{bmatrix} 0 & -1 & 0 \\ 1 & 0 & 0 \\ 0 & 0 & 1 \end{bmatrix}$$

This leads to a common stiffness matrix in which the linear components of the flexure stiffness are interchanged i.e.

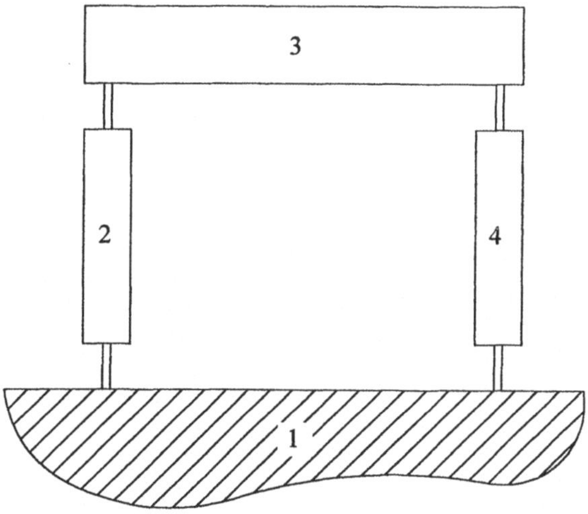

Figure 5.29: A simple linear spring ($s = 0.05$ m, $l'' = 0.035$ m, $l = 0.04$ m, $L = 0.005$ m, depth = 0.01 m, $h = 0.008$ m, width = 0.005 m, $t = 0.5$ mm, $E = 210$ GPa, ρ =8000 kg m^3). Note that, from equation (4.6), l'' has been chosen as a point at a distance of $L/3$ from the base to a corresponding point measured from the upper linear motion platform

$$[\mathbf{K_e}] = \begin{bmatrix} k_y & 0 & 0 \\ 0 & k_x & 0 \\ 0 & 0 & k_\theta \end{bmatrix}$$

Derivation of the linear transformation matrices is not as convenient. The necessary geometric parameters for this transformation are the length of the line, l connecting the origins of the auxiliary and principal local coordinates and the angle, θ_{jk}, between the x-axis and this line connecting the origins of the two coordinates (measured positive anti-clockwise). These are given in table 5.5.4.2 below

Table 5.5.4.2: Geometric parameters for the linear flexure

Parameter	Value (rad)	Parameter	Value (m)
$\theta_{21} = \theta_{41}$	$-\pi/2$	$l_{21} = l_{23} = l_{41} = l_{43}$	0.0225
$\theta_{23} = \theta_{43}$	$\pi/2$	$l_{32} = l_{34}$	0.025
θ_{32}	π		
θ_{34}	0		

Systematically inserting the angular values of table 5.5.4.2 into equation (5.124) yields the components of force acting on the three links

$$\{\mathbf{F}_2\} = \begin{bmatrix} -2k_y & 0 & -k_y(l_{21}-l_{23}) \\ 0 & -2k_x & 0 \\ -k_y(l_{21}-l_{23}) & 0 & -2k_\theta - k_y(l_{21}^2 + l_{23}^2) \end{bmatrix} \begin{Bmatrix} x_2 \\ y_2 \\ \theta_2 \end{Bmatrix} + \begin{bmatrix} k_y & 0 & 0 \\ 0 & k_x & -l_{32}k_x \\ -l_{23}k_y & 0 & k_\theta \end{bmatrix} \begin{Bmatrix} x_3 \\ y_3 \\ \theta_3 \end{Bmatrix} = -[\mathbf{k}_1]\{\mathbf{x}\}_2 - [\mathbf{k}_2]\{\mathbf{x}\}_3$$

$$\{\mathbf{F}_3\} = \begin{bmatrix} k_y & 0 & -l_{23}k_x \\ 0 & k_x & 0 \\ 0 & l_{32}k_y & k_\theta \end{bmatrix} \begin{Bmatrix} x_2 \\ y_2 \\ \theta_2 \end{Bmatrix} \begin{bmatrix} -2k_y & 0 & 0 \\ 0 & -2k_x & -k_x(l_{34}-l_{32}) \\ 0 & -k_x(l_{34}-l_{32}) & -2k_\theta - k_x(l_{34}^2 + l_{32}^2) \end{bmatrix} \begin{Bmatrix} x_3 \\ y_3 \\ \theta_3 \end{Bmatrix} + \begin{bmatrix} k_y & 0 & -l_{43}k_x \\ 0 & k_x & 0 \\ 0 & l_{34}k_y & k_\theta \end{bmatrix} \begin{Bmatrix} x_4 \\ y_4 \\ \theta_4 \end{Bmatrix} = -[\mathbf{k}_3]\{\mathbf{x}\}_2 - [\mathbf{k}_4]\{\mathbf{x}\}_3 - [\mathbf{k}_5]\{\mathbf{x}\}_4$$

$$\{\mathbf{F}_2\} = \begin{bmatrix} k_y & 0 & 0 \\ 0 & k_x & -l_{34}k_x \\ -l_{43}k_y & 0 & k_\theta \end{bmatrix} \begin{Bmatrix} x_3 \\ y_3 \\ \theta_3 \end{Bmatrix} + \begin{bmatrix} -2k_y & 0 & -k_y(l_{41}-l_{43}) \\ 0 & -2k_x & 0 \\ -k_y(l_{41}-l_{43}) & 0 & -2k_\theta - k_y(l_{41}^2 + l_{43}^2) \end{bmatrix} \begin{Bmatrix} x_4 \\ y_4 \\ \theta_4 \end{Bmatrix} = -[\mathbf{k}_6]\{\mathbf{x}\}_3 - [\mathbf{k}_7]\{\mathbf{x}\}_4$$

These forces can be combined to give the 9 x 9 stiffness matrix corresponding to equation 5.129

$$[\mathbf{K}] = \begin{bmatrix} \mathbf{k}_1 & \mathbf{k}_2 & \mathbf{0} \\ \mathbf{k}_3 & \mathbf{k}_4 & \mathbf{k}_5 \\ \mathbf{0} & \mathbf{k}_6 & \mathbf{k}_7 \end{bmatrix}$$

Of less complexity is the mass matrix given by

$$[\mathbf{M}] = \begin{bmatrix} m_2 & 0 & 0 & 0 & 0 & 0 & 0 & 0 & 0 \\ 0 & m_2 & 0 & 0 & 0 & 0 & 0 & 0 & 0 \\ 0 & 0 & I_2 & 0 & 0 & 0 & 0 & 0 & 0 \\ 0 & 0 & 0 & m_3 & 0 & 0 & 0 & 0 & 0 \\ 0 & 0 & 0 & 0 & m_3 & 0 & 0 & 0 & 0 \\ 0 & 0 & 0 & 0 & 0 & I_3 & 0 & 0 & 0 \\ 0 & 0 & 0 & 0 & 0 & 0 & m_4 & 0 & 0 \\ 0 & 0 & 0 & 0 & 0 & 0 & 0 & m_4 & 0 \\ 0 & 0 & 0 & 0 & 0 & 0 & 0 & 0 & I_4 \end{bmatrix}$$

Inserting the values for this case study into the above matrices, an eigenvalue and vector analysis produces 9 eigenvalues (the squares of the natural frequencies) with 9 eigenvectors associated with each eigenvalue. The results from this analysis are given in tables 5.5.4.3 and 5.5.4.4 below

Table 5.5.4.3: Natural frequencies based on square root of the eigenvalues for the simple linear spring in units of cycles per second or Hertz

Mode	Value (Hz)
ω_1	74.88
ω_2	2978
ω_3	3237
ω_4	7752
ω_5	7861
ω_6	11666
ω_7	29775
ω_8	31582
ω_9	32908

From these tables, it is possible to visualize the respective modes of this mechanism. In the above table, every third row corresponds to rotation of the links while the first of every three rows represents horizontal displacement. Most striking is the value of the lowest mode natural frequency, 75 Hz, in comparison to all other modes, which are in excess of 3 kHz. As would be expected this mode corresponds to parallelogram motion of the flexure between the upper platform and base. Assuming that the hinges are perfectly rigid in all but the rotational degree of freedom, the fundamental mode natural frequency can be approximated from the equation

$$\omega_n = \left(\frac{1}{2\pi l''}\right)\sqrt{\frac{4k_\theta}{m_3 + 2I_{pivot}/l''^2}} = 76.47 \qquad \text{(Hz)}$$

Table 5.5.4.4: Eigenvectors corresponding to the eigenvalue modes of the previous table

Mode 1	Mode 2	Mode 3	Mode 4	Mode 5	Mode 6	Mode 7	Mode 8	Mode 9
-0.0162	-0.7071	0.0722	0	-0.0001	0	0	0	0
0	0	0	0	0	0.0068	-0.7071	0.0990	0.0541
0.7066	0	0.7032	-0.7071	0.7071	-0.0016	0	0.0001	-0.0001
-0.0319	0	-0.0262	0	0.0009	0	0	0	0
0	0	0	0	0	0.0365	0	0	0.0009
0	0	0	0	0	0.9986	0	0.9902	0.9968
-0.0162	0.7071	0.0722	0	-0.0001	0	0	0	0
0	0	0	0	0	0.0363	0.7071	-0.0990	-0.0584
0.7066	0	0.7032	0.7071	0.7071	-0.0016	0	-0.0001	-0.0001

Artificially increasing the linear stiffness terms of the hinges by a factor of 100 increases the fundamental mode frequency to a value 75.02 Hz indicating a relatively close agreement between the simple and complete planar analysis.

The second mode can be visualized as equal and opposite horizontal (x-axis) displacements of the two support legs. For this mode the natural frequency can be approximated from the equation

$$\omega_2 = \left(\frac{1}{2\pi}\right)\sqrt{\frac{2k_y}{m_2}} = 2978$$

To the expected precision of the above analyses, the predicted frequencies from the two models are the same. The reader is left to ponder the remaining 7 modes.

5.5.5 Comments on general planar analysis

In the preceding analysis, it was found that the fundamental natural frequency was of considerably lower value than the higher modes. Clearly, the dynamic performance of the mechanism is predominantly governed by this lowest, or gravest, mode. As a consequence, it is reasonable in many cases to ignore all but the lowest mode and consider the hinges as 'ideal'. This reduces the degrees of freedom of each joint by two (which now becomes a 'hinge'), in many cases enabling a simple analytic solution for the lowest modes using Lagrange's equation or Rayleigh's method.

It would be erroneous to assume that all modes of interest can be assessed from this simple planar analysis. As already shown in section 5.1.1, in many designs, the flexure mechanism may possess relatively low natural frequencies in directions perpendicular to the plane of interest. Fortunately, because of the orthogonal geometry of most mechanisms (they are commonly manufactured using machines with orthogonal slideways and spindles) it is not necessary to resort to full three-dimensional models. In these cases, it is possible to view the mechanism from different, orthogonal planes and assess the natural frequencies using any of the preceding analyses. In practice, if there are natural modes at significant frequencies (either near to the required bandwidth or to vibration frequencies transmitted to the structure) these will invariably become excited and are likely to couple to any directions of interest. Ignoring higher modes may result in problems if there are excitation forces that coincide with one of the frequencies. In this case, it is possible (and more likely as the modes become increasingly large) that the high frequency excitation will couple vibrations to the lower frequency modes. To see this, consider the ratio of two numbers computed to a precision of, for example, 1%. As the ratio of the difference between the two numbers becomes greater than one hundred the computed value will always appear as an integer. Even for mechanisms of modest complexity, if the mode is out of the plane of interest, there is almost always some unwanted coupling. Considering an isolated rod, which must constitute the simplest flexure, separate excitation and the interaction of perpendicular longitudinal and lateral modes has confounded experimentalists for centuries. Both of these important points were clear to Rayleigh who wrote

> *..in actual experiments with bars which are neither uniform in material nor accurately cylindrical in figure it is often found impossible to excite longitudinal or torsional vibrations without the accompaniment of some measure of lateral motion. In bars of ordinary dimensions the gravest lateral motion is far graver than the gravest longitudinal or torsional motion, and consequently it will generally happen that the principal tone of either of the latter kinds agrees more or less perfectly in pitch with some overtone of the former kind. Under such circumstances the regular modes of vibrations become unstable, and a small irregularity may produce a great effect. The difficulty of exciting purely longitudinal vibrations in a bar is similar to that of getting a string to vibrate in one plane.*

In all of the previous analysis, the links were assumed to be perfectly rigid bodies. Real links will have a finite compliance and for more precise estimates it is recommended that this be appropriately apportioned to the flexure elements connecting links. If, for example, a link of compliance c_1 is connected by two flexures of compliance's c_2 & c_3, it is recommended that the compliance's of the flexures be modified to c_2' & c_3' according to

$$c_2' = c_2 + \frac{c_2}{c_2 + c_3} c_1$$
$$c_3' = c_3 + \frac{c_3}{c_2 + c_3} c_1 \quad (5.133)$$

In terms of stiffness values, the above becomes

$$k_2' = \frac{k_2 k_1 (k_2 + k_3)}{k_1 (k_2 + k_3) + 1}$$
$$k_3' = \frac{k_3 k_1 (k_2 + k_3)}{k_1 (k_2 + k_3) + 1} \quad (5.134)$$

It can be readily shown that, setting the link compliance to zero (or stiffness to infinity) corresponds to the previous rigid body model.

5.6 General dynamics of flexures

Construction of a generalized three dimensional model represents an extension of the planar analysis of section 5.5. Having derived the appropriate transformations, the procedure for solution of the matrices is the same. Such transformations, however, are not straightforward and great care should be exercised in their development. Inclusion of such an analysis would take us too far afield in a text of this length and must await further exposition, possibly in a second volume. The necessary transformations are discussed in considerable depth in the texts of Arya, 1990, Haug, 1989, Goldstein, 1980, Pars, 1965 and many other texts concerning kinematics of rigid bodies. However, to help the interested reader, the necessary transformations are mentioned here for completeness. The first transformation necessary is that of relating elements of the original stiffness matrix in terms of the orientation of the local coordinate system. Assuming that superposition applies for both linear and angular components of stiffness, this is most easily performed through successive rotations about three axes using Euler angles. The first rotation is about the *z*-axis of magnitude ϕ and the appropriate coordinate transformation corresponding to that of equation (5.116) is

$$[\mathbf{D}] = \begin{bmatrix} \cos(\phi) & \sin(\phi) & 0 & 0 & 0 & 0 \\ -\sin(\phi) & \cos(\phi) & 0 & 0 & 0 & 0 \\ 0 & 0 & 1 & 0 & 0 & 0 \\ 0 & 0 & 0 & \cos(\phi) & \sin(\phi) & 0 \\ 0 & 0 & 0 & -\sin(\phi) & \cos(\phi) & 0 \\ 0 & 0 & 0 & 0 & 0 & 1 \end{bmatrix} \quad (5.135)$$

After this rotation, the new coordinate system comprises the original set with a rotation in the xy plane. By definition of the *x-convention*, the second rotation θ is then taken about the x-axis of this new coordinate set and is effected by the matrix transformation

$$[\mathbf{C}] = \begin{bmatrix} 1 & 0 & 0 & 0 & 0 & 0 \\ 0 & \cos(\theta) & \sin(\theta) & 0 & 0 & 0 \\ 0 & -\sin(\theta) & \cos(\theta) & 0 & 0 & 0 \\ 0 & 0 & 0 & 1 & 0 & 0 \\ 0 & 0 & 0 & 0 & \cos(\theta) & \sin(\theta) \\ 0 & 0 & 0 & 0 & -\sin(\theta) & \cos(\theta) \end{bmatrix} \tag{5.136}$$

Finally the transformation from one coordinate system to the other is completed by a third rotation, ψ about the z-axis of this second intermediate coordinate set produced from the matrix

$$[\mathbf{B}] = \begin{bmatrix} \cos(\psi) & \sin(\psi) & 0 & 0 & 0 & 0 \\ -\sin(\psi) & \cos(\psi) & 0 & 0 & 0 & 0 \\ 0 & 0 & 1 & 0 & 0 & 0 \\ 0 & 0 & 0 & \cos(\psi) & \sin(\psi) & 0 \\ 0 & 0 & 0 & -\sin(\psi) & \cos(\psi) & 0 \\ 0 & 0 & 0 & 0 & 0 & 1 \end{bmatrix} \tag{5.137}$$

The complete coordinate transformation is a linear combination of these three rotations. Comparing with equations (5.117) and (5.119) the appropriate transformation matrix becomes

$$[\mathbf{A}_1] = [\mathbf{B}][\mathbf{C}][\mathbf{D}] \tag{5.138}$$

Assuming that the principle of superposition applies in this case, the stiffness matrix is chosen to be simple diagonal with the linear stiffness components in the first three cells of the diagonal and the rotational components in the three remaining.

Determination of the linear transformation matrices requires some additional definition of terms. In three dimensions, the relevant length of the link can be reduced to three components projected onto the three planes of the local coordinate system. For example, the length of the projection of the link of true length l onto the xy plane is denoted l_z and can be readily computed from the equation

$$l_z = l\left(\cos^2(\alpha) + \cos^2(\beta)\right)^{1/2} = l\left(1 - \cos^2(\gamma)\right)^{1/2} \tag{5.139a}$$

Projected lengths on the other two planes are readily computed from

$$l_x = l(1-\cos^2(\alpha))^{1/2}$$
$$l_z = l(1-\cos^2(\gamma))^{1/2} \qquad (5.139b)$$

The angles α, β and γ are the *direction cosines* of the line from the local to the auxiliary local coordinate system given by

$$\cos\alpha = \frac{x}{l}$$
$$\cos\beta = \frac{y}{l} \qquad (5.140)$$
$$\cos\gamma = \frac{z}{l}$$

Values for x, y & z are derived from the vector connecting the principal and auxiliary local coordinates with the principal local coordinate system as the origin (i.e. $\mathbf{l} = x\mathbf{i} + y\mathbf{j} + z\mathbf{k}$). To determine the forces in a Cartesian frame, it is necessary to derive a transformation matrix in terms of linear translations in the direction of the principal local coordinates of the link plus rotations about each of the three axes. Consequently, the displacement vector of an individual link must be of the form

$$\{\delta\mathbf{x}\} = \{\delta x \quad \delta y \quad \delta y \quad \delta\theta_x \quad \delta\theta_y \quad \delta\theta_z\}^T \qquad (5.141)$$

In terms of this coordinate set, the linear transformation matrix is of the form

$$[\mathbf{A}_3]^{\mathrm{T}} = \begin{bmatrix} 1 & 0 & 0 & 0 & -l_y\sin(\theta_{y0}) & -l_z\sin(\theta_{z0}) \\ 0 & 1 & 0 & -l_x\sin(\theta_{x0}) & 0 & l_z\cos(\theta_{z0}) \\ 0 & 0 & 1 & l_x\cos(\theta_{x0}) & l_y\cos(\theta_{y0}) & 0 \\ 0 & 0 & 0 & 1 & 0 & 0 \\ 0 & 0 & 0 & 0 & 1 & 0 \\ 0 & 0 & 0 & 0 & 0 & 1 \end{bmatrix} \qquad (5.141)$$

Eliminating columns and rows 3, 4 & 5 reduces to the transpose of the transformation matrix of equation (5.120). The initial angles of the projections of the line connecting the local coordinates are given by

$$\tan\theta_{x0} = \frac{z}{y}$$

$$\tan\theta_{y0} = \frac{x}{z} \qquad (5.142)$$

$$\tan\theta_{z0} = \frac{x}{y}$$

Solutions for the three dimensional mechanism proceed similarly to the planar case discussed in the preceding section.

5.7 Sources of interesting flexure mechanisms

A full discussion of the novel flexure designs that have been presented in the literature over the last few decades would contain sufficient material for the production of many volumes of literature. However, these include a large number of novel implementations of lever mechanisms and methods for combining links and joints to produce displacements in desired directions. Most of these can be appreciated at a glance and therefore the reader will be amply rewarded after perusing these references.

Alemanni M., Mana G., Pedrotti G., Strona P.P. and Zosi G., 1986, On the construction of a zerodur translation device for x-ray interferometric scanning, *Metrologia*, **22**, 55-63. One of few examples utilizing elliptic hinges in this case manufactured from a single monolithic block of the low thermal expansion glass ceramic Zerodur™. X-ray interferometry represents one of the more demanding applications for flexure mechanisms requiring rectilinear motion accurate picometer levels with parasitic rotations below tens of nano-radians. Translation distances of tens of millimeters are under investigation.

Bamford R.M. and Glaser R.J., 1997, Induced-strain actuator with mechanical amplification, *NASA Tech. Briefs*, Jan., 60. An entertaining page.

Becker P., Seyfried P. and Siegert H., 1982, The lattice parameter of highly pure single silicon crystals, *Z. Phys. B. -Condensed Matter*, **48**, 17-21. An interesting variant on the double compound linear spring with a high mobility to enable electromagnetic correction of parasitic errors.

Deslattes R.D., 1969, Optical and x-ray interferometry of a silicon lattice spacing, *Appl. Phys. Letts.*, **15**, 386-388. A levered, notch type rectilinear spring of extraordinary rectilinear characteristics.

Furukawa E., Muranaka Y., Takenouchi Y., Moriya T. and Andoh T., 1992, Development of a torsional ground motion pickup, *Int. J. JSPE*, **26**(2), 152-157. Some novel lever mechanisms and an alternative analytic approach to that adopted in chapter 7 of this book.

Furukawa E., Mizuno M. and Terada K., 1991, A magnifying mechanism for use on piezo-driven mechanisms, *Bull. Japan Soc. Prec. Engg*, **25**(4), 315 - 320

Furukawa E., Mizuno M. and Hojo T., 1994, A twin-type piezo-driven translation mechanism, *Bull. Japan Soc. Prec. Engg*, **28**(1), 70 -75

Hart M., 1968, An ångstrom ruler, *Brit. J. Appl. Phys. (J. Phys. D)*, **1**, 1405-1408. The first paper presenting the concept of x-ray interferometry for metrological applications. The flexure used in this study was produced from a single crystal of silicon by making two pairs of orthogonal cuts.

Howells M.R., 1995, Design strategies for monolithic adjustable radius mirrors, *Optical Engineering*, **34**(2), 410-417

Jones R.V., 1987, *Instruments and experiences*, Wiley & Sons, London. Containing a compilation of papers by this author, this is an excellent introduction to flexure applications with extensive discussion of practical implementation. Also contains many references plus an attempt to provide a brief historic perspective on the developments of this field

Kyusojin A. and Sagawa D., 1988, Development of linear and rotary movement mechanism by using flexible strips, *Bull. Japan Soc. Of Prec. Engg.*, **22**(4), 309-314. Some examples of flexures combined to produce hinges with defined instantaneous centers.

Nashimura K., 1991, A spring guided micropositioner with linearized subnanometer resolution, *Rev. Sci. Instrum.*, **62**(8), 2004-2007.

Ryu J.W., Gweon D-G. and Moon K.S., 1997, Optimal design of a flexure hinge based $XY\theta$ wafer stage, *Precision Engineering*, **21**(1), 18-28. An interesting analysis of lost motion in levered flexure systems (this is discussed in chapter 7)

Sydenham P.H., 1981, Mechanical design of instruments. 5: Putting elasticity to use (Part A), *Measurement and Control*, **14**, 179-185, (part B), *Ibid*, 219-227, - Elastic design of fine mechanism in instruments, *J. Phys. E: Sci. Instrum.*, **17**, 922-930. These references contain many examples of flexure elements plus an extensive list of references.

Teague E.C., Young R.D., Scire F. and Gilsinn D., 1988, Para-flex stage for microtopographic mapping, *Rev. Sci. Instrum.*, **59**(1), 67-73. Example of application of a variant of the hinge shown in figure 5.13(b).

Xu W. and King T., Flexure hinges for piezo-actuator displacement amplifiers: Flexibility, accuracy and stress considerations, *Precision Engineering*, **19**(1), 4-10.

References

Arya A.P., 1990, *Introduction to Classical Mechanics*, Allyn and Bacon, Boston, Ma. A careful reader should be able to correct a small transcribing error in equation (13.115) and a typographical mistake in (13.117) of this edition.

Bevan T., 1948, *The theory of machines*, Longmans, Greem and Co., Lond.

Goldstein H., 1980, *Classical Mechanics*, Addison-Wesley, Reading, Ma., chapters 4 & 5. Note that Goldstein uses the *x-convention* in the sequence of rotations of Euler angles. Because the rotations are not commutative, one must use these transformations with great care.

Haug E.J., 1989, *Computer aided Kinematics and dynamics of Mechanical Systems: Volume I: Basic Methods*, Allyn and Bacon, Boston, Ma.

Keown R.A., 1921, *Mechanism*, McGraw-Hill Book Co., NY.

Mabie H.H. and Reinholtz C.F., 1987, *Mechanisms and Dynamics of Machinery*, John Wiley & Sons, NY.

Pars L.A., 1965, *A Treatise on Analytical Dynamics*, Heinmann, London, chpt. VII. Note that Pars uses the *y-convention* in the sequence of rotations of Euler angles.

Plainevaux J.E., 1956, Etude de deformations d'une lame de suspension elastique, *Nuovo Cimento,*. **4**(5), 922-928 (in French).

Plainevaux J.E., 1956, Mouvement de tangage d'une suspension elementaire sur lames elastiques, *Nuovo Cimento*, **4**(5), 1133-1141 (in French).

Rayleigh J.W.S.,1894, *The Theory of Sound*, vol. I, Dover Publications, section 149, chapter VII .

Smith S.T. and Chetwynd D.G, 1990, Optimisation of a magnet/coil force actuator and its application to linear spring mechanisms, *Proc. Inst. Mech. Engrs.*, **204**(C4), 243-253.

Smith S.T., Chetwynd D.G. and Harb S., 1994, A simple two-axis ultraprecision actuator, *Rev. Sci. Instrum.*, **65**(4), 910-917.

Thorpe II, A.G., 1953, Flexure pivots - Design formula and charts, *Product Engineering*, Feb., 192 - 200.

Timoshenko S.P. and Goodier J.N., 1970, *Theory of Elasticity*, 3rd Ed., McGraw-Hill, NY.

6 Hinges of rotational symmetry

6.0 Overview

This chapter discusses the design of flexure elements that are constructed from solids of revolution to provide compliance in defined axes for the purposes of precise motion control. Because of the ease of manufacture, such mechanisms are often formed from cylindrical tubes and circular disks. Due to their omnipresence in many instrument designs, a brief discussion of the bellows and coil spring as flexure elements are included. Not surprisingly, notch and leaf type hinges can also be integrated into rotary couplings and two such designs are presented towards the end of this chapter.

6.1 Introduction

Design of circularly symmetric compliant elements is not new. Established markets already exist that provide various springs and elastic couplings that may be readily used for applications where transmission of linear translations or rotations are desired. This chapter draws from the wealth of information already developed for such mechanisms. In many cases it is desired that a flexible coupling be provided between two rigid bodies having rotational symmetry. In fact, for most motor driven shaft assemblies, a flexible coupling is necessary to accommodate the small misalignments inevitable in any assembly. In addition to misalignments, the ends of the shafts may also require some relative motion. Probably the most familiar couplings are the universal (or Hooke) joints and constant velocity couplings used to transmit rotary motion from engines to the driven wheels on automobiles. Although less familiar, there is a not-less-common requirement for couplings to connect between many types of machines such as electric motors to gearboxes and, subsequently, gearboxes to machinery.

Reflecting the enormous range of applications in which electric and other rotary motors are used, a bewildering array of mechanisms are available to provide the requisite freedoms and power transmission requirements, see for example Shigley 1989, Neale *et al.*, 1991. Fortunately, the emphasis of this chapter is towards applications that best utilize the advantages of flexure based designs i.e. couplings requiring smooth, friction and wear-free motion, minimal hysteresis and well defined distortions. The latter of these requirements immediately rules out the enormous variety of couplings linked by intermediate compliant elements such as; rubber inserts (spider, Barwell couplings), thin metallic strips (chain coupling) or loose fitting rigid connections (gear and chain couplings). Of interest within the context of flexure design, this chapter will discuss the following coupling geometry's

I. Coil springs.
II. Disk couplings.
III. Intermediate disk and coil spring designs: The rotary hinge.
IV. Flexible membranes.
V. Notch and leaf-type hinges applied to rotational couplings.
VI. The bellows.

6.2 Coil springs

Because of its high compliance in all degrees of freedom, the coil spring, when used to connect two rigid bodies, is not ideally suited as a flexure for precise motion control. It is included in this book because it is simple to analyse and identify the reasons why this is used as an energy storage device, a more complete analysis can be found in the book of Wahl, 1963. In its simplest form, a coil spring comprises a single wire formed into a helix. Resistance to an applied load is as a consequence of this applied force being transmitted along the wire from one end to the other. Because the load passes through the center of the spring, this must be supported by a torsional reaction of equal magnitude throughout the wire. There will, of course, be a vertical shear force in the wire providing an additional distortion. However, this may be ignored for most practical designs. For a coil spring comprising a thin circular wire of diameter d the torque, T, due to an axial applied load, F, at any position along the wire is given by

$$T = FR = \frac{GJ\theta}{L} \tag{6.1}$$

where R is the radius of the coil.

The expressions for the total angular twist, θ, and length of spring, L, for inclusion in the torsion equation on the right hand side of (6.1) are given by

$$\begin{aligned} L &\approx 2\pi Rn \\ \theta &\approx \frac{\delta}{R} \\ J &= \frac{\pi r^4}{2} \end{aligned} \tag{6.2}$$

where n is the number of turns of the helix, R is the nominal radius of the spring and δ is the total deflection between the two ends of the spring

Substituting equations (6.2) into (6.1) and rearranging gives

$$T = FR = \frac{Gr^4\delta}{4nR^2} \tag{6.3}$$

More generally, it is the stiffness, k, of the spring that is of interest. From (6.3) the stiffness can immediately be written in terms of radii or diameters

$$k = \frac{F}{\delta} = \frac{Gr^4}{4nR^3} = \frac{Gd^4}{8nD^3} \tag{6.4}$$

where d and D correspond to the wire and helix diameters respectively.

Based on this simple model, the maximum shear stress occuring at the surface of the wire can be obtained from the bending equation (2.100)

$$\tau = \frac{Tr}{J} = \frac{2FRr}{\pi r^4} = \frac{2FR}{\pi r^3} \tag{6.5}$$

As might be expected for such a common machine element, extensive studies of coil springs have been presented over many years. Probably the most notable compilation of this body of work is provided by Wahl, 1963. A more complete solution for the compliance and maximum stress in a coil spring includes the pitch angle, α, and poisson's ratio, ν, of the wire. For a spring of circular wire, the compliance can be calculated from, Young, 1989,

$$\frac{\delta}{F} = \frac{4nR^3}{Gr^4}\left[1 - \frac{3}{16}\left(\frac{r}{R}\right)^2 + \frac{3+\nu}{2(1+\nu)}(\tan\alpha)^2\right] \tag{6.6}$$

The maximum shear stress in the circular wire is given by

$$\tau = \frac{2FR}{\pi r^3}\left[1 + \frac{5}{4}\frac{r}{R} + \frac{7}{8}\left(\frac{r}{R}\right)^2\right] \tag{6.7}$$

Correspondingly for a coil spring of *square* wire ($R > 3b$)

$$\begin{aligned} \frac{\delta}{F} &= \frac{1}{k} = \frac{2.79R^3 n}{Ga^4} \\ \tau &= \frac{4.8FR}{8a^3}\left[1 + \frac{1.2a}{R} + 0.56\left(\frac{a}{R}\right)^2 + 0.5\left(\frac{a}{R}\right)^3\right] \end{aligned} \tag{6.8}$$

To see why coils springs are used in such a broad range of applications, it is informative to investigate various spring geometry's in terms of utilization of material. A straight rod subject to an axial applied load provides the simplest spring element to analyze and represents the most efficient geometry. At the limit of performance, the whole volume of the spring will be subject to the same maximum stress. Consequently, such a design may be considered to represent 100% material utilization as defined by the ratio of average stress in the volume of the spring divided by the total maximum stress-volume product. Although such a geometry might be reasonable for materials that can sustain high strains (i.e. rubber and other elastomers), for metallic, ionic and covalent solids, such a spring would be impracticably long if appreciable displacements were required. Other simple spring geometry's include the coil spring, a simple cantilever subject to a bending moment at the free end, a simple cantilever subject to a force at the free end in a direction perpendicular to the neutral axis and a leaf type flexure constrained to produce an 's' shaped deflection curve as discussed in section 4.1. For the coil spring made from wire of circular section, the shear stress in the plane of the wire section is a linear function of the radius from the center of the wire and at maximum deformation is given by

$$\tau = \frac{r}{r_m}\tau_{\max} \tag{6.9}$$

By definition, the materials utilization, u, for this spring geometry is

$$u = \frac{\iiint_{\mathrm{vol}} \tau d(\mathrm{vol})}{(\mathrm{vol})\tau_{\max}} = \frac{\frac{\tau_{\max}}{r_m}\int_0^{r_m} 2\pi r L r dr}{\pi r_m^2 L \tau_{\max}} = \frac{2}{3} \tag{6.10}$$

Material utilization can be readily evaluated for the other geometry's. As an example, the simple cantilever beam with a bending moment applied at the end will have a constant stress along the axis with a linear variation from maximum compressive on one face to an equal, tensile stress on the other. Clearly, the average stress and, therefore, materials utilization is 50% of maximum as measured across the face and constant along the axis. For an applied force at the free end of the beam there will be a linear variation of bending moment starting at zero at the free end to a maximum value at the support. For these simple geometry's the material utilization is tabulated overleaf. Clearly, the coil spring represents one of the best geometry's if we desire to utilize as much material as possible. They also provide relatively large displacements for a given strain, occupy a small volume and are, almost, rotationally symmetric. Clearly, the coil spring is an efficient and useful strain energy storage element.

Table 6.2.1: Materials utilization for different spring geometry's

Spring geometry	Materials utilization (%)
Rod subject to axial load	100
Coil spring wound from circular wire	67
Cantilever subject to moment at free end	50
Cantilever subject to force at free end	25
Leaf type flexure undergoing 's' shaped deformation	25

With a substantial knowledge of coil spring performance and materials selection accumulated over many decades, it is somewhat of a disappointment to realize that they are far from ideal for use as a flexure. The main reasons for this are

- The compliance in all other degrees of freedom is very high. Consequently, this does not provide 'guidance'. Equivalently stated, the coil spring often provides too many freedoms.
- With compression of the spring the diameter of the helix increases.
- As the coil spring is freely compressed the two ends will rotate resulting in a coupling of linear translation and axial rotation, or twisting, of the spring.

6.3 The disc coupling (freedoms θ_x, θ_z, x)

Connecting two shafts via a thin disc can readily produce an inexpensive coupling. In practice there are two common configurations for such a design, see figure 6.1. Most common for applications involving high pressure sealed couplings is the connection from the center of a disk to the outside rim, Wolff, 1951. For this coupling one pipe is joined to a hub in the center of the disc, or diaphragm. The second connects to the outer rim, figure 6.1(a), and will be referred to as an *inner to outer rim* disk coupling. In the second, both shafts are connected symmetrically about the circumference, figure 6.1(b). Henceforth, this is called an *outer rim* disc coupling. Design equations for each of these will be discussed in turn. Because this coupling is designed to provide one axial and two angular freedoms while transmitting a torque it is easy to confuse the terms describing the various components of stiffness. Considering this coupling as a simple idealized joint connecting two rigid bodies (in this case connecting the ends of two shafts), a simplified model of this mechanism is shown in figure 6.2. For most applications, the disc coupling is required to transmit torque while permitting 'free' translation along the axis and two rotations perpendicular. Consequently, these are considered to be three degree of freedom couplings. In

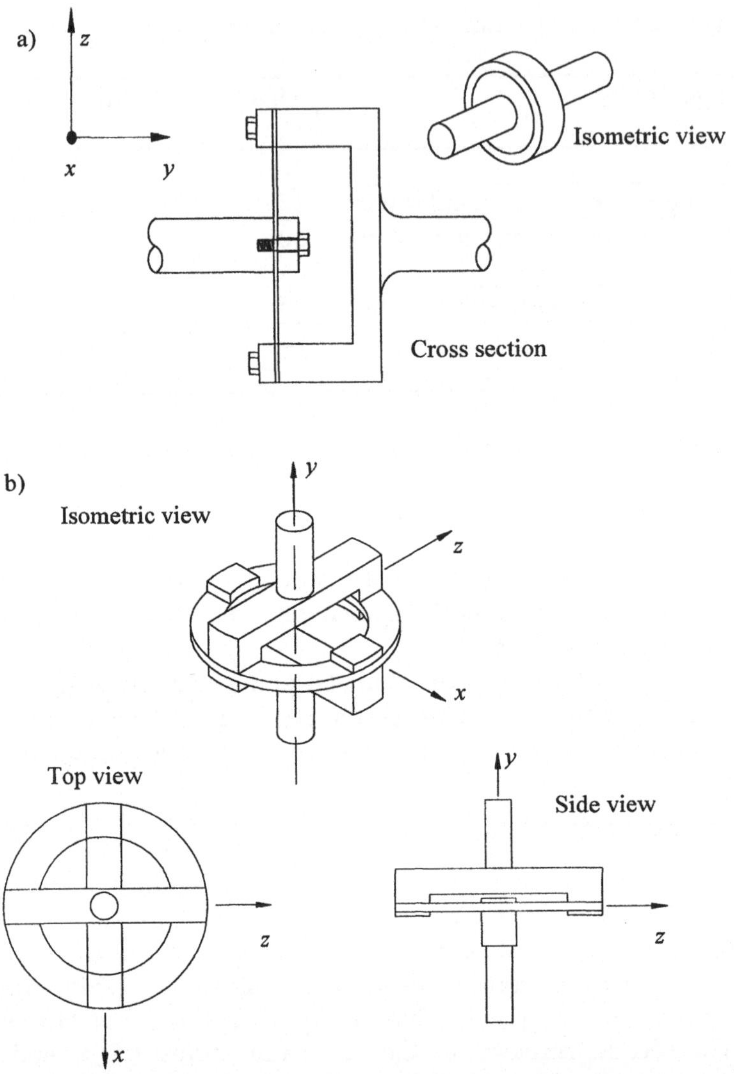

Figure 6.1: Two types of disk coupling, a) the inner to outer edge coupling, b) the outer edge disk coupling

the following two sections, equations for the stiffness provided by the coupling between shafts in relevant coordinate directions are presented. For clarity, the following terminology will be used to describe individual components of stiffness

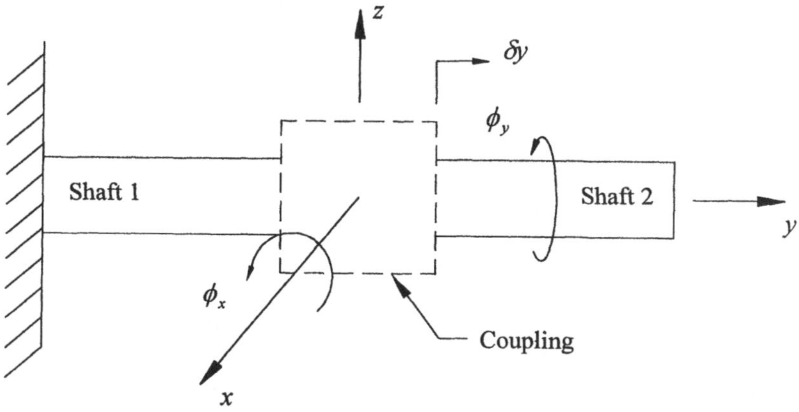

Figure 6.2: A schematic representation of a coupling (the disc coupling in this case) modeled as a joint connecting two rigid shafts

1. 'Axial stiffness', $k_{\delta_y F_y}$, relates the relative motion of the shafts, δ_y towards or away from each other as a consequence of an axial force F_y.

2. In most applications, disc couplings are used to transmit torque from one shaft to another. The ability to faithfully transmit rotation can be assessed from the 'torsional stiffness', this being the ratio of the applied torque, M_y, to the relative angular motion, ϕ_y (i.e. $k_{\phi_y M_y} = M_y / \phi_y$) between the two shafts.

3. Correspondingly, for a large number of designs, misalignment of the axes of two shafts is notoriously difficult to eliminate. Again the compliance of this coupling to accommodate these misalignments is a common concern. Hence the 'angular stiffness' (which, by symmetry, is constant about the x - z plane), refers to the ratio of the applied torque perpendicular to the shaft axis, M_x or M_z, to the subsequent angular rotations ϕ_x or ϕ_z respectively i.e. $k_{\phi_x M_x} = k_{\phi_z M_z} = M_x / \phi_x$.

6.3.1 The inner to outer rim disk coupling

Of primary interest for this coupling is its ability to transmit angular rotation about the axes of the shafts. This is the torsional stiffness of the coupling for which an approximate value can be computed from, Cherukuri, 1997,

$$\frac{M_y}{\theta_y} = \frac{4\pi G t a^2 b^2}{(b+a)(b-a)} \tag{6.11}$$

where a and b are the inside and outside radii respectively, t is the disc thickness and G is the modulus of rigidity, see section 2.4.3.

A torque of magnitude, M_y, will produce pure shear, the contours forming concentric circles. At a radius r the shear stress is given by

$$\tau_{r\theta} = \frac{M_y}{2\pi t r^2}$$

$$\therefore \tag{6.12}$$

$$\tau_{\max} = \frac{M_y}{2\pi t a^2}$$

From von Mises failure criterion, the maximum torque that can be transmitted is given by

$$M_y = \frac{\sigma_Y 2\pi t a^2}{\sqrt{3}} \tag{6.13}$$

Care should be exercized here in that an upper case Y in the subscript of σ is used to denote the *yield* stress and not stress in the *y*-direction.

Considering only radial bending of the disk, the axial stiffness of this coupling is, Young, 1989,

$$k_{\delta_y F_y} = \frac{F_y}{\delta_y} = \frac{2\pi a D}{b^3}\left[\frac{C_1 C_2}{C_4} - C_3\right]^{-1} = \frac{2\pi a D}{b^3} C_5 \tag{6.14}$$

where

$$C_1 = \frac{1}{4}\left[1 - \left(\frac{a}{b}\right)^2\left(1 + 2\ln\left(\frac{b}{a}\right)\right)\right]$$

$$C_2 = \frac{a}{4b}\left[\left(\frac{a}{b}\right)^2 - 1 + 2\ln\left(\frac{b}{a}\right)\right]$$

$$C_3 = \frac{a}{4b}\left\{\left[\left(\frac{a}{b}\right)^2 + 1\right]\ln\left(\frac{b}{a}\right) + \left(\frac{a}{b}\right)^2 - 1\right\} \tag{6.15}$$

$$C_4 = \frac{1}{2}\left[1 - \left(\frac{a}{b}\right)^2\right]$$

Stresses in the disc can be computed (i.e. $\sigma = 6M / 2\pi r t^2$) from the maximum bending moment which occurs at the inner edge of the disc and is given by

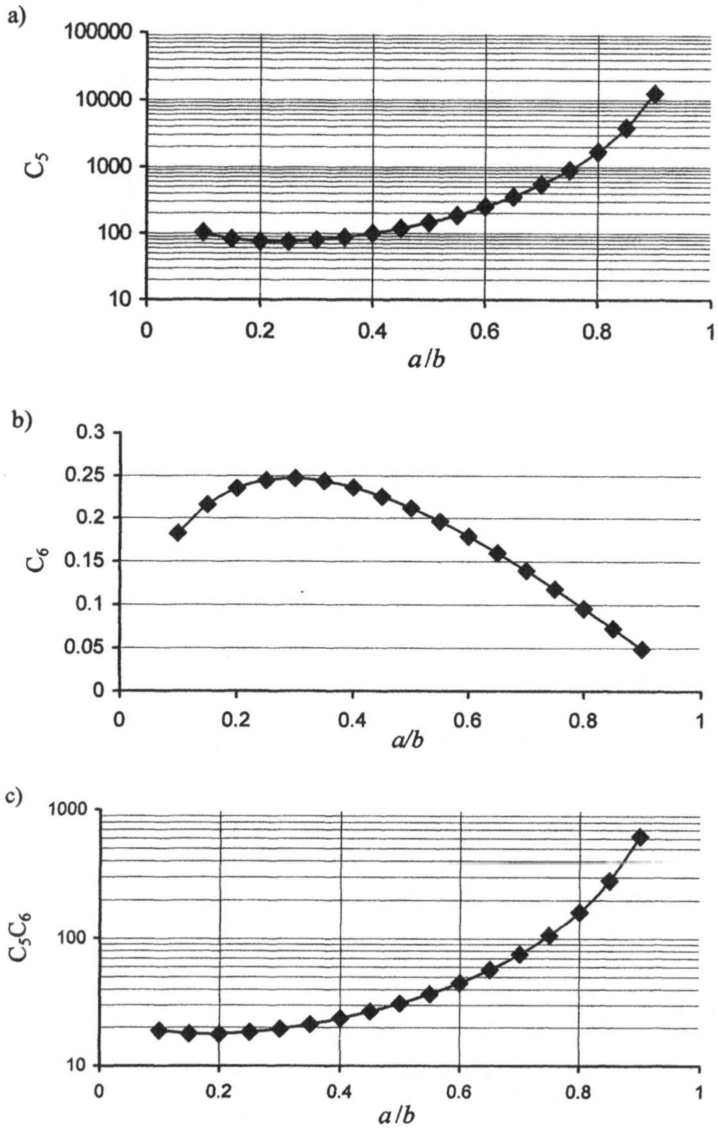

Figure 6.3: Geometry factors for determination of the deflection and moments for an inner to outer disc coupling subject to an axial load, a) the deflection constant C_5, b) the moment constant C_6, c) the product $C_5\,C_6$

$$M_{\max} = F_y b \frac{C_2}{C_4} = F_y b C_6 \tag{6.16}$$

From which

$$\sigma_{\max} = F_y b \frac{C_2}{C_4} = \frac{Et^2 \delta_y}{2(1-\upsilon^2)ab^2} C_5 C_6 \tag{6.17}$$

The constants C_5 and C_6 and their product are plotted for ratios a/b ranging from 0.1 to 0.9 in figures 6.3(a - c). From this figure, it can be seen that, for ratios of a/b between 0.1 and 0.35, the product of these two factors is nearly constant at a value of around 20. In this case, equation (6.17) can be rewritten as

$$\sigma_{\max} \approx \frac{10Et^2 \delta_y}{(1-\upsilon^2)ab^2} \tag{6.18}$$

The two other freedoms of this coupling have identical stiffness corresponding to angular deflections perpendicular to the coupling axis. A side view of this disc coupling indicating modeling parameters is shown in figure 6.4. It can be shown that the maximum stress for a given angular deflection again occurs at the inside edge and can be obtained from

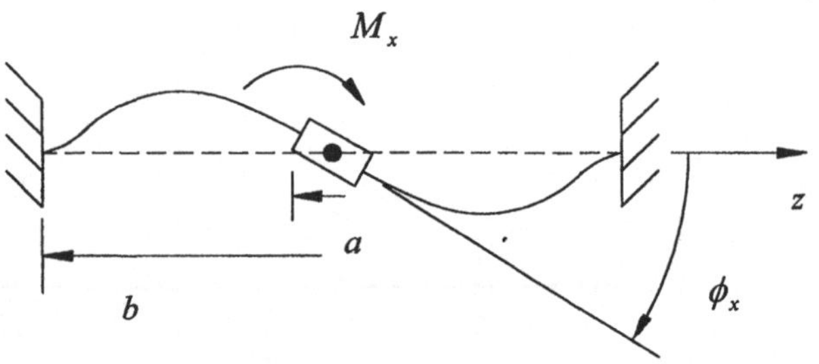

Figure 6.4: Cross section of a clamped circular disc with a moment applied to a central annulus of radius a

$$\sigma_{\max} = \alpha \frac{Et}{b} \phi_x \tag{6.19}$$

Correspondingly, the moment required to induce an angular deflection ϕ can be obtained from the equation

$$\frac{M_x}{\phi_x} = \frac{M_z}{\phi_z} = \beta E t^3 \tag{6.20}$$

The stress and angular stiffness factors, α and β, are plotted in figure 6.5 for the ratio a/b ranging from 0.1 to 0.8.

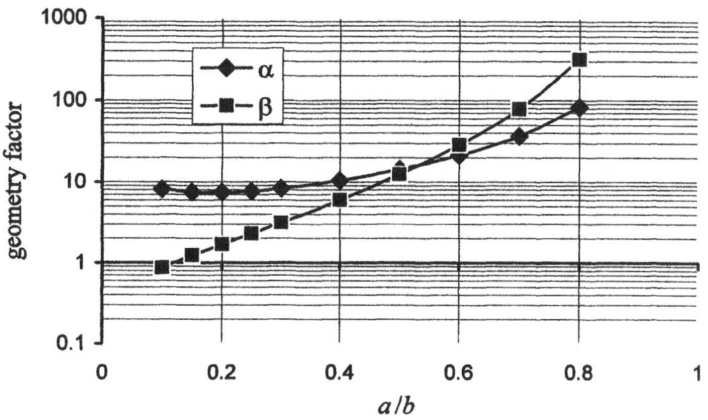

Figure 6.5: Geometric factors for prediction of maximum bending stress and angular deflections α of equation (6.19) β of equation (6.20)

6.3.2 The outer rim disc coupling

Analysis of the out rim disc coupling is less straightforward. This section assesses the stiffness of the coupling in the three freedoms that correspond to; angular misalignment that requires two rotational freedoms and axial motion requiring a linear freedom in the axial direction. Because this type of coupling is used for the transmission of torque, analysis of torsional stiffness is also presented. For an assessment of the equations developed in this section see Woody and Smith, 1999.

An approximate expression for the bending stiffness of a disc type coupling can be derived, provided that the following assumptions are made,

1. The width of the disc (i.e. difference between inside and outside radius) is small in comparison to its radius so that straight beam theories can be applied. In almost all cases the thickness of the 'beam' will also be substantially less than its width.
2. Displacements due to shear induced by loads normal to the beam axis can be ignored.
3. The distribution and magnitude of external loads is not affected by displacements of the beam.
4. Superposition is applicable.
5. Twisting about the axis of the disc induces no bending. Twisting can be visualized by cutting the circular disc and straightening it out so that it becomes a straight beam. The twisting couple is applied about the longitudinal axis of this beam. When bent back to a circular arc, the twisting moment is always tangential to the axis.

Under these conditions, the equation governing deflections of a circular ring due to symmetrically applied bending, M_θ, and twisting moments, T_θ, is given by, Levy, 1962

$$\frac{EI}{R^2}\left(\frac{\partial^3 y}{\partial\theta^3}+\frac{\partial y}{\partial\theta}\right)=\frac{EI}{R^2}\left(y'''+y'\right)=-\frac{\partial M_\theta}{\partial\theta}+\lambda T_\theta \tag{6.21}$$

The primes indicate partial differentiation with respect to θ, I is the second moment of area about the neutral axis of bending and λ is the ratio of the bending to torsion rigidity given by

$$\lambda=\frac{EI}{GJ} \tag{6.22}$$

For a beam of rectangular section, this can be computed from equations (6.57) and (6.61) of the following section. From the expressions for J, it can be shown that the above ratio is equal to $1+\upsilon$ for a disc of circular cross section section (i.e. a circular ring, made from a wire of circular section, an unusual design), $(1+\upsilon)/2$ for a thin plate ($J = bt^3/3$) and up to approximately 1.66 for a square cross-section made from a material having a Poisson's ratio of 0.3. Consequently, values ranging from 2/3 to 5/ will be used in the following analyses. Solutions to equation (6.21) must satisfy three boundary conditions given by the symmetric condition (in this case) of zero slope and deflection at the supports plus the requirement for prescribed slope or deflection at $\theta=90°$. Ultimately, for rotation of the axes of the disc coupling the slope at the fixed points are constrained so that the surfaces of the disc, where clamped, always forms a straight line perpendicular to the axis of the clamp. This rather complex boundary condition can be obtained from the combination of individual solutions for twisting and an applied normal force at $\theta=90°$ as shown in figure 6.6.

Having obtained an expression for the displacement, the twist of the beam β (rad) can be derived from

$$\frac{EI}{R}\beta=M_\theta-\frac{EI}{R^2}y'' \tag{6.24}$$

As a first step to the solution of this problem, consider the case of the circular ring subject to a twisting moment, $2T$, at the center (i.e. $\theta=90°$). From this, because the loads will be equally supported by the two quadrants, the subsequent bending and twisting moments in one quadrant are

$$\begin{aligned}M_\theta&=T\cos(\theta)\\T_\theta&=T\sin(\theta)\end{aligned} \tag{6.25}$$

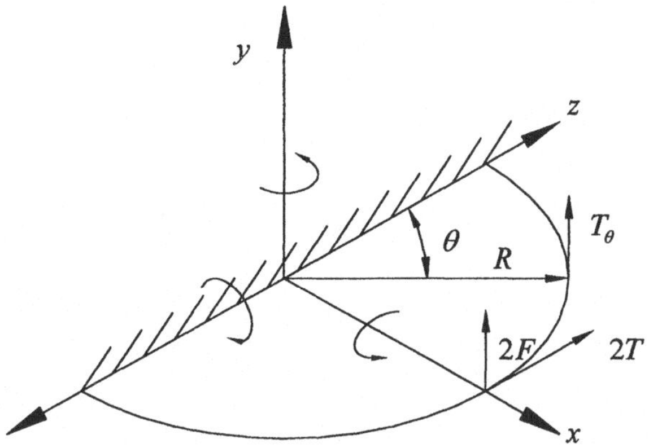

Figure 6.6: Parametric representation of a clamped semi-circular ring

Substituting (6.25) into (6.21) gives

$$\frac{EI}{R^2}(y''' + y') = T(1+\lambda)\sin(\theta)$$
$$y''' + y' = b\sin(\theta) \tag{6.26}$$

where

$$b = \frac{T(1+\lambda)R^2}{EI} \tag{6.27}$$

A general solution to this linear, differential equation can be obtained from the complementary function, y_{cf}, that provides a solution to the homogeneous component of equation (6.26) plus a particular integral, y_{pi}, that necessarily contains a non-coincident term in $\sin(\theta)$

$$y_{cf} = A\cos(\theta) + B\sin(\theta) + C$$
$$y_{pi} = (D + E\theta)\sin(\theta) \tag{6.28}$$

Differentiating the second of equations (6.28) gives

$$y'_{pi} = D\cos(\theta) + E\sin(\theta) + E\theta\cos(\theta)$$
$$y''_{pi} = -D\sin(\theta) + 2E\cos(\theta) - E\theta\sin(\theta) \tag{6.29}$$
$$y'''_{pi} = -D\cos(\theta) - 3E\sin(\theta) - E\theta\cos(\theta)$$

Substituting (6.29) into (6.26) yields

$$
\begin{aligned}
&2E\sin(\theta) = b\sin(\theta) \\
&\therefore \\
&E = \frac{b}{2}
\end{aligned}
\tag{6.30}
$$

Hence the complete solution can be obtained from the sum of the two terms in (6.28) from which

$$
y = A\cos(\theta) + B\sin(\theta) + C + \frac{b}{2}\theta\sin(\theta) \tag{6.31}
$$

The boundary conditions for this case are

$$
y\big|_{\theta=0} = y'\big|_{\theta=0} = y'\big|_{\theta=\pi/2} = 0 \tag{6.32}
$$

From the first of these $A = -C$ and equation (6.31) reduces to

$$
\begin{aligned}
&y = C(1-\cos(\theta)) + B\sin(\theta) + \frac{b}{2}\theta\sin(\theta) \\
&y' = C\sin(\theta) + B\cos(\theta) + \frac{b}{2}\sin(\theta) + \frac{b}{2}\theta\cos(\theta)
\end{aligned}
\tag{6.33}
$$

The second of the three boundary conditions result in the equations

$$
\begin{aligned}
&y'(0) = B = 0 \\
&y'(\pi/2) = C + \frac{b}{2} = 0
\end{aligned}
\tag{6.34}
$$

Substituting (6.34) into (6.33), the general solution for the displacement of the disc is given by

$$
\frac{2y}{b} = \cos(\theta) + \theta\sin(\theta) - 1 \tag{6.35}
$$

It can be shown that this corresponds to case 5 of the examples given in the paper of Levy, 1962 and, henceforth, results will be based on equations generated in this paper. Plotting the displacement y as a function of θ corresponds to a view of the displacement of the ring as observed from the origin of the coordinate system, figure 6.7(a). To view the displacement from the side as a function of the distance from the center (i.e. viewing the deflected beam as projected onto the x - y plane) it is necessary to plot (6.35) as a function of x/R as shown in figure 6.7(b). The slope of the disc due to bending as a function of the distance x from the center (i.e. once again, as would be observed from the projection of the neutral axis onto the the x - y plane) is given by

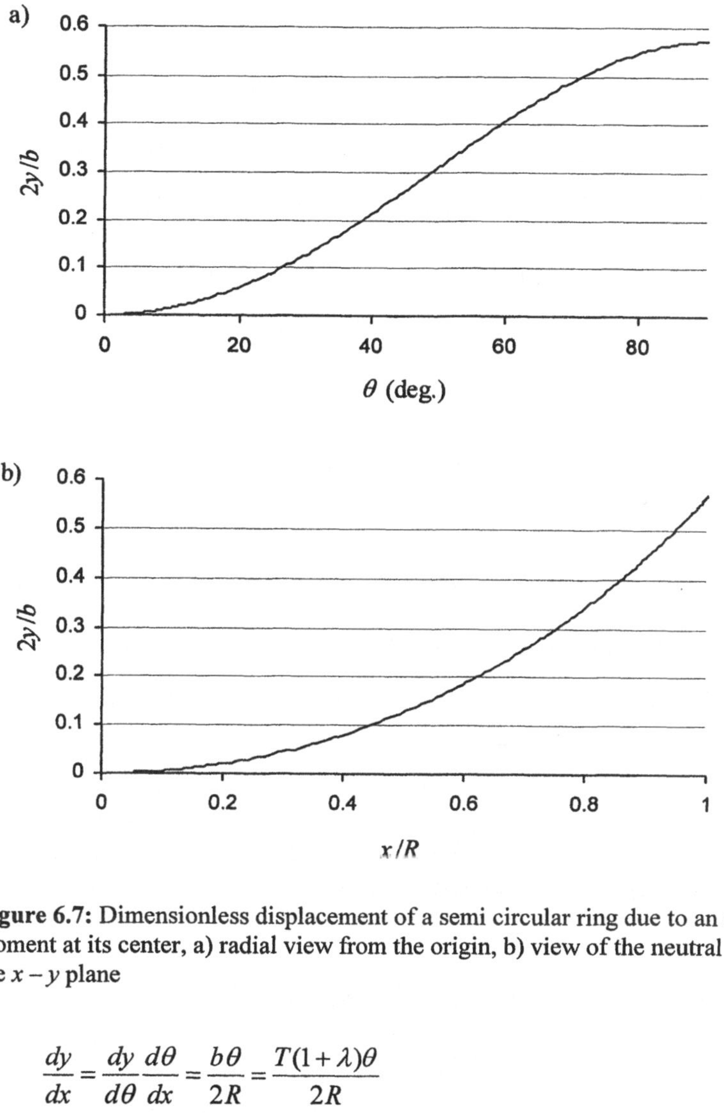

Figure 6.7: Dimensionless displacement of a semi circular ring due to an applied twisting moment at its center, a) radial view from the origin, b) view of the neutral axis projected onto the $x-y$ plane

$$\frac{dy}{dx}=\frac{dy}{d\theta}\frac{d\theta}{dx}=\frac{b\theta}{2R}=\frac{T(1+\lambda)\theta}{2R} \tag{6.36}$$

The twist of the beam is given by

$$\beta_T=\frac{TR}{2EI}[(\cos(\theta)+\theta\sin(\theta))+\lambda(\theta\sin(\theta)-\cos(\theta))] \tag{6.37}$$

Figure 6.8 shows plots of the function in parentheses for three values of the parameter λ. Examination of this graph reveals one of the potential errors associated with the assumptions of this model. At the fixed end a finite twist is predicted, the magnitude of which is related to the deviation of λ from unity. In

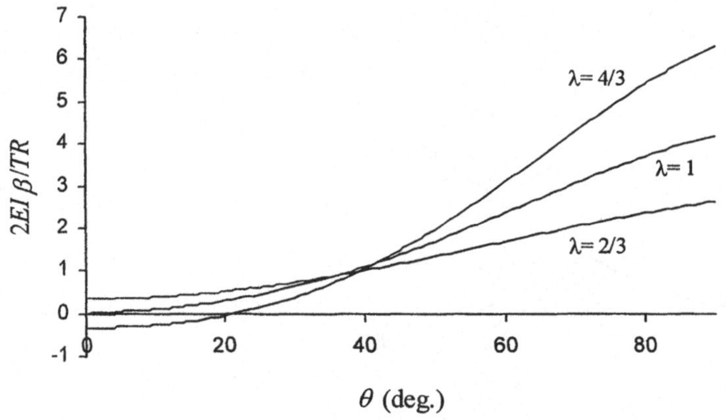

Figure 6.8: Twisting along the beam as a function of the angular position θ for values of $\lambda = 2/3$, 1 & 4/3

practice, twisting moments will be, at least partially, constrained at these supportsand is a potential source of error. Such a complication is ignored in the proceeding analysis.

From Levy, 1962, the displacement, y, of the semicircular section due to a normal force of magnitude $2F$ is

$$\frac{4EI}{FR^3}y = \begin{bmatrix} 2\theta\sin\theta + (\pi-2)(1-\cos\theta) - 2\sin\theta + 2\theta\cos\theta \\ -\lambda((6-\pi)(1-\cos\theta) + 6\sin\theta - 4\theta - 2\theta(\cos\theta + \sin\theta)) \end{bmatrix} \tag{6.38}$$

This is plotted in figure 6.9 for a range of values for λ. Proceeding as for the previous example, the slope of this curve in x is

$$\frac{4EI}{FR^2}\frac{dy}{dx} = \begin{bmatrix} 2\theta\cos\theta + (\pi-2\theta)\sin\theta \\ -\lambda((4-\pi+2\theta)\sin\theta + (4-2\theta)\cos\theta - 4) \end{bmatrix}\frac{1}{\cos\theta} \tag{6.39}$$

The twist due to this bending moment is given by

$$\beta_M = \frac{FR^2}{4EI}(1+\lambda)[2\theta\sin\theta + (\pi-2)\cos\theta - 2\sin\theta + 2\theta\cos\theta] \tag{6.40}$$

For the purpose of matching the boundary condition at the point of applied load, it is necessary to determine the expressions for the slopes and displacements at the center of the semi circular ring. The relevant equations are tabulated below

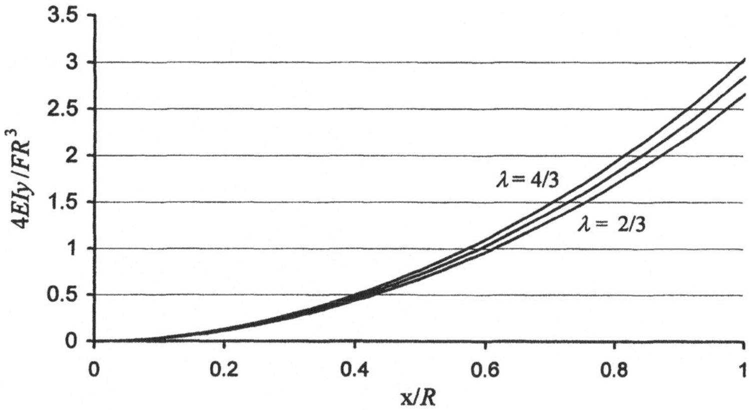

Figure 6.9: Displacement of a semicircular beam due to a normal force applied at the center

Table 6.3.2.1: Equations of displacement, slope and twist in one quadrant of a circular ring

Parameter	Equation at $\pi/2$	Equation no.
y torsion	$\frac{TR^2}{4EI}(1+\lambda)(\pi-2)=D_T T$	6.35
$\frac{dy}{dx}$ torsion	$\frac{T(1+\lambda)R\pi}{4EI}=C_1 T$	6.36
β_T	$\frac{TR}{4EI}\pi(1+\lambda)=C_2 T$	6.37
y bending	$\frac{FR^3}{4EI}((2\pi-4)+\lambda(4\pi-12))=D_B F$	6.38
$\frac{dy}{dx}$ bending	$\frac{FR^2}{4EI}(\pi-\lambda(4-\pi))=C_3 F$	6.39
β_M	$\frac{FR^2}{4EI}(1+\lambda)(\pi-2)=C_4 F$	6.40

It is informative to compare equation (6.38) with the expression derived by Timoshenko, 1941, using Castigliano's equation shown overleaf

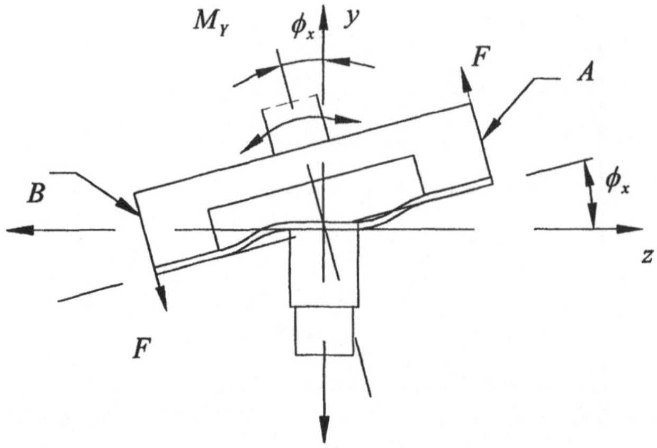

Figure 6.10: Side view of coupling indicating distortion due to applied bending moment (for the purposes of finite element modeling the bending moment can be generated by applying equal and opposite forces, *F*, on faces A and B)

$$\frac{2EI\,y|_{\theta=\pi/2}}{(2F)R^3} = 0.57 + \lambda(0.142) \qquad 6.38$$

$$= 0.47 + \lambda(0.037) \qquad \text{Timoshenko}$$

6.3.2.1 Angular stiffness

A constraint imposed by the disc coupling is that of common slopes at the surfaces of the ring at the point of clamping. Also, the center lines of the cross section must also pass through the pivot point, see figure 6.10. In this model the displacement and angle of the surface at the clamp has been assigned the symbols y and ϕ_x respectively. Clearly, these will comprise the sum of the effects of both 'torsion' and 'bending' due to a normal force which, if known, can be computed from the formulae given in table 6.3.2.1 above. Using the expressions in this table, the parameters for the model of figure 6.10 yield the equations

$$\begin{aligned} y &= y_B - y_T = D_B F - D_T T \\ \phi &= \phi_B - \phi_T = C_3 F + C_4 F - C_1 T - C_2 T = C_B F - C_T T \end{aligned} \qquad (6.41)$$

The boundary condition of common slopes coincident with the pivot provides the condition

$$\frac{y}{R} = \phi \qquad (6.42)$$

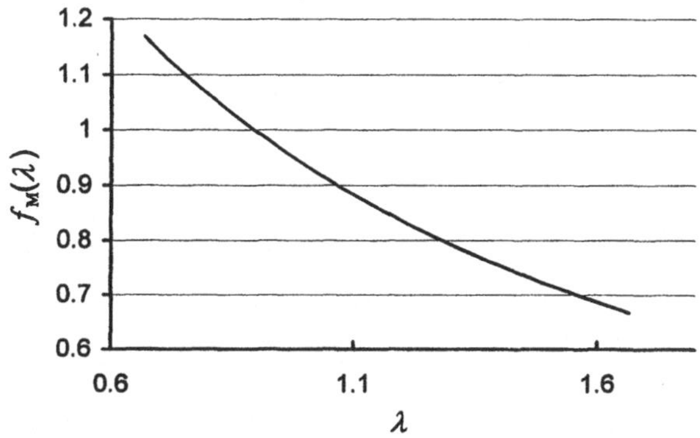

Figure 6.11: Stiffness factor $f_M(\lambda)$ for the bending of the disk coupling

It is important to remember that, in this section, ϕ refers to rotation about either the x or z axes. Substituting (6.41) into (6.42), it is possible to eliminate the twisting couple applied to the ring from

$$T = \frac{(D_B - C_B R)}{(D_T - C_T R)} F = \frac{a}{b} F \tag{6.43}$$

Additionally, it is known that the applied bending moment is simply the sum of all moments generated in the four quadrants of the disk. Consequently, the force can be expressed in terms of the applied moment in terms of the relationship

$$M = 4FR + 4T = 4F\left(R + \frac{a}{b}\right) \tag{6.44}$$

Substituting (6.44) into the second of (6.41) yields

$$\begin{aligned}\frac{M}{\phi} = k_{M\phi} &= \frac{16EI}{R}\left(\frac{(\pi+4) + \lambda(8-\pi)}{(2\pi^2 - 4\pi - 4) + \lambda(8\pi^2 - 18\pi - 16) + \lambda^2(6\pi^2 - 14\pi - 12)}\right)\\ &= \frac{16EI}{R} f_M(\lambda)\end{aligned} \tag{6.45}$$

The functional relation $f_M(\lambda)$ is shown plotted figure 6.11. For common value of λ this is usually close to 1. In practice, it is the ratio of the bending to torsional stiffness that is of concern to the designer and this is discussed at the end of section 6.3.2.3.

6.3.2.2 Axial stiffness

The axial stiffness of the disc type flexure can be obtained from the boundary condition

$$\phi\big|_{\theta=\pi/2} = 0 \tag{6.46}$$

From this, the displacement, y, due to an applied force, F^* (= 4F), is given by

$$\begin{aligned} y &= \frac{F^* R^3}{4EI}\frac{1}{3\pi}\left(\left(\pi^2 - \pi - 2\right) + \lambda\left(3\pi^2 - 7\pi - 6\right)\right) \\ &\approx \frac{F^* R^3}{12EI}\left(1.5 + 0.5\lambda\right) \end{aligned} \tag{6.47}$$

An equation for predicting an important case in which the axial stiffness of a coupling comprized from segments of included angle less than 90 degrees (i.e. $\theta < \pi/2$) is developed in section 6.4.

Figure 6.12: Parametric model of a single segment of a leaf spring subject to in-plane loads

6.3.2.3 Torsional stiffness

To determine the torsional stiffness, it is necessary to evaluate distortions due to forces and moments acting in the plane of the disk only. Again, the disc coupling can be modeled as comprising the sum of four arcuate segments, each of included angle ψ, fixed at one end and subject to combined forces and moments at the other. Displacement of the 'free' end of the segment can be derived from the superposition of the effects of an axial and radial force plus an applied moment. Both the forces, F_ψ, F_R, and applied moment, M, result in in-plane distortion of the segment, figure 6.12. From symmetry, it is assumed that the deflection of the 'free' end is in the direction of the axis. The displacements, u_ψ, u_R, of the free end and subsequent angular deflection, ϑ_B, can be calculated using Castigliano's theorem from the integral energy equation, Young, 1989

$$U = \int_0^\psi \frac{\lfloor F_R R\sin(\theta) + F_\psi R(1-\cos(\theta)) + M \rfloor}{2EI} R d\theta \tag{6.48}$$

where, in this case, I corresponds to the second moment of area about the neutral axis which is *normal* to the plane of the disc (i.e. $I = t(R_o - R_i)^3/12$).

Correspondingly, the displacements and angular deflections due to the applied forces and moment are given by

$$\begin{aligned}
u_\psi &= \frac{\partial U}{\partial F_\psi} = \frac{R^2}{2EI}\begin{bmatrix} M(\psi - 2\sin(\psi)) + F_R R(1+\cos^2(\psi) - 2\cos(\psi)) \\ ..+ F_\psi R(3\psi + \cos(\psi)\sin(\psi) - 4\sin(\psi)) \end{bmatrix} \\
u_R &= \frac{\partial U}{\partial F_R} = \frac{R^2}{2EI}\begin{bmatrix} 2M(1-\cos(\psi)) + F_R R(\psi - \cos(\psi)\sin(\psi)) \\ ..+ F_\psi R(1+\cos^2(\psi) - 2\cos(\psi)) \end{bmatrix} \\
\vartheta &= \frac{\partial U}{\partial M} = \frac{R}{EI}\lfloor F_R R(1-\cos(\psi)) + M\psi + F_\psi R(\psi - \sin(\psi)) \rfloor
\end{aligned} \tag{6.49}$$

Subjecting the segment to the condition of zero angular and radial deflection at the free end yields

$$\begin{aligned}
F_R &= F_\psi \frac{\sin(\psi)[2\cos(\psi) - 2 + \sin(\psi)\psi]}{2\sin^2(\psi) + 4(\cos(\psi)-1) - \cos(\psi)\sin(\psi)\psi + \psi^2} \\
M &= F_\psi R \frac{3\sin^2(\psi) + 4(\cos(\psi)-1) - \cos(\psi)\sin(\psi)\psi + \psi^2 - \sin(\psi)\psi}{2\sin^2(\psi) + 4(\cos(\psi)-1) - \cos(\psi)\sin(\psi)\psi + \psi^2}
\end{aligned} \tag{6.50}$$

Substituting equations (6.50) into (6.49), the axial deflection, u_ψ, of the beam becomes

$$u_\psi = \frac{F_\psi R^3}{2EI}\left[\frac{4\sin(\psi)\cos(\psi) + 5\psi - \psi^3 - 4(\sin(\psi) + \cos(\psi)\psi) - \cos^2(\psi)\psi}{2(1+\cos^2(\psi)) + \cos(\psi)\sin(\psi)\psi - 4\cos(\psi) - \psi^2}\right] \tag{6.51}$$

For a coupling comprised from 4 segments, the applied axial toque, T, is $T = 4FR$. Defining the angular rotation, $\phi_y = u_\psi / R$, about the central axis, the angular stiffness, $k_{T\phi_y}$, of the disk coupling is

$$k_{T\phi_y} = \frac{8EI}{R}\left(\frac{2(1+\cos^2(\psi)) + \cos(\psi)\sin(\psi)\psi - 4\cos(\psi) - \psi^2}{4\sin(\psi)\cos(\psi) + 5\psi - \psi^3 - 4(\sin(\psi) + \cos(\psi)\psi) - \cos^2(\psi)\psi}\right) \tag{6.52}$$

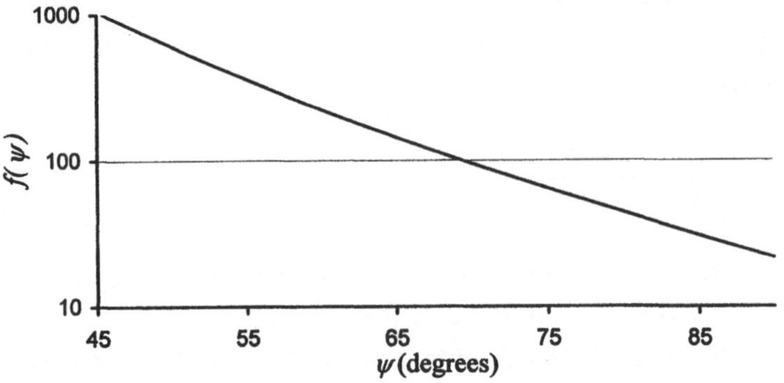

Figure 6.13: Factor $f(\psi) = k_{T\theta}R / 8EI$ for the torsional stiffness of a disc coupling

The expression in parentheses of equation (6.52) for the torsional stiffness as a function of the included angle of the segment is plotted in figure 6.13. In practice, additional clamping, boundary, shear and other compliances can *significantly* reduce this effective stiffness, Woody and Smith, 1999. For a disc coupling consisting of four quadrants, equation (6.52) reduces to

$$k_{T\phi_y} = \frac{16EI}{R}\left(\frac{8-\pi^2}{20\pi - 32 - \pi^3}\right) \tag{6.53}$$

Equation (6.53) applies to the torsional stiffness of the hinge of the following section. In practice, it is unlikely that a disk coupling comprised from arcuate segments of included angle 90° will be produced. However, viewed as a coupling to provide specific constraints and freedoms, it is more relevant to consider the *ratio* of stiffness in different coordinate directions. Particularly relevant to the disk coupling is the ratio of torsional to angular stiffness. Dividing equations (6.53) by (6.45) yields

$$\frac{k_{T\phi_y}}{k_{M\phi}} = \frac{(R_o - R_i)^2(8-\pi^2)}{t^2\left(20\pi - 32 - \pi^3\right) f_M(\lambda)} \approx \frac{11(R_o - R_i)^2}{t^2}$$

In most designs, this ratio will range from 50 to 500 or more and increases as the square of the ratio of the breadth to thickness of the disk. For couplings of included angle less than 90° this is ratio not expected to be significantly different.

6.4 Rotationally symmetric leaf type hinge (axial stiffness of the disc coupling revisited)

In many ways similar to the disk coupling of the previous section, this type of flexure is also used to provide three freedoms, one translation and two rotations.

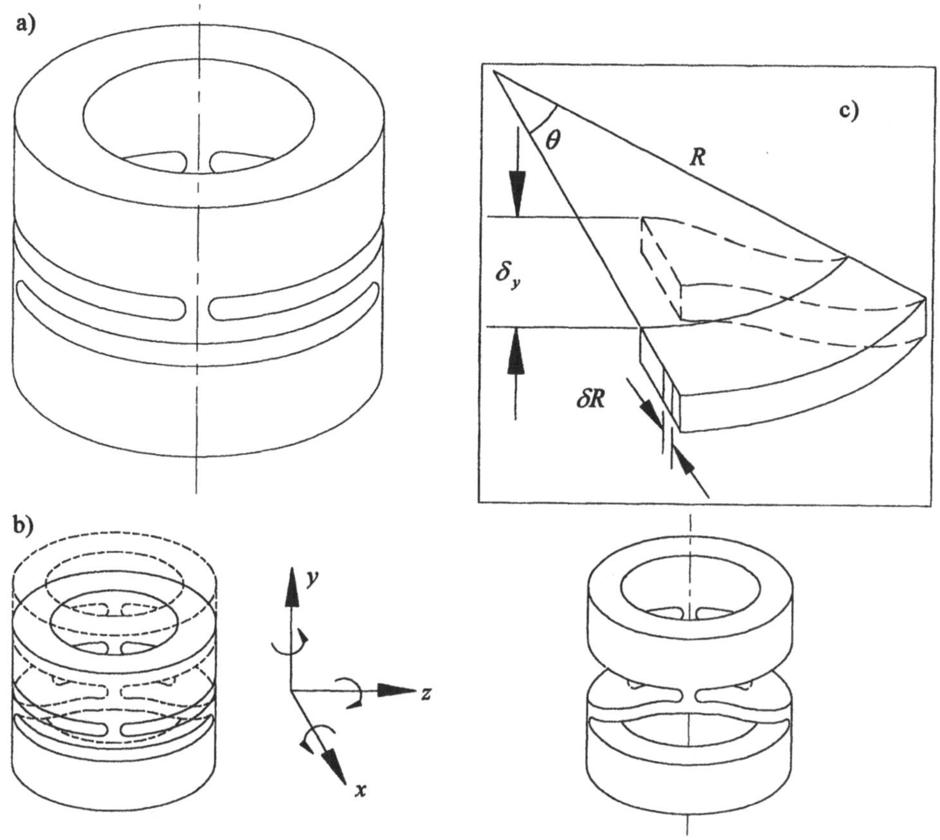

Figure 6.14: The rotationally symmetric hinge, a) isometric view, b) exaggerated axial distortion, c) modeling parameters for deriving axial stiffness equations

Although, in principle, the same technique could be employed for the prediction of axial stiffness of this flexure, satisfaction of the boundary conditions for the sectors leads to considerable mathematical complexity. An alternative analysis is presented in this section.

Figures 6.14(a & b) show both an isometric view of such a flexure and a exaggeratedly distorted flexure after a linear distortion in the axial, or z-axis, direction. In its simplest form, this flexure can be easily manufactured by making four slits in a tube. First, two slits are cut into either side of the tube directly opposite one another. This can be readily achieved by slitting the tube in a milling machine using a slitting saw. With the tube vertical, the first cut would make a horizontal slit perpendicular to its axis the depth of this being less than half the outside diameter of the tube. A second identical slit is made on the other side of the tube after which two small bridges remain that prevent the tube from separating into two pieces. Following this, the tube is then rotated about its axis by 90 degrees and the slitting saw moved down a distance corresponding to the thickness of the saw plus the desired flexure depth. Repeating the same two slits at this depth will then produce the complete flexure. It is important that the upper slit is made first so that the final machining forces are not supported by the thin flexure (it is left to the reader to see why this is so). Other manufacturing methods such as electro-discharge machining or fabrication do not impose a sequential manufacturing requirement, see chapter 8. Once machined, this flexure is simply a monolithic version of the disc coupling discussed above with each segment necessarily having an included angle, ψ, of less than 90 degrees.

6.4.1 Axial stiffness and maximum stress calculations for the rotationally symmetric leaf type flexure system

To derive a suitable formula for the axial stiffness of a rotationally symmetric hinge it is possible to consider the full flexure consisting of the superposition of two cantilever arches subject to a force, F_y, at the ends that is normal to the plane of curvature, see figure 6.14(c). Under these conditions the deflection at the end of the cantilever can be determined by integrating the bending moment and torsion equations and assuming that superposition holds, see Seely and Smith, 1932. Under these conditions, the moment M_x and torque M_z at a position specified by the polar coordinate ψ can be obtained from the equations

$$\begin{aligned} M_x &= F_y R_m \sin\psi \\ M_z &= F_y R_m (1-\cos\psi) \end{aligned} \tag{6.55}$$

Using a techniques called the 'dummy load method', the end deflection δ_y of the arched beam can be determined from the equation

$$\begin{aligned} \delta_y &= \int_0^{\psi} \frac{F_y R_m^3 \sin^2\psi}{EI} d\psi + \int_0^{\psi} \frac{F_y R_m^3 (1-\cos\psi)^2}{GJ} d\psi \\ &= \frac{F_y R_m^3}{EI}\left[\left(\frac{\psi}{2} - \frac{\sin(2\psi)}{4}\right) + \frac{2(1+\upsilon)I}{J}\left(\frac{3\psi}{2} - 2\sin\psi + \frac{\sin(2\psi)}{4}\right)\right] \end{aligned} \tag{6.56}$$

where R_m is the mean radius of the beam and ψ is the angle subtended by cantilever.

From equation (2.104) of chapter 2, the polar second moment of area for a beam of rectangular section is given by the equation

$$
\begin{aligned}
J &= \frac{(R_o - R_i)d^3}{3}\left[1 - \frac{192}{\pi^5}\frac{d}{(R_o - R_i)}\sum_{n=1,3..}^{\infty}\frac{1}{n^5}\tanh\left(\frac{n\pi(R_o - R_i)}{2d}\right)\right] \\
&= 4I\left[1 - \frac{192}{\pi^5}\frac{\varepsilon}{(1-\gamma)}\sum_{n=1,3..}^{\infty}\frac{1}{n^5}\tanh\left(\frac{n\pi(1-\gamma)}{2\varepsilon}\right)\right] \\
&= 4If_J(\varepsilon,\gamma)
\end{aligned}
\tag{6.57}
$$

where γ and ε are the dimensionless ratios given by

$$\gamma = \frac{R_i}{R_o} \tag{6.58}$$

$$\varepsilon = \frac{d}{R_o} \tag{6.59}$$

It can be readily shown that

$$0 < \gamma < 1 \tag{6.60}$$

From (6.57), the ratio of bending to torsional rigidity (λ of the previous section) is

$$\lambda = \frac{2(1+\upsilon)I}{J} = \frac{(1+\upsilon)}{2f_J(\varepsilon,\gamma)} \tag{6.61}$$

Although these values have been arbitrarily chosen, it would be expected that γ and ε would typically be within the range

$$
\begin{aligned}
0.75 < \gamma < 0.95 \\
0.01 < \varepsilon < 0.1
\end{aligned}
$$

Noting that

$$R_m^3 = \left(\frac{R_o + R_i}{2}\right)^3 = \frac{R_o^3}{8}\left(1 + 3\gamma + 3\gamma^2 + \gamma^3\right)$$

equation (6.56) can be rearranged to provide the linear stiffness of the curved beam given by

$$k_{\delta_y F_y} = \frac{4E(1-\gamma)d\varepsilon^2}{3(1 + 3\gamma + 3\gamma^2 + \gamma^3)}\left[\left(\psi - \frac{\sin(2\psi)}{2}\right) + \frac{(1+\upsilon)}{f_J(\varepsilon,\gamma)}\left(\frac{3\psi}{2} - 2\sin\psi + \frac{\sin(2\psi)}{4}\right)\right]^{-1} \tag{6.62}$$

Connecting two curved cantilever beams together at their ends will produce a curved spring represented by a cord of total angle 2ψ with clamped ends subject to a linear translation of the supports in a vertical, or y axis, direction perpendicular beam axis. Clearly equation (6.62) corresponds to the y axis stiffness of the rotational flexure subtending a total angle θ. The stiffness of this new configuration will be equivalent to half of that for each cantilever. Consequently, the stiffness of the complete flexure is given by the equation

$$k_{\delta_y F_y} = \frac{2E(1-\gamma)d\varepsilon^2}{3(1+3\gamma+3\gamma^2+\gamma^3)}\left[\left(\psi - \frac{\sin(2\psi)}{2}\right) + \frac{(1+\upsilon)}{f_J(\varepsilon,\gamma)}\left(\frac{3\psi}{2} - 2\sin\theta + \frac{\sin(2\psi)}{4}\right)\right]^{-1}$$
$$= \frac{2E(1-\gamma)d\varepsilon^2}{3(1+3\gamma+3\gamma^2+\gamma^3)}\left[\left(\frac{\theta}{2} - \frac{\sin(\theta)}{2}\right) + 2\lambda\left(\frac{3\theta}{4} - 2\sin\left(\frac{\theta}{2}\right) + \frac{\sin(\theta)}{4}\right)\right]^{-1} \tag{6.63}$$

An alternative, and considerably simplified, approximate formula for the stiffness of the flexure can be obtained by conceptually splitting it into radial strips. Assuming that each strip will distort along its axis in the same way as a simple leaf flexure undergoing an 's' shaped distortion the elemental stiffness of each strip can be obtained from equation (4.9) of chapter 4 i.e.

$$dk_{\delta_y F_y} = \frac{12EI}{L^3} = \frac{12Ed^3dR}{12(R\theta)^3} = \frac{Ed^3dR}{(R\theta)^3} \tag{6.64}$$

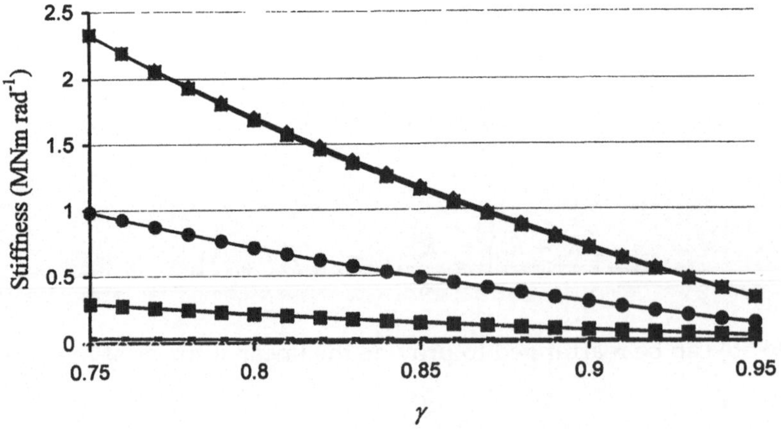

Figure 6.15: Stiffness of a single arcuate element using both the full and simple theoretical models of equations (6.63) and (6.66). $E = 210$ GPa, $R_o = 1$, $\nu = 0.3$ values of ε correspond to 0.04, 0.03, 0.02 and 0.01 with the larger values corresponding to high stiffness

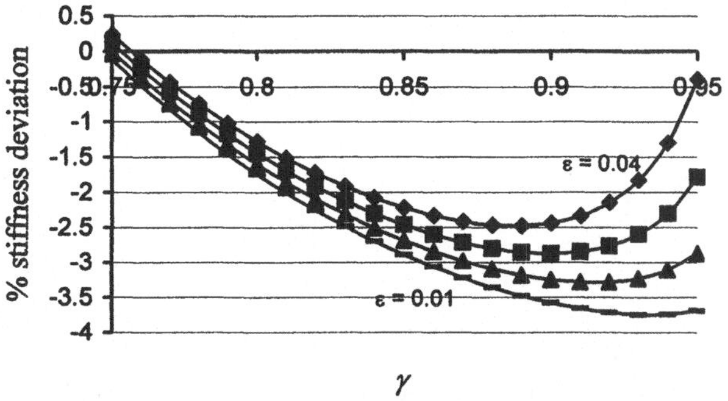

Figure 6.16: Percentage deviation between the simple and full theoretical models for the plots shown in figure 6.15

Integrating (6.64) yields the total stiffness, *k*, of the flexure

$$k_{\delta_y F_y} = \frac{Ed^3}{\theta^3}\int_{R_i}^{R_o}\frac{dR}{R^3} = \frac{Ed^3}{2\theta^3}\left(\frac{1}{R_i^2} - \frac{1}{R_o^2}\right) \tag{6.65}$$

In terms of the dimensionless parameters ε and γ this can be expressed by the equation

$$k_{\delta_y F_y} = \frac{Ed}{2\theta^3}\left(\frac{\varepsilon}{\gamma}\right)^2\left(1-\gamma^2\right) \tag{6.66}$$

Equations (6.65) and (6.66) have been plotted as a function of the variable γ of the range 0.75 to 0.95 with values of ε ranging from 0.01 to 0.1, figure 6.15 with the percentage stiffness deviation between these two equations plotted in figure 6.16. Both graphs have been computed using an included angle of $\theta = 75$ degrees. For symmetry, in minimum of four hinges per flexure is necessary and therefore this is considered to represent an upper bound on the angles likely to be encountered with this particular hinge design. It can be seen that the error for all plotted values of γ does not exceed 4%. For higher values of γ the error tend to deviate to greater than this limit with rapid deviations occuring earlier for the higher values of ε. Considering that the range of values considered in this analysis represents the limits of design commonly used, these may be considered upper limits.

6.4.2 Simplified equation for maximum stress

Using the simple model of equation (6.66), the maximum stress is found to occur at each end of the flexure at the inside edge and can be determined from the equation

$$\sigma = \frac{6Ed}{2R^2\theta^2}\delta_y \tag{6.67}$$

or

$$\begin{aligned}\sigma_{\max} &= \frac{6Ed}{2R_i^2\theta^2}\delta_y \\ &= \frac{6E\varepsilon}{2R_o\gamma^2\theta^2}\delta_y\end{aligned} \tag{6.68}$$

The range of validity of this equation has yet to be experimentally determined. However, finite element methods provide a complimentary method for comparison. The results of such a study are outlined in the following section.

6.4.3 Assessment of approximate equations using finite element models

Ultimately, equations must be verified experimentally. In the absence of such information, finite elements represent a reasonable check on the validity of mathematical models. In this analysis, finite element models were generated and computed. In all analyses solid pyramidal elements were used with nodal points midway along each line. Although these elements provide reliable results for stiffness calculations, they are less suited for solutions for determination of stress. As a consequence considerably more deviation between stresses predicted from simple beam bending and finite element analysis should be expected. A simple curved beam was used with one end fixed and the other constrained with a freedom in the direction of traverse only. Force was applied as a point load at the central node on the face of the moving end of the beam. As a result of this, there was a stress concentration at this node. Consequently, the stress was determined from the maximum value at the fixed end. No significant differences were found in a comparison the results from an alternative model in which a traction force

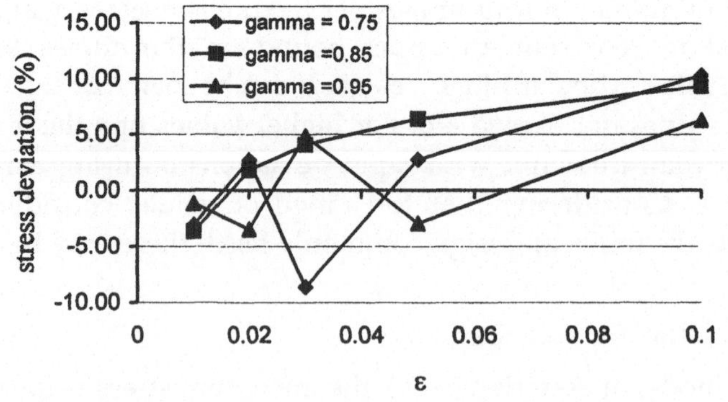

Figure 6.17: Stress deviation between simple formula and results from finite element analysis ($\theta = 45$ degrees)

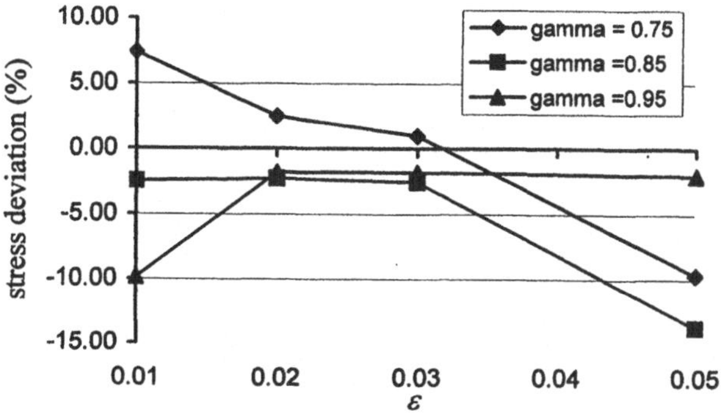

Figure 6.18: Stress deviation between simple formula and results from finite element analysis ($\theta = 60$ degrees)

was applied to the moving face. Deflections were measured at the moving face with variations across the face due to strains being insignificant in comparison to the macroscopic displacements of the elements.

The closeness of the finite element model to the stresses and deformations of a real flexure is always difficult to assess. In practice, it is more general to consider accuracy as the extent to which the model corresponds to an analysis with an infinite number of elements or the 'exact' solution based on continuum models incorporating the same boundary conditions. For all of the geometries analyzed, the stress was concentrated predominantly in the inside region of the hinge at the end suppports. As a consequence, finite element meshes were automatically generated with the ends of the beams partitioned in these regions to contain a fine mesh having at least four elements in the radial and thickness directions while the central region of the hinge was reduced to fewer elements. In general, for such meshes, results appeared to converge to within a few percent for meshes having more than a thousand elements. For some geometries,

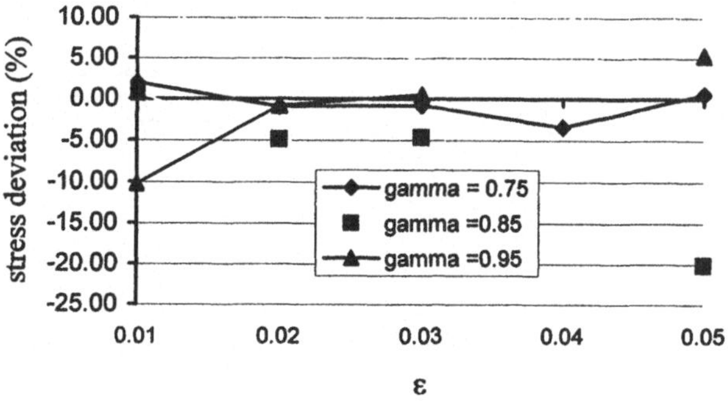

Figure 6.19: Stress deviation between simple formula and results from finite element analysis ($\theta = 75$ degrees)

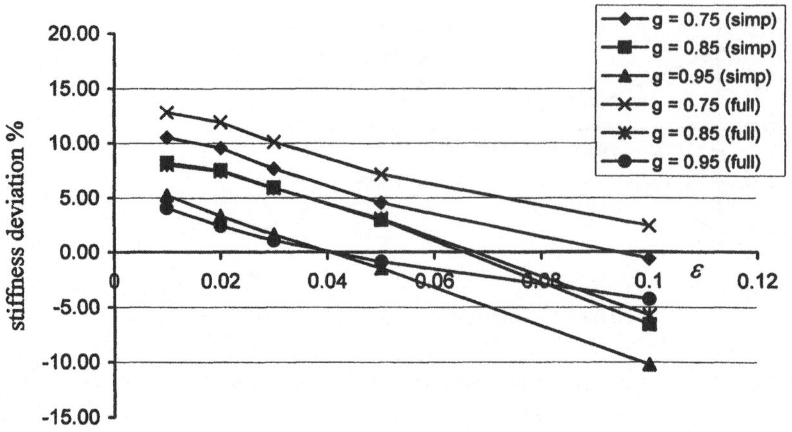

Figure 6.20: Stiffness deviation between simple and full analytic models and results from finite element analysis ($\theta = 45$ degrees)

pathalogical results appeared in which case the number of elements was increased significantly. In all cases the deviations between model and finite element analysis are derived from the equation

$$\text{Fractional deviation} = \left(1 - \frac{\text{results from finite element analysis}}{\text{values calculated from formulae}}\right) \tag{6.69}$$

Percentage deviations are simply obtained by multiplying equation (6.69) by 100. Based on the above equation, negative errors correspond to the values based on finite elements being higher than those predicted from the theoretical models. Assuming that the finite element models correpsond to the 'true' value

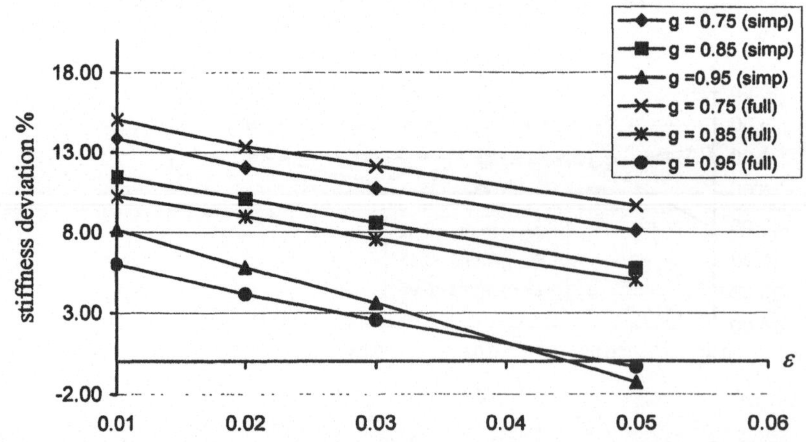

Figure 6.21: Stiffness deviation between simple and full analytic models and results from finite element analysis ($\theta = 60$ degrees)

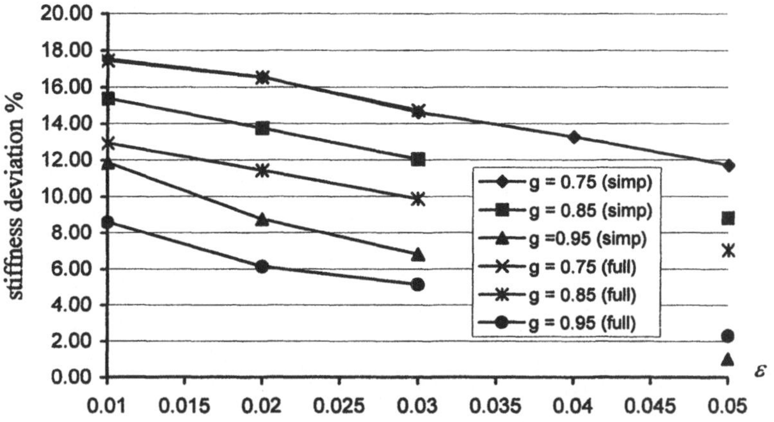

Figure 6.22: Stiffness deviation between simple and full analytic models and results from finite element analysis ($\theta = 75$ degrees)

then a negative error would correspond to an 'unsafe' design. However, the 'true' values for such a hinge system are not known. Consequently, the deviations of this analysis represent the magnitude of the safety margin that should be incorporated into such a design. Results of this analysis are plotted in figures 6.17 to 6.22. Figures 6.17 - to 6.19 show the stress deviations between those predicted from the simple formula (6.68) and those computed using finite element analysis for include hinge angles, θ, of 45, 60 and 75 degrees respectively. Figures 6.20 to 6.22 show the corresponding deviation between stiffness predicted from simple and full theoretical models and finite element results. It must also be emphasized that the equations should not be used for geometries outside of those assessed here without further verification.

Approximate equations for the angular stiffness of this flexure can be obtained from the preceding section for the limited case of a individual segments of the flexure comprising arcuate sectors with a total angle of 90 degrees.

6.5 The bellows as a flexure element

Few can have gone through life without encountering the bellows being used for a flexible coupling. Sometimes called gators, examples include; the seal connecting a manual shift gearbox to the gear-stick, the grease retaining seal across constant velocity couplings in most vehicles and often being used by blacksmiths in days of yore (and, as a consequence, becoming a regular backdrop in movies set in medieval times). A little closer observation and these can be seen on many pipelines to accommodate thermal and seismic motion and also in instrumentation for transmission of rotation from motor drives to micrometer spindles, etc. In many cases, when used to accommodate motion of pipelines under high pressure or carrying hazardous materials, their reliable function

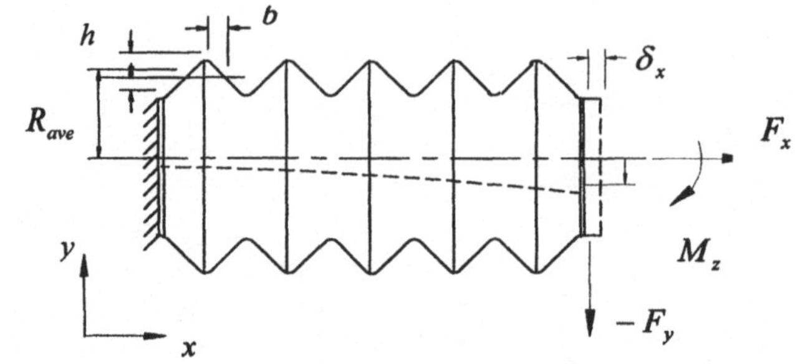

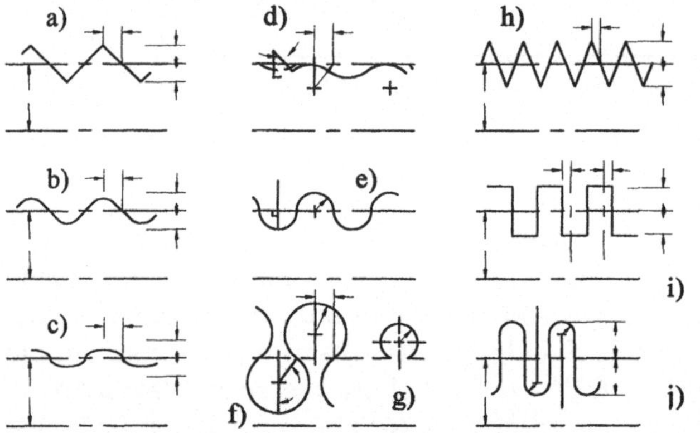

Figure 6.23: Ten different bellows configurations indicating mathematical modeling parameters, a) triangular, b) parabolic or sinusoidal, c) elliptical, d) 's'-shaped, e) semi-circular, f) 's'-shaped, g) single omega, h) crimped plates, I) rectangular, j) 'u'-shaped

becomes a safety issue. Consequently, there is extensive literature on the bellows including the effects of pressure and fatigue life estimation, Hamada and Tsuda, 1997. Recent reviews by Wilson, 1984, and Becht, 1986, contain a wealth of information plus extensive reviews of the literature on this subject. In practice, the instrument designer is likely to choose a particular bellows from those commercially available. In this section, only the important conclusions relevant to flexures will be discussed. Consequently, this section should not be taken as a reference for bellows design as much as a guide towards important considerations in the choice of bellows for integration into a particular design.

Like the simple coil spring, the bellows has a relatively high compliance in all freedoms except for torsion about its longitudinal or, *z*-axis. It is possible to produce a high ratio of torsion to the other five stiffness values with a bellows design and there are a large number of different bellows geometry's, each possessing intrinsically different merits and limitations. Figure 6.23 taken from Wilson, 1984, shows ten different cross sections commonly in use. The goal of the bellows is to provide a compliant link between two rigid bodies and, therefore, it is only the relative displacements and rotations between each end that is of importance. Consequently, this coupling can be modeled as being fixed at one end (i.e. a cantilever) after which the displacement of the free end represent the relative motion between the two rigid bodies that it connects. One other form of distortion is that due to misalignment between the two shaft axes. This will cause the bellows to undergo an 's' shaped distortion as shown in figure 6.24.

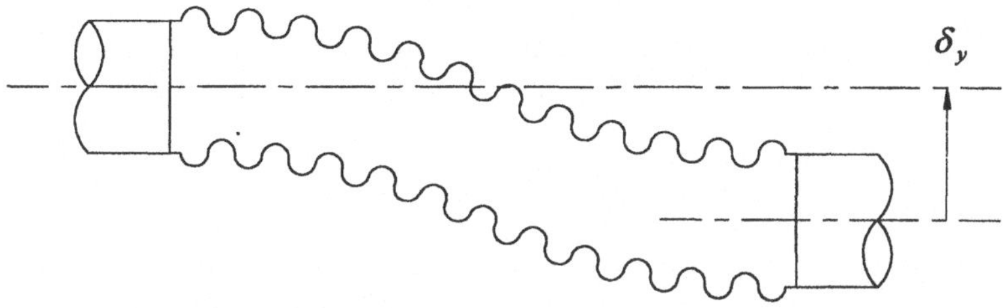

Figure 6.24: 'S' shaped distortion of a bellows coupling due to parallel, mis-aligned shafts

Not surprisingly, because any particular bellows consists of repeated cycles of an axial meander shape (sometimes called a convolution), all can be mathematically modeled in terms of the same dimensionless groups. Therefore, conclusions from one analysis are generally applicable across the range of designs. The 'U' shaped bellows is most commonly used in instrumentation[1]. Unfortunately, analysis of this geometry is complex and has, so far, only been carried out using finite element methods, Hamada *et al.*, 1976. Comparison between these finite element results and the experimental data of Berliner and Vikhman, 1976, deviated by a factor of almost two. Deviations of better than ±15 % were observed between these experimental values and those based on the theoretical model of Haringx, 1952. It is the object of this section to provide the reader with a 'feel' for the relatively compliance's of bellows type flexures. Consequently, because of its relative simplicity, analysis of the rectangular bellows will be chosen for illustration purposes. Although commonly used for connecting sealed piping, it will be assumed throughout the following discussions that the bellows are not pressurized (for those interested in this issue, there exists substantial literature, some of which is referenced in the works cited here). Most of the following analysis has been extracted from the review paper of James Wilson, 1984. For the simple cantilever mode displacements, Donnell, 1932 showed that, by substituting an equivalent elastic modulus E' for E, simple beam bending formulae could be applied. For a bellows of N convolutions, average radius R_{ave}, the length L and mean cross sectional area A are given by

[1] These are commercially available in a variety of forms from Servometer® Corporation, Cedar Grove, NJ 07009-1291. This company also produces an excellent design guide.

$$L = 4Nb$$
$$A = 2\pi R_{ave}t \tag{6.70}$$

Based on these equations, the axial stiffness of the bellows is given by

$$K_{\delta_x F_x} = \frac{E' A}{L} = \frac{E' 2\pi R_{ave} t}{4Nb} \tag{6.71}$$

Similarly, the angular stiffness can be obtained from the equation

$$\frac{M_y}{\theta_y} = \frac{M_z}{\theta_z} = K_{\theta_y M_y} = \frac{E' I}{L} = \frac{E' \pi R_{ave}^3 t}{4Nb} \tag{6.72}$$

Consequently, for any given design it is only necessary to determine the parameter E' that corresponds to a particular compliance. Fortunately, for predicting the torsional stiffness of a bellows, reasonable approximations can be derived directly from the torsion equation, see chapter 3. Some manufacturers refer to this as 'wind up' and has the units of angular rotation per unit torque (e.g. rad N m^{-1} or arc sec per inch oz, etc.).

6.5.1 Torsional stiffness of the rectangular bellows

The rectangular bellows may be considered to consist of pairs of cylinders of length $2b$ connected by annular disks, or washers. Assuming that the connecting disk has a stiffness considerably higher than that of the cylinders, the total compliance of the bellows can be calculated for the sum of compliance's over all cylindrical elements, N. Based on these assumptions the equation for the bellows compliance is

$$\frac{1}{K_{\theta_x M_x}} = \frac{1}{K_i} + \frac{1}{K_o} \tag{6.73}$$

where $K_{\theta_x M_x}$ is the stiffness of the bellows while the subscripts i and o represent the inner and outer cylinders respectively. Although simple to include, the compliance of the connecting disk is ignored.

Treating the bellows as a thin shell of thickness t, a reasonable assumption, equation 6.71 can be expressed in the form

$$\frac{1}{K_{\theta_x M_x}} = \frac{2b}{\pi G t}\left[\frac{1}{R_i^3} + \frac{1}{R_o^3}\right] \tag{6.73}$$

Defining the dimensionless parameter

$$\eta = \frac{R_i}{R_o} \tag{6.75}$$

equation (6.73) can be written in the equivalent form

$$K_{\theta_x M_x} = \frac{\pi G t}{2b}\left[\frac{R_o^3 R_i^3}{R_o^3 + R_i^3}\right] = \frac{\pi G t R_o^3}{2b}\left[\frac{\eta^3}{1+\eta^3}\right] \tag{6.76}$$

Theoretically values of η can range from 0 to 1. In practice, this is rarely less than 0.5 and typically greater than 0.6 for applications in which the torsion stiffness is important. It is interesting to contrast this calculation with one in which the mean radius R_{ave} is used and the bellows considered as a simple thin tube. Based on this assumption the torsion stiffness can be written immediately as

$$\widetilde{K}_{\theta_x M_x} = \frac{G\pi t R_{ave}^3 t}{4b} \tag{6.77}$$

The average radius is given by

$$R_{ave} = \frac{R_o + R_i}{2} \tag{6.78}$$

from which

$$\begin{aligned} R_{ave}^3 &= \frac{1}{8}\left[R_o^3 + 3R_o^2 R_i + 3R_o^2 R_i^2 + R_i^3\right] \\ &= \frac{R_o^3}{8}\left[1 + 3\eta + 3\eta^2 + \eta^3\right] \end{aligned} \tag{6.79}$$

Substituting equation (6.79) into (6.77) gives

$$\widetilde{K}_{\theta_x M_x} = \frac{G\pi R_o^3 t}{32b}\left[1 + 3\eta + 3\eta^2 + \eta^3\right] \tag{6.80}$$

The difference between these two equations can be seen from the ratio of the stiffness values

$$\frac{K_{\theta_x M_x}}{\widetilde{K}_{\theta_x M_x}} = \frac{16\eta^3}{\left(1+\eta^3\right)\left(1+\eta\right)^3} \tag{6.81}$$

Figure 6.25 shows this ratio as η is varied from 0.5 to 1. It can be seen that for $\eta > 0.72$ the difference is always less than 15%. Typically, a bellows will contain

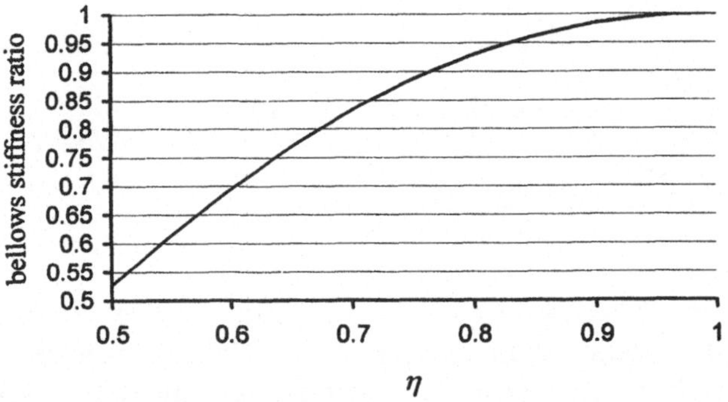

Figure 6.25: Torsional stiffness ratio of equation (6.81)

five or more convolutions. However, it is not necessary to know the number of convolutions to predict the torsional stiffness. This is apparent by assuming that instead of the meander cross-section, all of the inner radii cylinders and all of the outer ones are joined together. The torsional stiffness will remain unchanged while the cross-section will resemble that of the single convolution already analyzed. Consequently, it is possible to rewrite the formula for the single convolution in terms of the total bellows length, L (= $4b$), and this should be valid for any number of convolutions (remembering, of course, that the shell thickness of the bellows has been ignored). Based on this assumption, the stiffness of a rectangular bellows based on the average radius is given by

$$K_{\theta_x M_x} \approx \frac{G\pi t R_o^3 t}{8L}\left[1+3\eta+3\eta^2+\eta^3\right] \tag{6.82}$$

Of practical interest is a comparison of the stiffness of a typical bellows over the range of geometry commercially available. Figure 6.26 shows a comparison between theoretical predictions based on the average radius of a rectangular flexure, equation (6.82), and 16 different flexure geometry's for nickel based 'U' shaped bellows commercially available from the Servometer® Corporation. Throughout the range of available geometry the deviations between these two values are always less than 10%. This illustrates a more generally accepted assumption that, *for most types of bellows*, the torsional stiffness may reasonably be calculated using the torsion equation for thin shells based on the average radius of the bellows. It can be seen that there is a deviation of much less than ±10% over the complete geometric range, even though the equation is based on the assumption of a rectangular geometry!

6.5.2 Axial stiffness of the rectangular bellows

An expression for the axial stiffness of a rectangular bellows was first demonstrated by Haringx, 1952. A later finite element analysis by Hamada et al.,

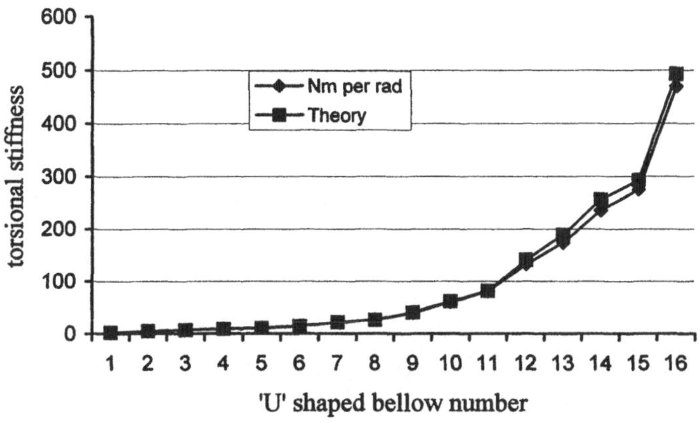

Figure 6.26: Comparison between predicted torsional stiffness of a 'U' shaped bellows, equation (6.82), and values obtained from the data sheets for an 'FC' series bellows manufactured by Servometer Inc.

1976, proved to be at variance with experimental results. However, based on its relative success, Haringx' theory will be presented in this section. Before presenting these results, it is worth mentioning that the relevant parameters for the rectangular bellows can be expressed in terms of the dimensionless groups

$$h/R_{ave} \qquad b/R_{ave} \qquad t/R_{ave} \tag{6.83}$$

where $h = (R_o - R_i)/2$ and represents the distance from the average to either the inside or outside radius. Another important dimensionless group is given by

$$\lambda = \frac{R_{ave} t}{b^2} \tag{6.84}$$

Consistent with these thin shell groups, the axial stiffness a single convolution of the rectangular bellows can be determined from equation (6.71) with E' given by

$$E' = \chi \frac{Et4\lambda (b/R_{ave})^3 (1-\eta^2)}{3R_{ave}(1+h/R_{ave})^2(1-\nu^2)\left[(1-\eta^2)^2 - (2\eta \ln \eta)^2\right]} \tag{6.85}$$

Unfortunately, this introduces a further two dimensionless groups χ and η given by

$$\eta = \frac{1 - h/R_{ave}}{1 + h/R_{ave}}$$
$$\chi = 1 + \frac{2\left(b/R_{ave}\right)}{1 + h/R_{ave}}\left(A/C + B/C\right) \tag{6.86}$$

It can be readily verified that the first of these is identical to the dimensionless parameter, equation (6.75), used in the previous analysis. The coefficients for the second of (6.86) can be obtained from

$$A = \left(1 - \eta^2 + 2\eta^2 \ln \eta\right)^2 \lambda_u$$
$$B = \eta\left(1 - \eta^2 + 2\ln \eta\right)^2 \lambda_i$$
$$C = \left(1 - \eta^2\right)\left[\left(1 - \eta^2\right)^2 - \left(2\eta \ln \eta\right)^2\right]$$
$$\lambda_u = \frac{\sinh 2Q_u + \sin 2Q_u}{2Q_u\left(\cosh 2Q_u + \cos 2Q_u\right)}$$
$$\lambda_i = \frac{\sinh 2Q_i + \sin 2Q_i}{2Q_i\left(\cosh 2Q_i + \cos 2Q_i\right)}$$
$$Q_u = \frac{\left[3\left(1 - \nu^2\right)\right]^{1/4}}{\left[\lambda\left(1 + h/R_{ave}\right)\right]^{1/2}}$$
$$Q_i = \frac{\left[3\left(1 - \nu^2\right)\right]^{1/4}}{\left[\lambda\left(1 - h/R_{ave}\right)\right]^{1/2}} \tag{6.87}$$

All but the those of an unusual mathematical inclination can derive any insight from such a convoluted series of equations. It is possible to get a 'feel' for the magnitude of this by calculating the ratio E'/E for a range of bellows sizes. For the bellows geometry's of the previous analysis the values of this parameter are shown in figure 6.27. It can be seen from this figure that the modified modulus is less than 0.1 % of the value based on a solid tubular beam model. This can be compared directly with the torsion equation (6.76) and interpreted as a decoupling between the torsional stiffness along the axis and rotational stiffness perpendicular. Alternatively, it can be said that torsion forces for a given relative angular motion between the two ends of the bellows are three orders of

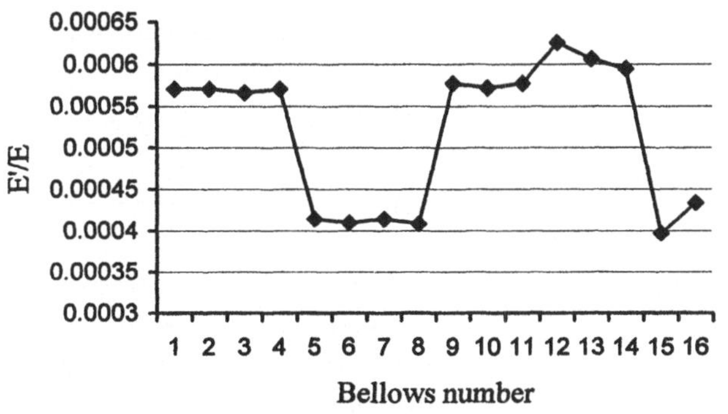

Figure 6.27: Elastic modulus scaling factor computed from equations

magnitude or more greater in the axial than the tangential direction for a typical bellows design.

In the axial direction, a direct comparison between linear and torsional compliances is difficult. However, the reduction in axial stiffness is again of the order of 1,000 or more than would be the case for a solid tube.

6.5.3 'S' Shaped distortion of the bellows

As mentioned in the introduction to this section, misalignments of the shafts at which the two ends of the bellows are connected will result in a 'S' shaped distortion. Superposition can be used and this configuration can be considered as the addition of two cantilevers of half the total length of the bellows. Assuming the original displacement at the ends of a cantilevered bellows of length L_s is given by

$$\delta_z = F\frac{L_s^3}{3E'I} \tag{6.88}$$

It is relatively straightforward to show that the stiffness of the bellows can be obtained from

$$K_{F_z\delta_z} = \frac{12E'I}{L^3} = \frac{12E'\pi t}{(4N)^3}\left(\frac{R_{ave}}{b}\right)^3 = \frac{3E'\pi t}{16N^3}\left(\frac{R_{ave}}{b}\right)^3 \tag{6.89}$$

Typically, this is considerably higher than the axial stiffness of the flexure. However, the small value of the modulus ratio in conjunction with the inverse cube power law in N generally dominates this expression to provide a relatively high compliance.

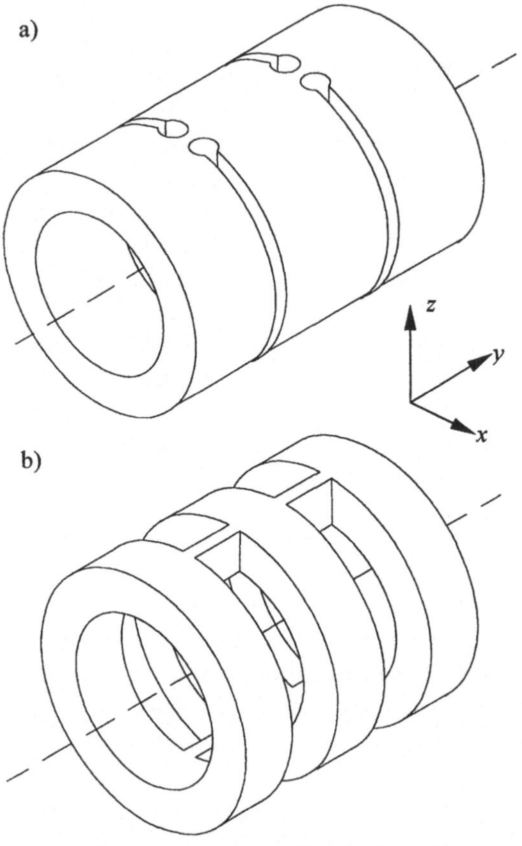

Figure 6.28: Two freedom cylindrical couplings, a) notch hinge coupling with constraint in y, z, θ_x and θ_y b) similar coupling utilizing leaf-type hinges

6.6 The notch and leaf type hinge applied to couplings of rotational symmetry

In some applications, it is desired to connect two shafts using a coupling of relatively high torsional and axial stiffness. Freedoms to accommodate angular and linear misalignments in a plane normal to the shaft axes can be provided by separating two ends of a cylinder using notch or leaf-type hinges. Typical examples of this are shown in figures 6.28(a and b). Each shaft connects to either end of the cylinder that constitutes the coupling. Both couplings in this figure are two freedom mechanisms. The first of these has been produced by drilling two, closely spaced pairs of holes through the cylinder in a direction perpendicular to its axis. Slots are then removed from the cylinder to leave notch hinges that provide, in this case, an angular freedom about the z-axis. Each hinge consists of two notch hinges either side of the cylinder with their axes being collinear. Consequently, these two hinges are considered to behave as a single hinge. For the coupling of figure 6.28(a), there are two such hinges each providing a single angular freedom. Consequently, this can be modeled as three rigid bodies series connected by single freedom joints. If one end of the cylinder is considered to be the fixed link, it is clear from the figure that the other end would be free to rotate about the z-axis as well as translating in x with the intermediate cylinder acting as a rigid link. The second coupling of this figure shows a mechanism of similar function. However, this coupling utilizes leaf type hinges instead of notches. Manufacture of such a coupling could proceed starting from a solid cylinder and cutting wide slots perpendicular to its axis. In many ways, this is a variant of the rotationally symmetric hinge discussed in section 6.4. In this case, however, the requisite compliance has been produced by letting the circumferential leaf springs become thick enough to be considered rigid while the material connecting the solid bodies has been thinned and lengthened to form a leaf spring the axis of which is parallel to that of the cylinder. Equations for the prediction of hinge compliances and stresses can be readily derived from those for the notch and leaf type hinges presented in chapter 4.

Couplings of other freedoms are readily produced by adding flexure elements or changing their orientation about the axis of the cylinder. For example, a coupling providing angular freedoms in θ_x and θ_z can be produced by arranging for the axes of the two flexure elements to be perpendicular to each other so that the hinge axes are parallel to the x and z axes respectively. In fact, such a coupling would result if a large hole were machined along the vertical axis of the two axis hinge shown in the center of the photograph of figure 4.42. The number of freedoms can be extended by adding hinges disposed at 90 degrees about and at intervals along the cylinder axis. Examples of such mechanisms can be found in the patents of Linley, 1976, and Agius, 1994.

6.7 Case study: A metrological three-axis translator for constant force profilometry

This case study has been extracted from the paper of Badami et al., 1996, and is chosen because it illustrates a variety of flexure applications as well as a simple drive coupling, the latter being a topic discussed in chapter 8.

Stylus and constant force transducers form the basis of a broad range of instruments for metrological characterization of fine surface features, see for example Thomas, 1982, Chetwynd and Smith, 1991, Sarid, 1994. Typically, stylus based instruments consist of a long range traverse mechanism for translation of the specimen relative to the probe, with the probe system mounted with its axis of measurement perpendicular to the traverse direction. In practice, it is the Abbé errors associated with the misalignment between the axis of the probe that interrogates the surface and the measurement axis of the transducer used to sense the displacement of the probe that dominate the achievable precision in profile height measurement. This case study presents a force probe system developed to measure fine optical surfaces with an accuracy of 1 nm and a resolution of 0.1 nm for vertical surface profile deviations of up to 3 mm. The heart of this profiler consists of a constant force probe similar to that developed by Smith and Howard, 1994, combined with a laser interferometer. The force probe consists of a simple cantilever flexure with a stylus attached at the free end. Forces are monitored by measuring deflection of the cantilever using capacitance gaging. Since the Abbe error is due to both the offset and the angle of tilt, control of either parameter can be used to minimize this error. The precision of alignment between the axis of the laser interferometer and the stylus probe is limited by manufacturing considerations. By controlling the angle of tilt, it is possible to minimize this error. To this end a novel, three-axis flexure arrangement has been designed and evaluated for active control of two rotations

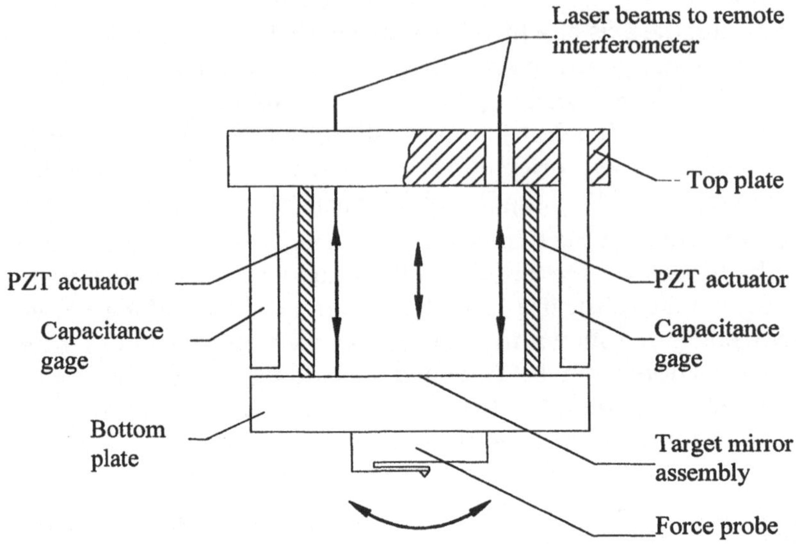

Figure 6.29: Relative dispositions of capacitance gages, interferometer and piezo-electric (PZT) actuators

over a range of 30 μrad while allowing a displacement of approximately 1.5μm. The tilt control mechanism may also be exploited to obtain an estimate of the Abbe offset by monitoring the displacements as measured by the laser and the probe while rocking the probe head through a known angle.

6.7.1 Principle of Operation

A schematic diagram indicating the relationship between the various parts of the device is shown in figure 6.29. The force probe is mounted to the target mirror assembly, which, in turn serves as a plane mirror for a remote interferometer. The mirror assembly also functions as a target for capacitance gages. The PZT actuators can be controlled individually in combination to provide tilt and translation of the force probe relative to the top plate. Since tilt in two axes is required, a minimum of three actuators and three capacitance gages are required.

The force probe assembly used to interrogate the surface is shown in figure 6.30. It consists of a stylus probe mounted on a glass cantilever. Electrodes deposited on the back of the cantilever and the top of the base form the plates of a capacitor. Deflections of the cantilever change the spacing between these plates, changing the capacitance, thus enabling the deflections of the cantilever to be monitored. This design is based on the work of Howard and Smith and is capable of sub-nanometer resolution over a 1 μm range.

In addition to providing tilt cancellation, the probe head also serves as a high-bandwidth actuator that is used to provide constant force stylus measurement. This is achieved by servo controlling the extension of the probe

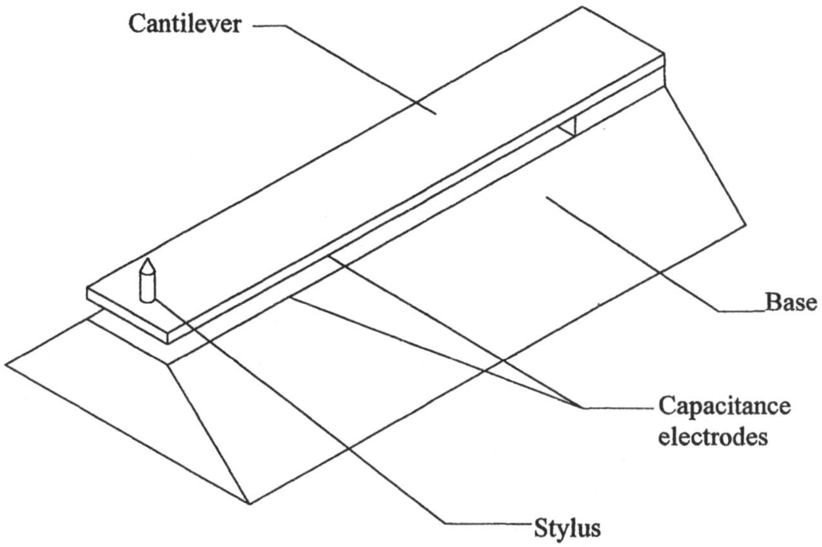

Figure 6.30: Schematic diagram of force probe

head to maintain constant cantilever deflection. The probe head in turn is to be mounted to a low-bandwidth, long-range vertical slide that provides the travel required by the large vertical profile deviations envisaged for its particular application. For absolute measurements, the position and tilt of the force probe must be measured relative to a metrology frame by interferometric methods. For this purpose, the rear surface of the probe is an optically flat reflector.

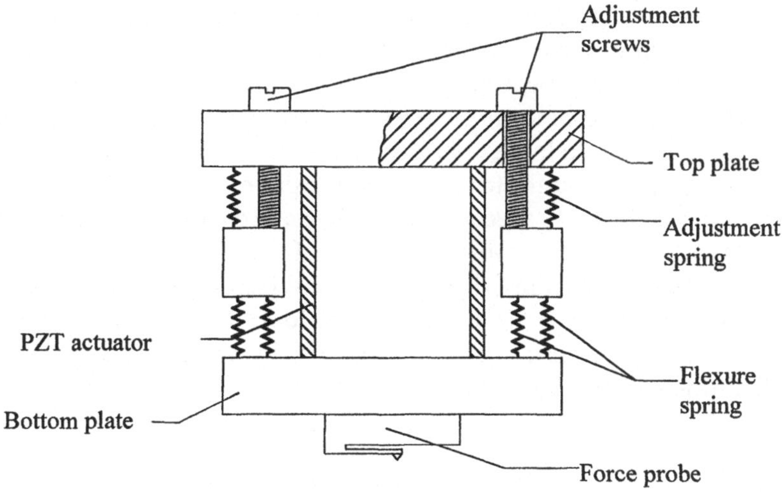

Figure 6.31: Schematic diagram of flexure arrangement

The force probe assembly is mounted to the probe head that consists of the actuators, flexures and capacitance displacement sensors. The flexure consists of two sets of three flexure elements (represented schematically in Figure 6.31 by the springs). The lower set of flexures provides the compliance for the PZT actuators to enable the required displacement and tilt. The upper set is used to adjust the pre-load on the actuators during assembly and allow for slight dimensional mismatch between the length of the flexure and that of the actuators. The adjustment screws control the deflections of the more compliant upper flexures that in turn control the pre-load on the actuators due to the deflection of the lower flexures. Once the pre-load is set, the controlling stiffness is that of the lower flexures. The deflections of the individual PZT actuators are monitored by capacitance gages, which can be used in combination to measure both displacement and tilt of the probe. Although the position of the force probe is monitored by a remote interferometer, the capacitance gages provide a more tightly coupled metrology loop that enhances the controllability of the entire system. In addition, the present design allows for the operation of the device as a stand-alone device to facilitate bench testing.

Figure 6.32 shows an exploded view of the complete assembly. The two sets of rotationally symmetric flexures (section 6.4) are part of a monolithic structure made out of aluminum, which also includes slots for the actuators. The flexures are equi-spaced around the circumference of a circle giving the monolithic flexure three-fold symmetry. Three PZT actuator segments are mounted on the top plate and pass through slots in the flexure. The actuators are made from a single piece of ceramic and exploit the Poisson's ratio effect to produce the desired fine motion. These actuators bear on the bottom plate via spherical contact surfaces that allow for the angular tilt. Capacitance gages mounted in the top plate monitor the displacement of each PZT with the bottom plate forming the target for the capacitance gages. The force probe is mounted to the bottom of the target mirror, which in turn is clamped to the lower surface of the bottom cap using a semi-kinematic clamping arrangement (not shown). Changes in length of the PZT actuators cause the bottom plate to tilt and translate relative to the top plate, the latter of which has a large aperture to allow the laser beams of a remote interferometer to pass through to the target mirror. While the aluminum structure of the probe head is susceptible to thermal changes, the force probe assembly and target mirror are constructed out of Zerodur™ and are directly referenced to the metrology frame through the interferometer, thus minimizing the thermal sensitivity of the main measurement loop. Monolithic construction has been employed to minimize joint losses and hysteresis and to obtain a compact, high stiffness structure.

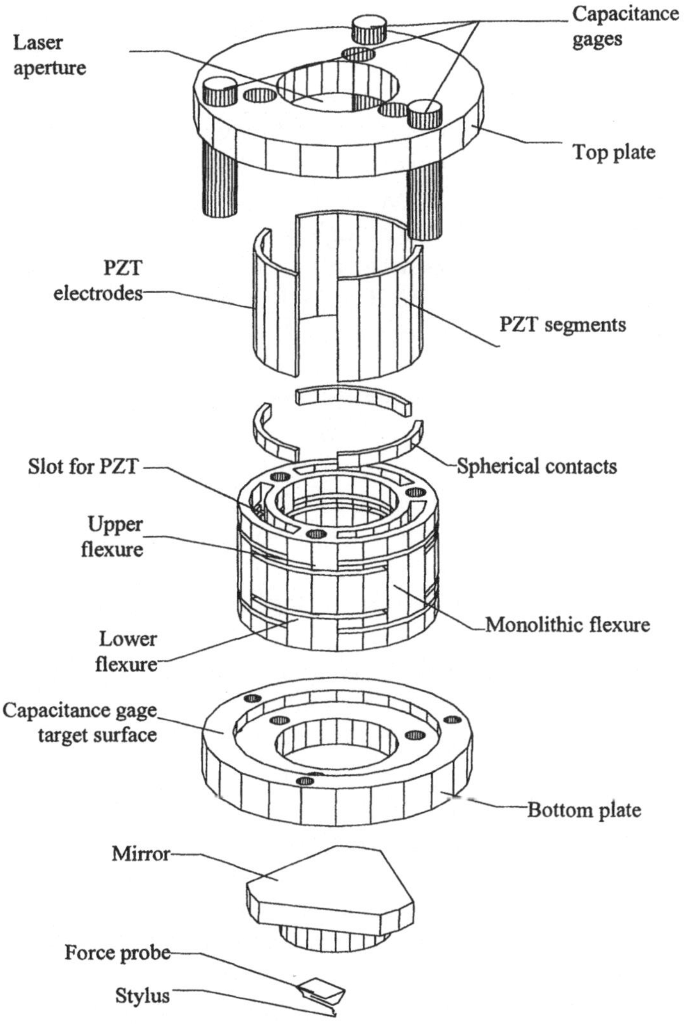

Figure 6.32: Exploded view of complete assembly

References

Aguis J.G., 1994, Constant velocity flexible coupling and connection with the drive and driven members, *USA patent US5299980*.

Badami V.G., Smith S.T. and Patterson S.R., 1996, A metrological three-axis translator and its application for constant force profilometry, *Proc. ASPE*, **14**, 391-395

Berliner Y.I. and Vikhman Y.L., 1976, The axial rigidity of of flexible elements of compensators with U- and Ω-shaped profiles of waves, *Chemistry and Petroleum Eng.*, **12**, 31.

Brecht IV C., 1986, Predicting bellows response by numerical and theoretical methods, *Trans. ASME: J. of Pressure Vessel Technology*, **108**, 334-341

Cherukuri H., 1997, personal communication.

Chetwynd, D. G., and Smith, S. T., 1991, High precision surface profilometry: From stylus to STM, in *From Instruments to Nanotechnology*, ed. Gardner, J. W., and Hingle, H. T., Gordon and Breach, London, 273-300.

Donnell L.H., 1932, The flexibility of corrugated pipes under longitudinal forces and bending, *ASME Trans. J. Appl. Mech.*, **APM-54-7**, 69

Hamada M. et al., 1976, Design diagrams and formulae for U-shaped bellows, *Int. J. of Pressure Vessel Technology*, **4**, 315

Hamada M. and Tsuda T., 1997, On design formulas of U-shaped bellows, *Trans. ASME: J. of Pressure Vessel Technology*, **119**, 127-131

Haringx J.A., 1952, Instability of bellows subjected to internal pressure, *Phillips Res. Rep.*, **7**, 189-196.

Howard, L. P., and Smith, S. T., 1994, A metrological constant force profiler, *Rev. Sci. Instrum.*, **65**(4), 892-902.

Neale M., Needham P. and Horrell R., 1991, *Couplings and Shaft Alignment*, Mechanical Engineering Publications, London

Juvinall R.C. and Marshek K.M., 1991, Fundamentals of Machine Component Design, 2nd ed., John Wiley and Sons Inc., NY.

Levy R., 1962, Displacements of circular rings with normal loads, *Proc. ASCE, J. Struct. Div.*, **88**(ST1), 23-54. Equation (1) in this paper has been erroneously typeset.

Linley F.M., 1976, Anti-backlash, self-aligning nut, *USA patent US3977269*.

Pilkey W.D., 1994, *Formulas for Stress, Strain and Structural Matrices*, John Wiley and Sons, Inc, NY.

Sarid, D., 1994, *Scanning force microscopy: with applications to electric, magnetic and atomic forces*, Oxford University Press, NY.

Seely F. B. and Smith J. O., 1932, *Advanced Mechanics of Materials*, J. Wiley and Sons, NY chpt. 15.

Shigley J.E., 1989, *Mechanical Engineering Design*, McGraw-Hill, NY.

Smith, S. T., and Howard, L. P., 1994, A precision, low force balance and its application to atomic force probe characterization, *Rev. Sci. Instrum.*, **65**(4), 903-903

Timoshenko S.P. and Woinowski-Kreiger S., 1959, *Theory of Plates and Shells*, McGraw-Hill Book Company, NY (NB Symbols for geometry factors of equations (6.17) and (6.18) correspond to the form given on page 289)

Timoshenko S.P., 1941, *Strength of Materials: Pt. II Advanced Theory and Problems*, 2nd ed., D. Van Nostrand, Inc., NY, chapter II.

Thomas, T. R., 1982, *Rough Surfaces*, Longman Press, London.

Wahl A.M., 1963, *Mechanical Springs*, 2nd ed., McGraw-Hill, NY.

Wilson J.F., 1984, Mechanics of bellows: a critical survey, *Int. J. Mech. Sci.*, **26**(11/12), 593-605

Woody S.C. and Smith S.T., 1999, Design and assessment of disk couplings, to be published.

Wolff P.H.W., 1951, The design of flexible disk misalignment couplings, *Proc. Inst. Mech. Engrs.*, **169**, 165-175

Young W.C., 1989, *Roark's Formulas for Stress and Strain*, McGraw-Hill Book Co., NY.

7 Levers

7.0 Overview

For a broad range of flexure applications, a displacement actuator is necessary to provide the drive forces. For a number of practical reasons, the most common drives are the feedscrew and piezoelectric actuators. Because these drives provide either long motion with limited resolution or short ranges with continuous control, it is often necessary to introduce some form of lever mechanism to amplify or attenuate the drive action. For example, displacements produced using micrometers often require attenuation while piezoelectric drives may have insufficient range thereby requiring motion amplification. Although the concept of a lever is simple, its practical implementation involves a number of compromises. Two types of lever are identified, these being the mechanical lever and soft-spring/stiff-spring attenuation. Following a discussion on the types of lever used in flexures, this chapter assesses the consequences and some issues. In particular, a method for analyzing the effects of 'lost motion', and influence on the dynamic performance are considered. A detailed analysis of an attenuation mechanism is presented to illustrate some important considerations in any flexure design, in this case, with emphasis on lost motion and associated modeling errors. Manufacture of the flexure in the case study also provides information relevant to discussions presented in chapter 8.

7.1 Introduction

Many flexure designs necessarily include some form of drive mechanism. Of all type available, these can be split into two categories these being contact or non-contact. Non-contact drives are electromagnetic, electrostatic or hydrostatic. Contacting drives all consist of actuators designed to impose a force or strain in a solid body that is some way connected between two points in the flexure. Contacting drives can be motor driven micrometers, friction driven pushers, piezoelectric, magneto-strictive, electro-strictive, hydraulic piston, pressurized diaphragms, Poisson's ratio, shape memory, bimetallic and thermal expansion (for a rather fun version of this latter drive see Snyder, 1993). Some of the important attributes of these mechanisms are discussed in chapter 8 while the interested reader is referred to Smith and Chetwynd, 1991 for a more detailed discussion. Non-contact drives impose a force on the flexure without significantly effecting stiffness or adding mass. Contact drives however can be considered to add, often, significant stiffness at their point of application. In general, the added mass of the actuator can be lumped into the 'stationary' link. For example, in the case of a micrometer the stationary link is the one to which the micrometer is rigidly secured while the moving link is attached to the anvil of the micrometer via some form of coupling. Consequently, the only additional mass on the moving link is that due only to the rigidly attached component of the coupling.

As mentioned in the summary, in many cases, either the range or resolution of a particular drive may be inadequate for a given application. Under such conditions, it is often necessary to introduce a lever mechanism either for the amplification or attenuation of the motion from the actuator to the flexure.

7.2 Mechanical levers

In its simplest form, the mechanical lever consists of a rigid body connecting the moving, or 'output', body to some reference, or 'base' link. Applying a displacement in a specific direction on this rigid body results in relative motion between the moving platform and base. Such a rigid lever mechanism is directly analogous to the electrical transformer. An alternative lever (in this case *attenuation*) effect can be achieved by imposing a compliant coupling between the drive and flexure. This can be suitably modeled as actuating a stiff spring

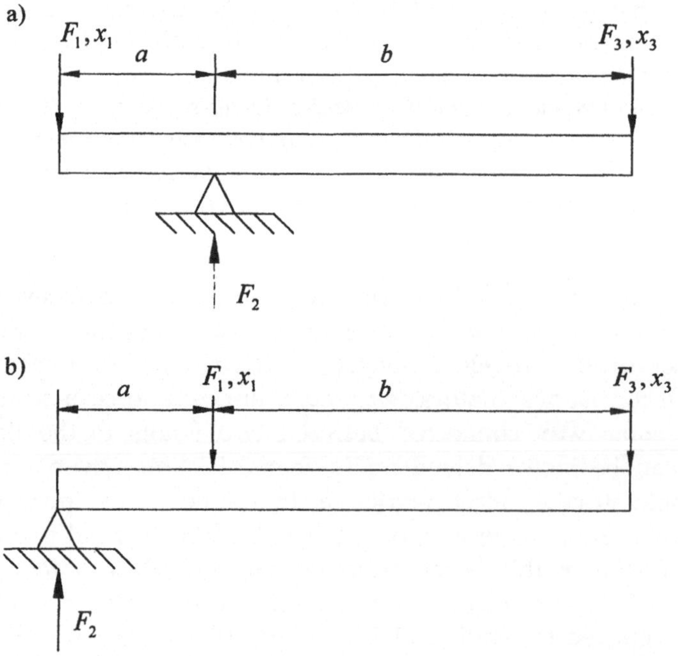

Figure 7.1: The simple, mechanical lever, a) pivot 2 between driver 1 and driven 3, b) pivot outside of driver and driven

through a soft one. This latter mechanism is directly analogous to the potential divider commonly employed in electrical circuits. Clearly, the displacement at the stiff spring will always be less than that of the actuator.

Before moving on, it should be noted that there are a large number of alternative lever mechanisms. In particular, the feedscrew, which can be considered a lever, probably represents one of the most common machine elements. Other lever mechanisms such as friction drives and the wedge may be advantageous for some applications, see for example Smith and Chetwynd, 1991.

7.2.1 The rigid lever

In its simplest form, the lever can be modeled as a rigid rod rotating about a pivot as shown in figure 7.1(a). Assuming that the input is applied at 1, the lever ratio, n, is defined as

$$n = \frac{\text{output displacement}}{\text{input displacement}} = \frac{b}{a} \tag{7.1}$$

Alternatives to this mechanism can be obtained by interchanging the input and output (in which case the lever ratio is inverted) or moving the pivot as shown in figure 7.1(b). For this configuration the lever ratio becomes

$$n = \frac{a+b}{a} \tag{7.2}$$

Fortunately, there is no restriction on the shape of the lever arm and in its more general form, the lever can be represented as shown in figure 7.2(a). Lengths of the lever arms are readily obtained based on the fact that the vector of rotation about the pivot is always at right angles to the applied forces

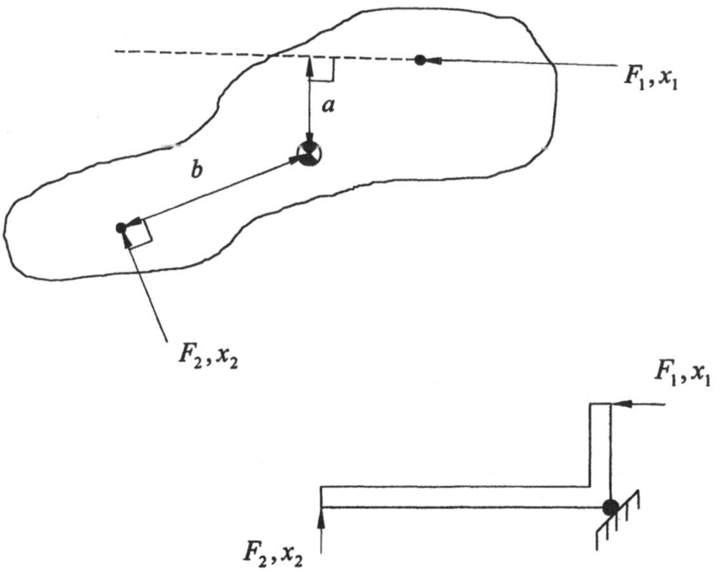

Figure 7.2: Levers of different shape, a) the general lever, b) lever arm with orthogonal axes between driver and driven.

for both input and output. It is often convenient to arrange that the input and output displacements are at right angles which, for small displacements, can be achieved using the link of figure 7.2(b). For a perfect lever, and assuming small displacements, the following may be deduced:

1. The work done at the input will be equal to that at the output
2. Assuming that the links are infinitely rigid, the angular rotation about the pivot of both the input and output links are equal
3. In static equilibrium, the vector of all three forces forms a closed loop.
4. The lever mechanism is reversible. That is, transposing the driver and driven inverts the lever ratio.
5. Motion between any two fixed points in a solid body must be in a direction perpendicular to an axis passing through the two points. This is *theorem I* of chapter 5.

From 7.2(a), it can be seen that, as the force vector is directed towards the pivot point, the lever ratio will increase asymptotically. This can be exploited in flexure mechanisms to achieve large amplifications from small displacements. In many applications, because of their smooth, continuous action, piezo-electric actuators are chosen. Because of their limited displacement range, a large variety of amplifying mechanisms have been implemented, see for example Furukawa *et al.*, 1990, 1991, 1994, Xu and King, 1996. In the following, it will be assumed that piezoelectric actuators requiring amplification are to be employed. Although this helps to illustrate some important design details, such an assumption is not necessary and driver and driven can be reversed in all of the levers to produce a reduction mechanism.

Figure 7.3 shows three piezo-electrically driven amplifying elements. The first of these figures represents what might be considered a conventional lever. In this figure, however, some further design details have been added. The hinges that constitute the coupling and pivot are included. Assuming that the applied force at the output acts in the direction shown it is obvious that the coupling will experience a compressive load while the pivot will be in tension. A notch hinge is chosen because of its relatively high compressive strength and resistance to buckling. For the pivot, a leaf type spring is considered adequate. It has already been shown in chapter 4 that the tensile force will induce an increase in the rotational stiffness of the flexure. From a simple bending moment analysis it can be shown that the forces in the pivot and coupling are

$$\begin{aligned} F_2 &= \frac{b}{a} F_3 = (n-1)F_3 \\ F_1 &= F_2 + F_3 = nF_3 \end{aligned} \tag{7.3}$$

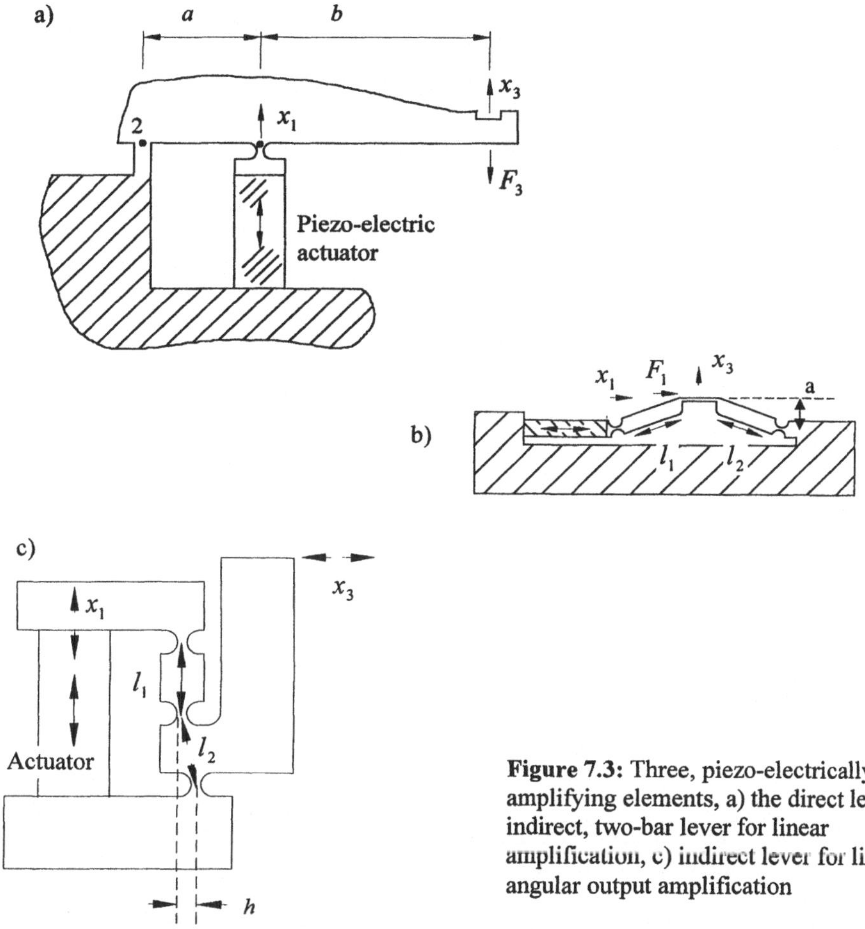

Figure 7.3: Three, piezo-electrically driven amplifying elements, a) the direct lever, b) indirect, two-bar lever for linear amplification, c) indirect lever for linear or angular output amplification

The subscripts 1 and 2 refer to the coupling and pivot respectively while the lever ratio n is given by

$$\frac{a+b}{a} \tag{7.4}$$

It is clear that the forces in the pivot increase in proportion to the lever ratio. Additionally, there will be a corresponding increase in bending moment about the lever arm. Bending of the lever arm will contribute to lost motion in the mechanism.

A more complex **two-arm** lever arrangement is shown in figure 7.3(b). In this, the lever ratio can be envisaged as that due to the rotation of an arm fixed to the pivot due to the forces from another free arm connected at some angle. In the particular example shown, it is obvious that the output displacement is due to the motion of the midpoint necessary as the spacing between opposite ends of the two rods changes. An equivalent mechanism with a piezo-electric drive is shown in figure 7.3(c). This last, and probably most common, mechanism can be modeled as shown in figure 7.4. Again, point 1 represents the input, 2 the pivot and point 3 is the output. In this particular model, the input displacement is parallel to the link connecting the pivot to the output.

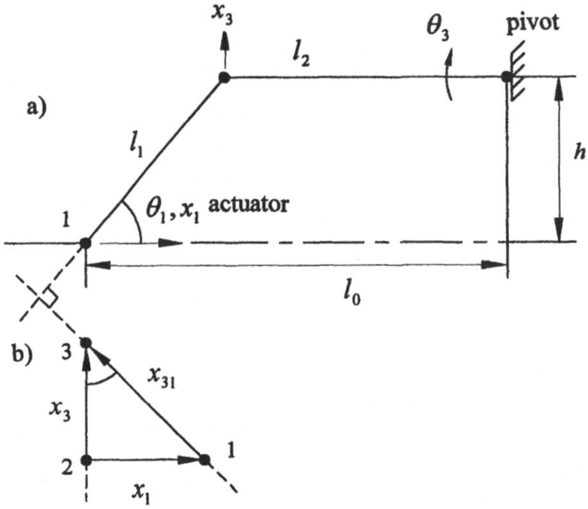

Figure 7.4: Parameters for a mathematical model of the two-arm levers shown in figures 7.3(b & c), a) mathematical model, b) displacement diagram

By far the simplest way to determine a relationship between the input and output is to use a 'displacement' diagram. The interested reader will find an alternative approach adopted by Furukawa and his coworkers. This is analogous to velocity diagrams often used to determine the instantaneous velocity in complex mechanisms. Obviously over short time intervals the instantaneous velocities and displacements will be directly proportional. Consequently, it is possible to construct a displacement diagram in the same way as one would a velocity diagram. The first step in the construction of this diagram is to define a 'stationary' point that coincides with the instantaneous position of the pivot, 2. The motion of the actuator, x_1, relative to the pivot can be immediately plotted. To determine the motion of point 3, the general directions of relative motions can be plotted. Firstly, it is known that the point 3 moves vertically with respect to the pivot. Consequently, a vertical line passing through point 2 is drawn on the diagram. Finally, because only rigid links are being considered, the motion of point 3 relative to point 1 is perpendicular to the link l_1. Consequently, the displacement line passing through point 1 can be drawn as shown. Clearly, because point 3 occupies a unique position in space (i.e. it can only be in one place) this must be located at the point where the two lines intersect. The length of the line connecting points 1 and 3 represents the relative displacement due to rotation of the link l_1. From this displacement diagram, provided that the small displacement assumption is valid, a number of simple input-output relationships may be readily derived.

Firstly, it is necessary to introduce some geometric relationships provided by continuity considerations. These are

$$l_1 \cos\theta_1 = l_0 - l_2 \tag{7.5}$$

$$l_1 \sin\theta_1 = h \tag{7.6}$$

Combining (7.5) and (7.6) to eliminate θ_1 gives

$$l_0 - l_2 = l_1\sqrt{1-\left(\frac{h}{l_1}\right)^2} \tag{7.7}$$

From the displacement diagram, the lever ratio represented by the relationship between input and output displacement is given by

$$\frac{x_3}{x_1} = n = \frac{1}{\tan\theta_1} = \text{ctn}\theta_1 = \frac{l_1}{h}\sqrt{1-\left(\frac{h}{l_1}\right)^2} \tag{7.8}$$

Clearly, for small angles, or as h reduces, the lever ratio can take on very large values. Alternatively, as θ_1 tends towards $\pi/2$ the lever ratio tends towards zero. The change in the ratio as a function of the initial angle can be simply obtained by differentiating (7.8) to give

$$\frac{dn}{d\theta_1} = -\text{cosec}^2\theta_1 = -\left(\frac{l_1}{h}\right)^2 \tag{7.9}$$

This may be considered a measure of the non-linearity of the ratio, and therefore, of the lever. This shows that, as the lever ratio increases linearly with a reduction in h, the non-linearity will increase as the inverse square.

It is interesting to note that this ratio is independent of the length of the link connected to the pivot. However, under some circumstances it might be desirable to provide a large rotation from a small displacement. Such an action can also be provided due to the rotation at the pivot of the mechanism in figure

7.4. Rotation of the other link is rarely of interest. In this instance, the lever ratio (having dimensions of rad m-1) is given by

$$\frac{\theta_3}{x_1} = \frac{x_3}{x_1 l_2} = \frac{l_1}{l_2 h}\sqrt{1-\left(\frac{h}{l_1}\right)^2} \tag{7.10}$$

The sensitivity of this ratio as a function of the angle is given by

$$-\frac{1}{l_2}\left(\frac{l_1}{h}\right)^2 \tag{7.11}$$

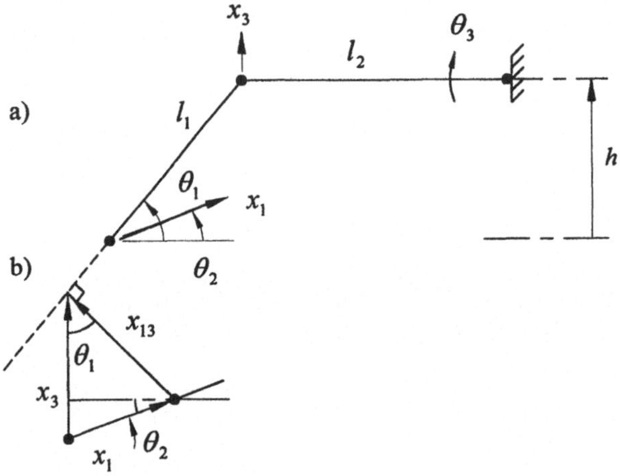

Figure 7.5: Parameters for a mathematical model of the generalized, two-arm lever, a) mathematical model, b) displacement diagram

Extension of the above analysis to the general, two-arm lever involves little additional complexity. In this case, the input can be directed in an arbitrary direction. The modified model for the generalized two-arm lever is shown in figure 7.5. Based on the displacement diagram, the lever ratio is readily obtained as

$$\frac{x_3}{x_1} = \sin\theta_2 + \frac{\cos\theta_2}{\tan\theta_1} \tag{7.12}$$

In terms of the output rotation, θ_3, the ratio becomes

$$\frac{\theta_3}{x_1} = \frac{1}{l_2}\left(\sin\theta_2 + \frac{\cos\theta_2}{\tan\theta_1}\right) \tag{7.13}$$

For a particular class of two-arm levers, it is arranged that the actuator action is co-linear with the axis of link l_1 (i.e. $\theta_1 = \theta_2$). In this particular case, the lever ratio becomes

$$\frac{x_3}{x_1} = \frac{1}{\sin\theta_1} \tag{7.14}$$

Again, a measure of the linearity of this ratio can be obtained from the derivative of (7.14) given by

$$\frac{d\,{}^{x_3}\!/_{x_1}}{d\theta_1} = \frac{-\cos\theta_1}{\sin^2\theta_1} \tag{7.15}$$

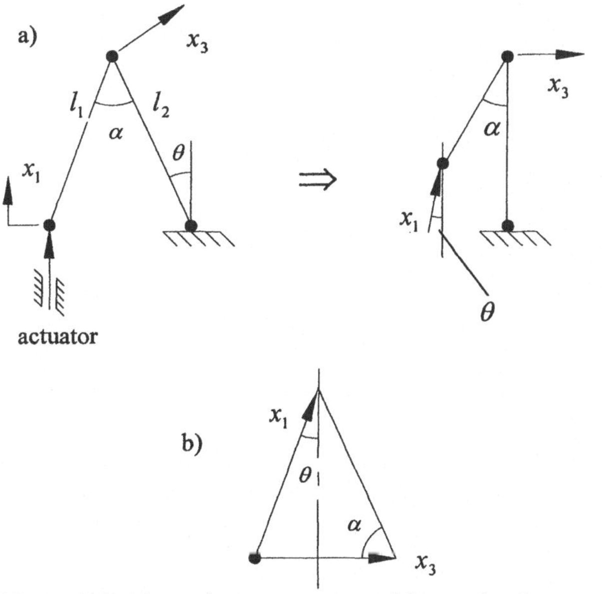

Figure 7.6: Alternative arrangement of the two bar lever, a) equivalent geometric models, b) displacement diagram

It can be seen that, as the angle reduces, non-linear effects increase as the inverse square and as a consequence may significantly reduce the available displacement range. Figure 7.6 illustrates a two bar lever that is equivalent to that of figure 7.5 with different angles between the links. The displacement diagram is included from which the lever ratio can be similarly determined as above.

7.2.2 Soft spring – stiff spring attenuation

Assuming that all pivots are infinitely rigid, all of the above mechanisms can be reversed to operate as either a displacement magnifying or attenuating mechanism. In some instances it may be necessary to introduce a large attenuation. In particular, relatively precise and inexpensive feedscrews can be obtained in the form of a micrometer spindles and precision ground ball screws. The resolution of these drives will always be limited by their size (determines manufacturing process), stiffness (pre-load forces introduce hysteresis in the form of oblique frictional tractions, Mindlin and Deresiewicz, 1953, or rolling resistance, Johnson, 1985) and manufacturing tolerances (continuously improving). A relatively simple way to increase the resolution of a long-range drive mechanism is to use a reduction mechanism. Many of the mechanisms outlined in the previous section are adequate for such a task. However, for high precision applications, the inherent non-linearity of these devices will have to be dealt with. An alternative is to attenuate the input using the soft spring - stiff spring concept illustrated in figure 7.7. Those readers familiar with electrical circuits will recognize this as analogous to a potential divider. Again, it is assumed that this mechanism is driven by a prescribed displacement, x_1, and this imposes a force, F, on an intermediate spring, k_1, and this same force is transmitted to the displacement flexure of stiffness k_2. The subsequent displacement of the flexure, x_3, is given by

$$x_3 = F/k_3 = \frac{k_1(x_1 - x_3)}{k_3}$$

$$\therefore \qquad (7.16)$$

$$\frac{x_3}{x_1} = \frac{k_1}{k_1 + k_3}$$

Use of this attenuation mechanism has the relative advantages that

1. It remains linear over relatively large displacement ranges
2. Extremely large attenuation ratios are possible.
3. A highly compliant flexure can also be used to provide precisely controlled, small forces (see case study III).

The only drawback to such a lever is

1. The high compliance of the intermediate spring between the drive and moving stage adds little to the dynamic performance while the relative advantages of coupling to a stiff drive are also lost.

A planar mechanism utilizing such a lever reduction is shown in figure 7.8. Other implementations are left to the designer's imagination. An interesting example can be found in the paper of Hart, 1968. In this paper, such a reduction mechanism is used to drive a single crystal silicon flexure with precise control of motion at sub-nanometer levels suing a simple micrometer. In this particular case, a reduction ratio of around 10,000:1 is achieved. Included in this paper is a novel method for obtaining fine angular adjustment using hanging chains to provide a torque.

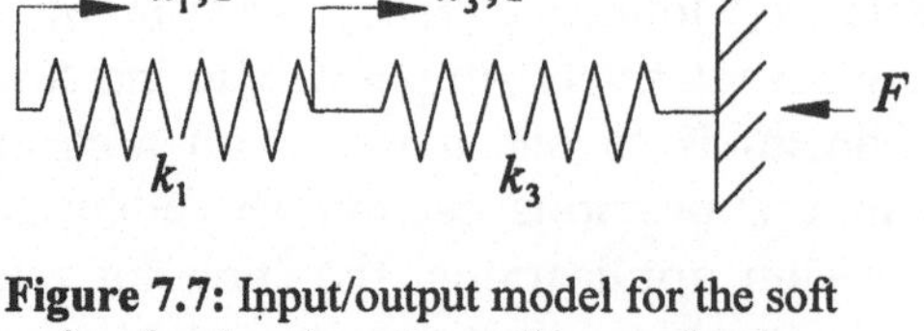

Figure 7.7: Input/output model for the soft spring, hard spring attenuation mechanism

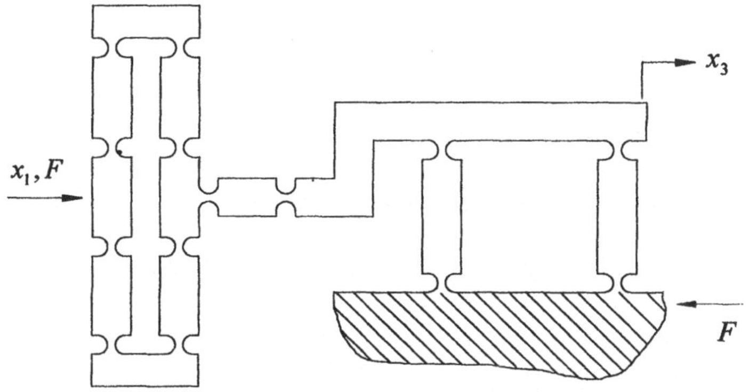

Figure 7.8: A planar mechanism utilizing a soft-spring, hard-spring attenuation mechanism

Torsion bars may provide a simple high compliance intermediate flexure, particularly for mechanisms requiring fine angular positioning. In its simplest form this consists of a long, thin rod attached to a rotary drive at one end and the flexure at the other. Figure 7.9 shows a simple, high-resolution angular adjustment mechanism. In this case, the stiff spring is achieved by using a similar, but larger diameter (and/or shorter length) rod. Stiffness values for each of the rods, k, can be readily computed from the torsion equation (2.100) while the lever ratio is easily shown to be

$$\frac{\theta_3}{\theta_1} = \frac{k_1}{k_1 + k_3} \tag{7.17}$$

Sometimes referred to as a filar suspension, examples of these can be found in the papers of Marsh, 1961, and case study II at the end of this chapter.

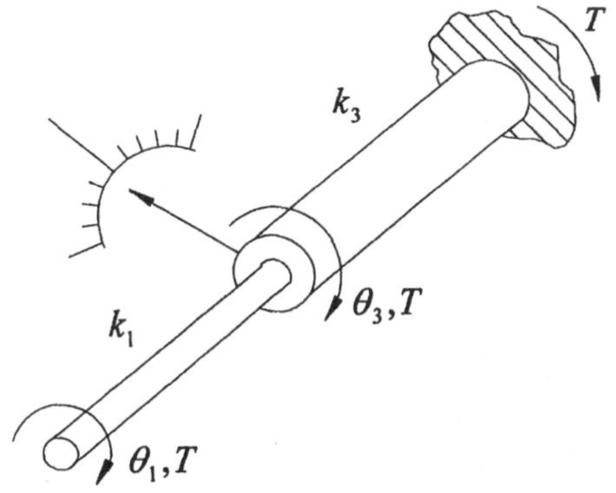

Figure 7.9: A simple angular adjustment mechanism

7.3 Lost motion

Piezoelectric or magneto/electro-strictive actuators are frequently used to drive flexures requiring translations ranging from tens of micrometers or more. A compact form of actuator consists of a stack of actuators connected in series. Unfortunately, for such a large range, the physical dimensions and number of elements in the stack become unwieldy. For a more compact device, considerable space can often be saved using a lever to amplify the motion. To appreciate the models of this mechanism, consider the simple notch type linear spring driven by a piezoelectric element as shown in figure 7.10. Three different configurations are shown. All three designs utilize one of the support legs as a lever arm. In figure 7.10(a), leverage is achieved by pushing the support leg near the stationary pivot. Analogously, it is possible to close a door by pushing near to the hinge without having to move ones hand a large distance (although considerably more force is required). For an infinitely rigid coupling between the actuator and support leg and also assuming that the notches act as perfect hinges, then the ratio, n, between displacement of the actuator and upper platform is simply

$$n = \frac{b+a}{a} \tag{7.18}$$

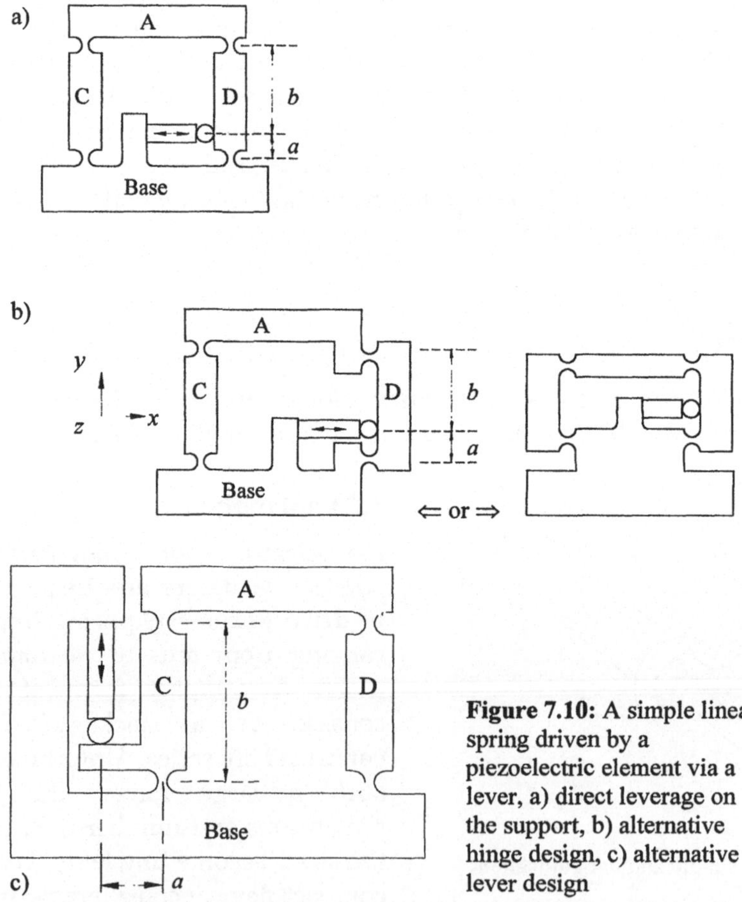

Figure 7.10: A simple linear spring driven by a piezoelectric element via a lever, a) direct leverage on the support, b) alternative hinge design, c) alternative lever design

Intuitively, it is apparent that there will be a large sideways, or lateral, force on the flexure. As a consequence it might be expected that this might introduce unacceptable distortion due to the compliance in this direction given by equation (4.109). A alternative design is shown in figure 7.10(b) in which the actuator will induce forces in the longitudinal direction of the notch. The benefit of this alternative design can be seen from ratio between lateral and longitudinal stiffness values given in equations (4.109) and (4.111) given by

$$\frac{k_{\delta_y F_y}}{k_{\delta_x F_x}} = \frac{8\beta^2\left(1-2\sqrt{\frac{\beta}{3}}\right)}{9\pi^2\left(1-\beta^2\right)} \tag{7.19}$$

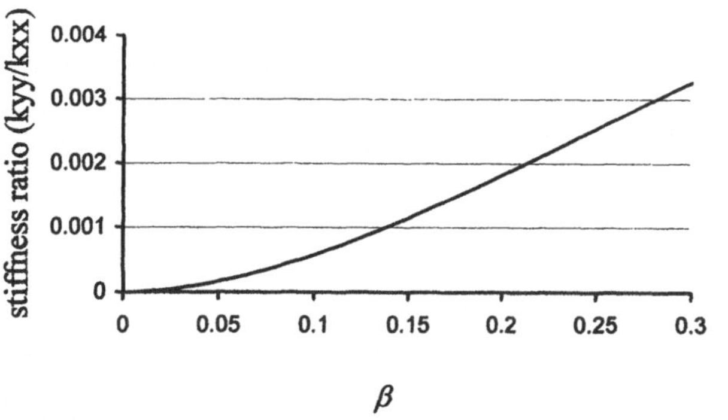

Figure 7.11: The ratio of lateral to longitudinal stiffness for a notch hinge

Equation (7.19) is plotted in figure 7.11. From this, it can be seen that, over a typical range of values for the dimensionless ratio β, the lateral stiffness is more than two orders of magnitude lower than that in the longitudinal direction. It will be shown shortly that compliance of this pivot is responsible for most of the lost motion within a lever system.

Finally, figure 7.10c) shows an alternative lever mechanism in which the actuator has been moved 'outside' of the flexure. In this case the lever ratio is given as $n = b/a$. Such a design is often favored when it is desired to use the space 'within' the flexure.

To assess the effects of coupling and pivot compliance on the lever ratio, it is possible to model the single degree of freedom system as shown in figure 7.12. In this, the flexure is modeled having the *same* stiffness, k_3, that it would have were it driven directly at the moving platform while the pivot and coupling are identified by the subscripts 2 and 1 respectively. For a perfect pivot and coupling, the relationship between a force applied at the coupling to the displacement at the moving platform can be obtained by equating the work done when a force is applied at the moving platform to that when a force is applied at the lever. From this, the following is readily obtained

$$F_1 = nk_3x_3 \tag{7.20}$$

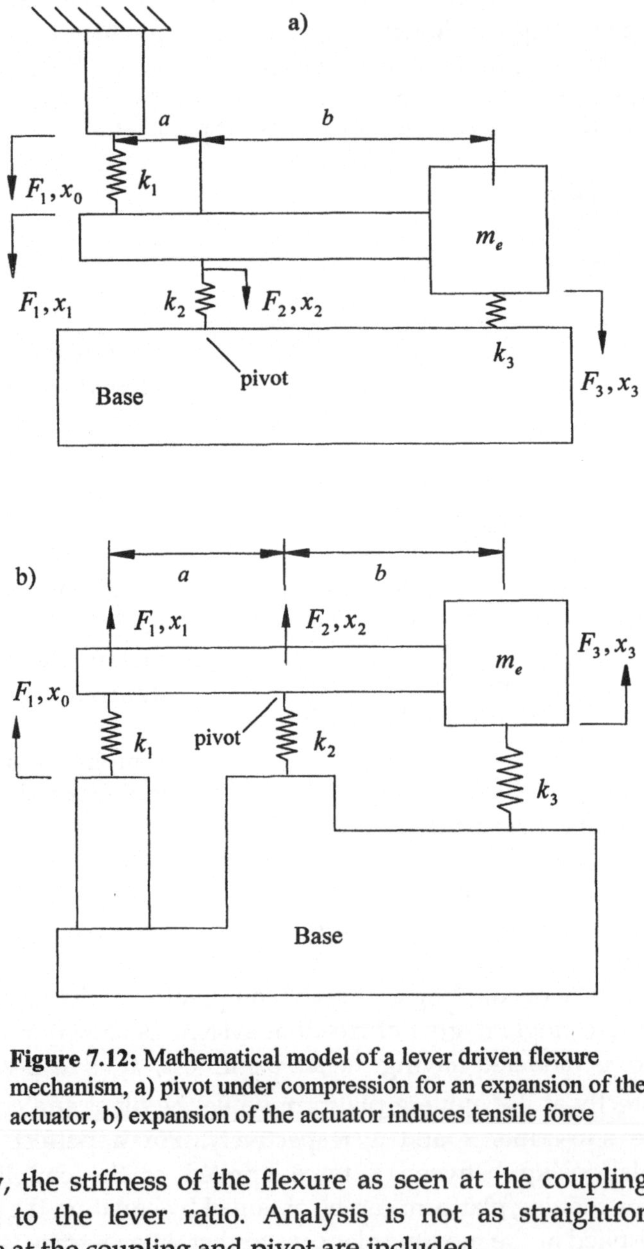

Figure 7.12: Mathematical model of a lever driven flexure mechanism, a) pivot under compression for an expansion of the actuator, b) expansion of the actuator induces tensile force

Clearly, the stiffness of the flexure as seen at the coupling varies in direct proportion to the lever ratio. Analysis is not as straightforward when the compliance at the coupling and pivot are included.

The objective of the lever is to provide some form of amplification or attenuation between the force or displacement at the actuator and that produced at the moving stage of the flexure. In terms of the model of figure 7.12 this consists of the desired ratio x_3 / x_0. From Newton's 3rd law it is known that, for the flexure to be in static equilibrium, the vector sum of forces must vanish i.e.

$$F_1 + F_2 + F_3 = 0 \tag{7.21}$$

It can also be seen that the spring forces applied to the flexure are

$$\begin{aligned} F_1 &= k_1(x_1 - x_0) \\ F_2 &= k_2 x_2 \\ F_3 &= k_3 x_3 \end{aligned} \tag{7.22}$$

Ignoring the angular stiffness of the springs, taking moments about the pivot gives

$$F_1 = \frac{b}{a} F_3 = nF_3 \tag{7.23}$$

Finally, because the flexure is assumed to be rigid, it is possible to derive two equal expressions for the angular deflection

$$\theta = \frac{x_1 - x_2}{a} = \frac{x_2 - x_3}{b} \tag{7.24}$$

An immediate outcome of equations (7.21-7.23) is that there will be a displacement in the pivot of magnitude

$$x_2 = -\frac{F_1(1 + 1/n)}{k_2} \tag{7.25}$$

Substitution of equations (7.21 – 7.23) into (7.24) gives

$$\begin{aligned} x_1 &= \frac{a(x_2 - x_3)}{b} + x_2 = \frac{F_2}{k_2}\left(1 + \frac{1}{n}\right) - \frac{F_3}{nk_3} \\ &= \frac{-F_1}{k_2}\left(1 + \frac{1}{n}\right)^2 - \frac{F_1}{n^2 k_3} \end{aligned} \tag{7.26}$$

From the first of (7.22), the actuator displacement is given by

$$x_0 = x_1 - \frac{F_1}{k_1} = -F_1\left(\frac{1}{k_1} + \frac{1}{k_2}\left(1 + \frac{1}{n}\right)^2 + \frac{1}{n^2 k_3}\right) \tag{7.27}$$

Finally, the relationship between output and input displacements is given by

$$\frac{x_3}{x_0} = \frac{-1}{n\left(\frac{k_3}{k_1} + \frac{k_3}{k_2}\left(1+\frac{1}{n}\right)^2 + \frac{1}{n^2}\right)} \tag{7.28}$$

It is informative to study some of the characteristics of this expression. To further simplify (7.27) the dimensionless stiffness ratios are replaced by the algebraic symbols α and β to yield

$$\frac{x_3}{x_0} = \frac{-n}{\left(n^2\alpha + n^2\beta\left(1+\frac{1}{n}\right)^2 + 1\right)} \tag{7.29}$$

Where α represents the stiffness ratio between the flexure and coupling and β the stiffness ratio between the flexure and pivot.

The fractional lost motion, f, is defined by

$$f = \frac{(-n) - \frac{x_3}{x_0}}{(-n)} = 1 + \frac{x_3}{nx_0}$$

$$= 1 - \frac{1}{\left(n^2\alpha + n^2\beta\left(1+\frac{1}{n}\right)^2 + 1\right)} = \frac{\alpha + \beta\left(1+\frac{1}{n}\right)^2}{\left(\alpha + \beta\left(1+\frac{1}{n}\right)^2 + \frac{1}{n^2}\right)} \tag{7.30}$$

As either the flexure stiffness tends to zero or both the coupling and pivot stiffness tend to infinity, the lever ratio tends towards the ideal value $-n$. For an ideal pivot, $\beta \Rightarrow 0$, equation (7.29) reduces to

$$\frac{x_3}{x_0} = \frac{-n}{(n^2\alpha + 1)} \tag{7.31}$$

while the fractional lost motion becomes

$$f = \frac{n^2\alpha}{(n^2\alpha + 1)} \tag{7.32}$$

Both of the above can be readily verified using a simplified static model.

For the case of a perfect coupling

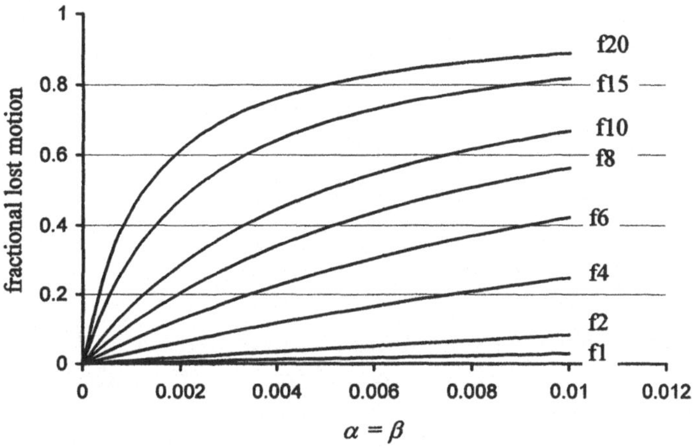

Figure 7.13: Fractional lost motion as a function of the ratios of stiffness between flexure and pivot. The stiffness of the actuator coupling has been made equal to that of the pivot. In the legend f2 represents the fractional lost motion for a flexure mechanism with a nominal lever $n = 2$.

$$\frac{x_3}{x_0} = \frac{-n}{\left(n^2\beta\left(1+\frac{1}{n}\right)^2+1\right)}$$

&

$$f = \frac{n^2\beta\left(1+\frac{1}{n}\right)^2}{\left(n^2\beta\left(1+\frac{1}{n}\right)^2+1\right)} \tag{7.33}$$

In both cases it can be seen that the fractional lost motion can only be reduced by minimizing the products $n^2\alpha$ or $n^2\beta$. Assuming that the stiffness of the pivot and couplings are identical, figure 7.13. For example, if an ideal lever ratio of 10 is assumed and 20 % lost motion is considered acceptable, the stiffness ratios for both pivot and coupling must be below 0.0012.

Surprisingly, the effect of cascading levers requires little additional complexity. At the outset, the stiffness of the flexure, k_3, was defined as the force required to produce a displacement. For a lever system, this is in direct proportion to the ideal lever ratio and independent of the stiffness of the pivot. Consequently, to determine the lost motion of the first lever, it is adequate to replace the stiffness of the new flexure to include that of the second lever. However, the output then becomes the displacement of the second lever.

Fortunately, this second lever can be modeled as a separate lever having an ideally stiff coupling. So, if two levers are cascaded and the first and second have ideal lever ratio's n_1 and n_2 the total fractional lost motion becomes

$$f_2 = \frac{n_2\alpha_1 + n_2\beta_1\left(1+\frac{1}{n_1}\right)^2}{\left(n_2\alpha_1 + n_2\beta_1\left(1+\frac{1}{n_1}\right)^2 + \frac{1}{n_1^2}\right)} \cdot \frac{n_2^2\beta_2\left(1+\frac{1}{n_2}\right)^2}{\left(n_2^2\beta_2\left(1+\frac{1}{n_2}\right)^2 + 1\right)} \tag{7.34}$$

If the stiffness ratios are the same as the previous analysis and the new lever ratios are split and equal to the square root of the ratio for a single lever (7.34) becomes

$$f_2 = \frac{\sqrt{n}\alpha + \sqrt{n}\beta\left(1+\frac{1}{\sqrt{n}}\right)^2}{\left(\sqrt{n}\alpha + \sqrt{n}\beta\left(1+\frac{1}{\sqrt{n}}\right)^2 + \frac{1}{n}\right)} \cdot \frac{n\beta\left(1+\frac{1}{\sqrt{n}}\right)^2}{\left(n\beta\left(1+\frac{1}{\sqrt{n}}\right)^2 + 1\right)} \tag{7.35}$$

Figure 7.14 shows the ratio of the fractional lost motion for a single lever of ideal ratio's 10 and 100 and two levers of ratio $\sqrt{10}$ and 10. As the stiffness ratios α and β are reduced, it is clear that a cascaded pair represents considerable improvement. However, this advantage diminishes as the ratio increases.

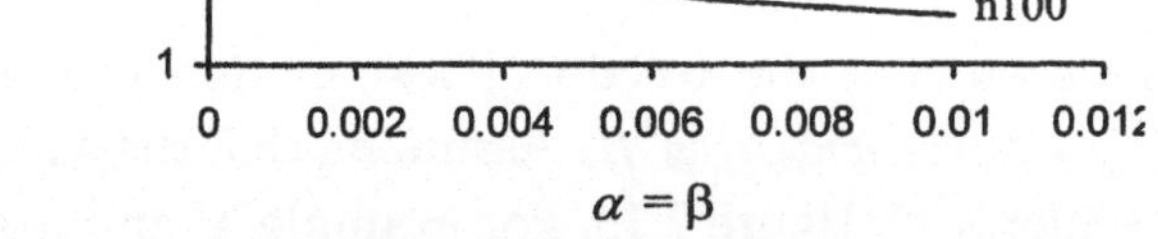

Figure 7.14: Ratio of fractional lost motion between a single lever and two levers in series

7.4 Effects of levers on flexure dynamics (influence of the drive)

In previous discussions, the influence of the driver upon the dynamic performance of the flexure has been ignored. In practice, with devices such as the feedscrew or piezo-electric actuators, the added stiffness of these drives can dominate over that of the flexure. However, the necessity of an intermediate coupling often results in a considerable reduction of the drive stiffness as 'seen' at the flexure. As mentioned in section 5.1.4, this drive stiffness will add to the potential energy storage and therefore to the effective stiffness of the flexure. This section considers the influence of an intermediate lever on the dynamic response of the flexure.

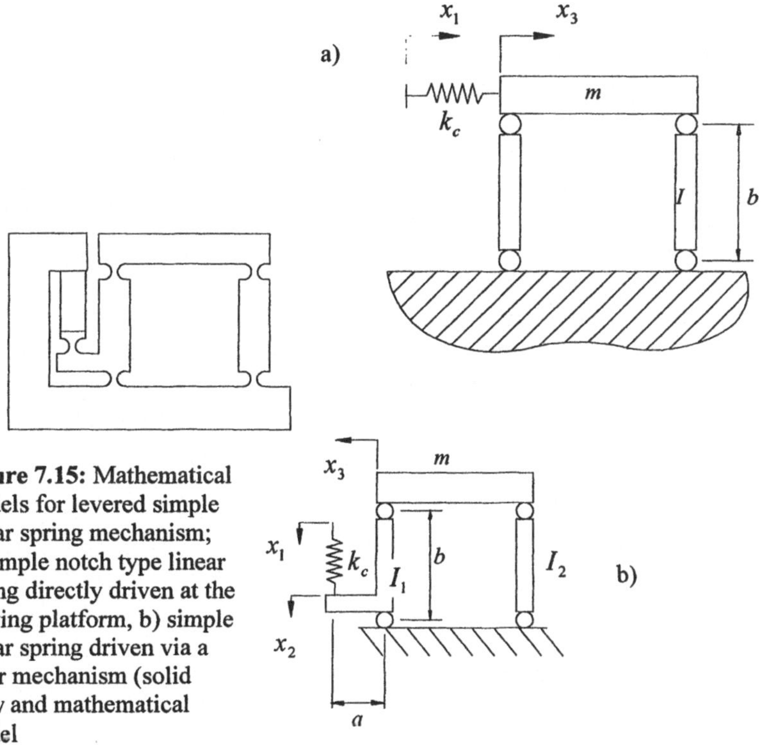

Figure 7.15: Mathematical models for levered simple linear spring mechanism; a) simple notch type linear spring directly driven at the moving platform, b) simple linear spring driven via a lever mechanism (solid view and mathematical model

To compare the influence of the lever on the dynamic response of the mechanism, it is necessary to determine the natural frequency of a mechanism directly driven. Consider the simple, notch type linear spring show in figure 7.15(a) in which the drive is directly attached to the moving platform via a coupling of effective stiffness k_c. Assuming the hinges to act at a point, it is relatively straightforward to show that the potential and kinetic energy of such a system is given by

$$T = \frac{1}{2} M\dot{x}_3^2 + I\frac{\dot{x}_3^2}{b^2} = \frac{1}{2}\left(M + \frac{2I}{b^2}\right)\dot{x}_3^2$$
$$V = \frac{1}{2} k_c (x_1 - x_3)^2 + \frac{1}{2}\left(4k_{\theta M_z}\right)\frac{x_3^2}{b^2} \tag{7.36}$$

The second moment of mass for each of the support legs corresponds to that for rotation about the stationary pivots. Lagrange's equation readily provides the equation governing motion for a prescribed input displacement, x_1

$$\left(M+\frac{2I}{b^2}\right)\ddot{x}_3+\left(\frac{4k_{\theta_z M_z}}{b^2}+k_c\right)x_3=k_c x_1 \tag{7.37}$$

From the homogeneous solution, the natural frequency is

$$\omega_n^2=\frac{\dfrac{4k_{\theta_z M_z}}{b^2}+k_c}{M+\dfrac{2I}{b^2}} \tag{7.38}$$

Clearly the coupling stiffness adds to the natural frequency of the flexure as mentioned in section 5.1.4 (see also equation 5.46)). Turning our attention now to levered mechanisms, of which three examples are shown in figure 7.10, the kinetic and potential energy equations become, see figure 7.15(b)

$$\begin{aligned}T&=\frac{1}{2}M\dot{x}_3^2+\frac{1}{2}(I_1+I_2)\frac{\dot{x}_3^2}{b^2}=\frac{1}{2}\left(M+\frac{2I_e}{b^2}\right)\dot{x}_3^2\\V&=\frac{1}{2}k_c\left(x_1-\frac{x_3}{n}\right)^2+\frac{1}{2}\left(4k_{\theta_z M_z}\right)\frac{x_3^2}{b^2}\end{aligned} \tag{7.39}$$

Lagrange's equation immediately gives

$$\left(M+\frac{2I_e}{b^2}\right)\ddot{x}_3+\frac{k_c}{n}\left(\frac{x_3}{n}-x_1\right)+\left(4k_{\theta_z M_z}\right)\frac{x_3}{b^2}=0 \tag{7.40}$$

Rearranging equation (7.40) and adding energy dissipation, the equation governing motion can be written in the form

$$\left(M+\frac{2I_e}{b^2}\right)\ddot{x}_3+\left(\frac{b_{12}}{n^2}+b_3\right)\dot{x}_3+\left(\frac{4k_{\theta_z M_z}}{b^2}+\frac{k_c}{n^2}\right)x_3=\frac{b_{12}}{n}\dot{x}_1+\frac{k_c}{n}x_1 \tag{7.41}$$

To determine the frequency response of this levered mechanism a prescribed input displacement is assumed to be of the form

$$x_1(t)=X_1e^{i\omega t} \tag{7.42}$$

Substituting (7.42) into (7.41), the general frequency response can be readily obtained as

$$H(i\omega)=\frac{X_3}{X_1}=\frac{i\omega\dfrac{b_{12}}{n}+\dfrac{k_c}{n}}{\left(\dfrac{4k_{\theta_z M_z}}{b^2}+\dfrac{k_c}{n^2}\right)-\omega^2\left(M+\dfrac{2I_e}{b^2}\right)+i\omega\left(\dfrac{b_{12}}{n^2}+b_3\right)}$$

$$=\frac{i\omega \mathrm{B}_2+\mathrm{K}_2}{\mathrm{K}_1-\omega^2\mathrm{M}+i\omega\mathrm{B}_1}=\frac{i\omega\dfrac{\mathrm{B}_2}{\mathrm{K}_1}+\dfrac{\mathrm{K}_2}{\mathrm{K}_1}}{1-\dfrac{\omega^2}{\omega_n^2}+i\omega\dfrac{\mathrm{B}_1}{\mathrm{K}_1}} \tag{7.43}$$

Although containing different terms, equation (7.43) is of the same form as that of a single stage vibration isolator discussed in section 3.6. From this equation, the undamped natural frequency of the lever driven mechanism is given by

$$\omega_n^2=\frac{\mathrm{K}_1}{\mathrm{M}}=\frac{\dfrac{4k_{\theta_z M_z}}{b^2}+\dfrac{k_c}{n^2}}{M+\dfrac{2I_e}{b^2}} \tag{7.44}$$

Comparison between equations (7.38) and (7.44) readily shows the influence of the lever on the dynamic performance of the mechanism. For an amplifying mechanism (i.e. $n(=b/a)>1$) there are two deleterious effects. Firstly, and probably most importantly, the effective stiffness of the coupling is reduced by the inverse square of the coupling. For ratio's greater than one, this will always result in a reduction of the natural frequency of the flexure and this is compounded by the increased inertia of the lever mechanism itself. Figure 7.16(a) over leaf shows the magnitude frequency responses of a levered flexure system.

At high frequencies, the response of the system is limited by the inertial term in equation (7.41). To assess the inertial effect on the drive, it is interesting to look at the relationship between the drive displacement and that produced at the input to the lever. In this case (7.41) can be rewritten in the form

$$n^2\left(M+\frac{2I_e}{b^2}\right)\ddot{x}_2+\left(b_{12}+n^2b_3\right)\dot{x}_2+\left(\frac{n^2 4k_{\theta_z M_z}}{b^2}+k_c\right)x_2=b_{12}\dot{x}_1+k_c x_1 \tag{7.45}$$

From this equation, it can be seen that the effective mass of the flexure as 'seen' by the drive increases with the square of the lever ratio.

It can be readily shown that the dynamic 'lost motion' is given by

$$f = \frac{n - \frac{X_3}{X_1}}{n} = 1 - \frac{1}{n^2}\frac{i\omega b_{12} + k_c}{\mathrm{K}_1 - \omega^2\mathrm{M} + i\omega\mathrm{B}_1} \tag{7.46}$$

a)

b)

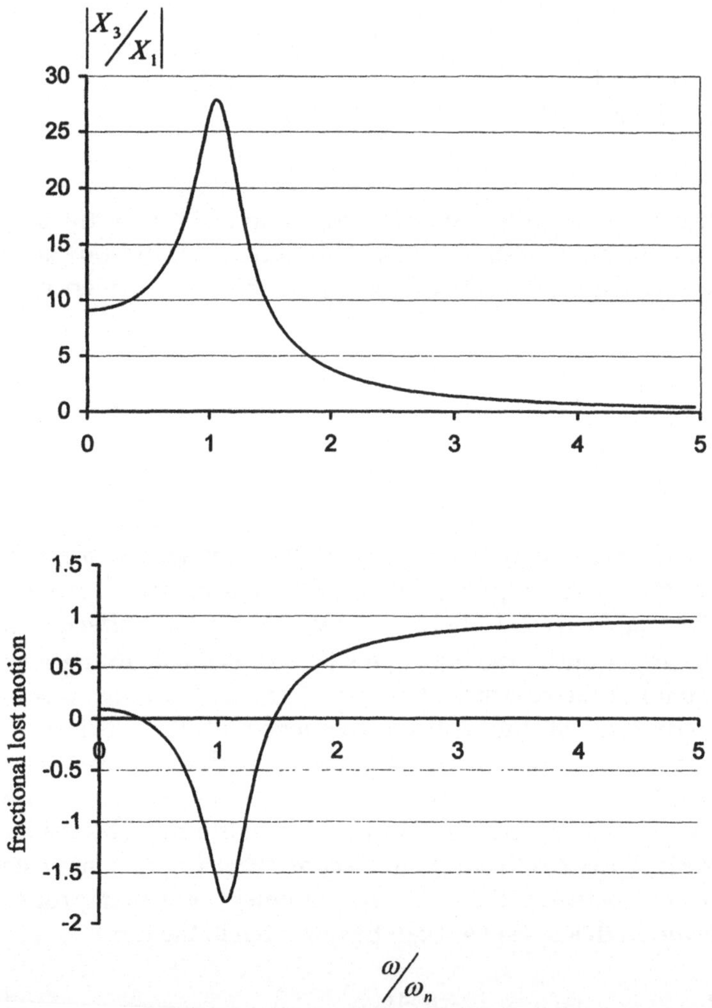

Figure 7.16: Dynamic response of a levered flexure, a) frequency response, b) fractional lost motion; k_c = 100 N m^{-1}, $k_{flexure}$ = 0.1 Nm^{-1}, M = 1 kg, b_{12} = 3 Ns m^{-1}, b_3 = 0.1 Ns m^{-1}, n = 10

It is informative to consider the above equation in terms of three major regions, these being low frequency, near resonance and high frequency. Each of these regions corresponds to stiffness, damping and mass dominated regimes respectively. For the first region, the 'static' lost motion can be obtained by setting $\omega \Rightarrow 0$ at which the above reduces to

$$f = \frac{n^2 K_1 - k_c}{n^2 K_1} = \frac{n^2 4k_{\theta_z M_z}}{n^2 4k_{\theta_z M_z} + b^2 k_c} \tag{7.47}$$

Being identical to equation (7.32), it can be seen that this tends to zero with large coupling or low flexure stiffness.

At high frequencies, the mass term will dominate causing the right term of (7.46) to tend to zero at which the fractional lost motion goes to 1.

The intermediate frequency near to resonance, at which damping has the dominant influence, is a little more complicated. Figure 7.16(b) shows the fractional lost motion corresponding to the frequency response plot of figure 7.16(a). Clearly, the choice of damping, as always, is a compromise. Such compromises are left to the reader.

7.5 Case study I: A fine adjustment mechanism for motion attenuation

In the design of an instrument for the measurement of hardness indentations, it was required that a small displacement sensor be positioned alongside the indenter to measure its depth of penetration. Figure 7.17(a) shows a solid view of the lever mechanism designed to enable fine adjustment of the sensor so that its zero point could be positioned adjacent to the tip of the hardness indenter. Figures 7.17 (b & c) are engineering drawings of a lever design indicating its dimensions. In the undistorted condition, the lever consists of a rigid bar with two leaf-type springs attached to its corners at one end. These springs comprise the lever pivot while it is the larger linear flexure that provides the motion of the sensor (mounted on the platform) relative to the base. A line diagram of the lever mechanism in the distorted condition is shown in figure 7.18. In fact, the lever mechanism in this design corresponds to the two-beam two degree of freedom flexure already discussed in section 4.4.2. In this case, it is the axial rigidity of the beams that is being exploited. One of the main design features is that the output motion of the stage is perpendicular to the lever adjustment enabling manual adjustment from, in this instrument, an accessible direction that does not obstruct the sensor or mounting mechanisms.

To translate the platform relative to the base, a force is applied to the lever at a height d as measured from the pivot point of the flexures, figure 7.18. In the distorted condition the lever flexures will provide an equal and opposite axial force to the base and moving platform. Idealistically, the lever flexure might be modeled as two 'perfect' pivots. Based on this assumption, the lever ratio, n, can be approximated from the equation

$$n = \frac{\text{dispalcement of platform}}{\text{displacement of lever}} = \frac{s}{d} \tag{7.48}$$

where the distance *d* based on the analysis of section 4.1.1 is measured from a point 1/3 of the length of the cantilever from the base to the point of application of the force and *s* is the separation of the neutral axes of the lever flexures.

a)

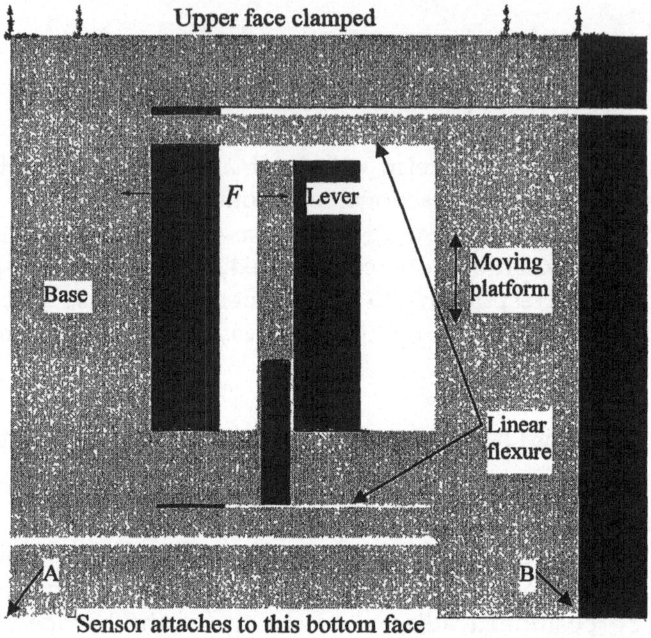

b)

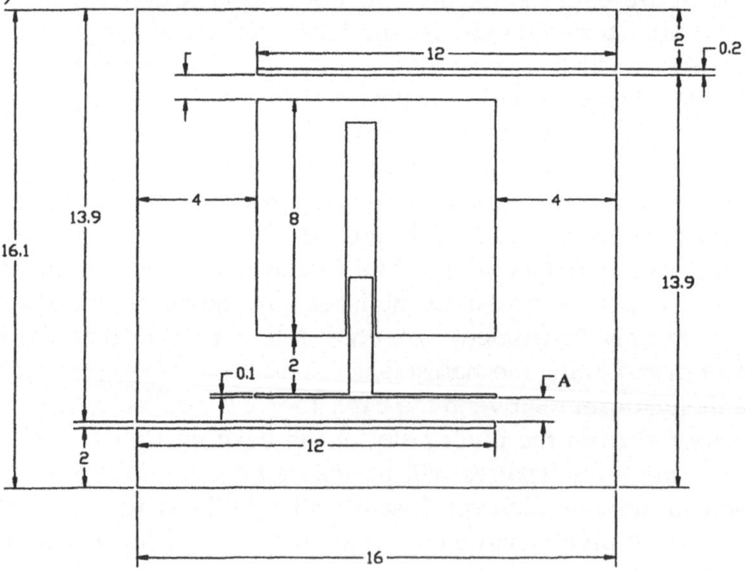

figure continues over

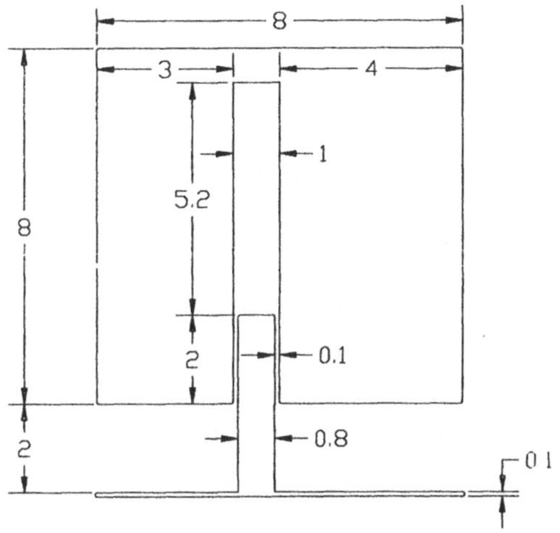

c)

Figure 7.17(cont'd): Lever mechanism for fine adjustment of hardness penetration depth sensing probe, a) solid view, b) major dimensions, c) exploded view of central portion showing dimensions of lever

For this design, a total deflection of the platform of 40-50 μm was required. To enhance the resolution of the screw adjuster, a lever ratio of 1:6 between input and output was considered adequate. In the final design, this flexure was manufactured using wire-electro-discharge machining, see chapter 8, from a 12 mm thick plate of bronze. Some of the relevant parameters for this design are given in Table 7.5.1 below. Equation (7.48) immediately gives an approximate lever ratio of 1:6.37. This is slightly larger than the desired value.

7.5.1 Finite element analysis of the lever flexure

The objective of this study is to assess the fidelity of this design. In this case the two parameters of interest are

1. Stiffness of the upper platform
2. Rectilinearity of motion of the lower face of the moving platform

Stiffness implies both dynamic response and immunity to vibration. Usually, the dynamic response is limited by the fundamental, or gravest, mode, natural frequency. Invariably this fundamental mode is proportional to the square root of the most compliant relative motion in the mechanism, this being the desired freedom. Consequently, the fastest response corresponds to the highest flexure stiffness that, in turn, is limited only by the maximum stress tolerated by the material of construction (where possible, minimizing mass by removing material where stresses are small and maximizing the ratio of elastic modulus to density should also be considered).

Initially, a design based on maximum permissible stresses is analyzed as an example of a flexure to provide a fast dynamic response. Based on this simple analysis, it is shown that the design will contain significant lost motion and undesirable distortions of the structural components (i.e. the 'rigid' links) of the flexure.

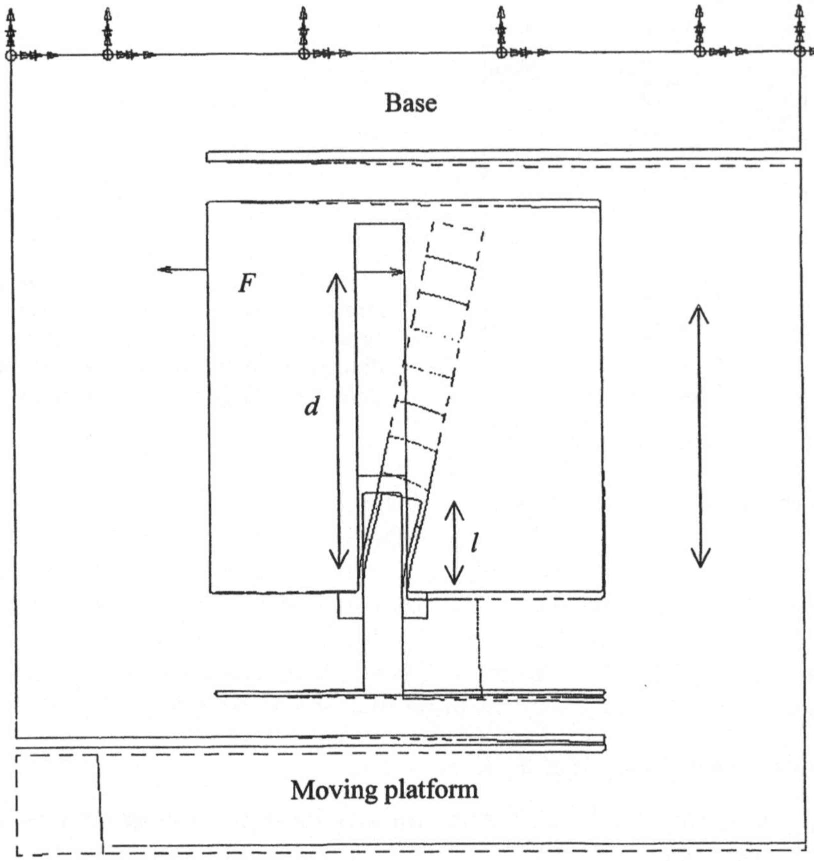

Figure 7.18: Line diagram of the distorted lever mechanism

Validity of the assumption that it is feasible to use simple beam bending equations can be evaluated from the dimensionless axial load parameter γ of section 4.1. In general, the simple beam bending equation is usually adequate if γ significantly less than one. This parameter depends on the axial force applied to the beam, which, in this design, corresponds to the force required to displace the translation spring. The stiffness of the translation spring is given by

$$k_f = \frac{24EI}{L^3} = 3.36 \times 10^6 \quad \text{(N m}^{-1}\text{)} \tag{7.49}$$

Based on this, the applied force for 40 μm output deflection of is 134 N. Being the axial load applied to the lever flexures, the dimensionless load parameter is

$$\gamma = \frac{l}{2}\sqrt{\frac{F_x}{EI}} = 1.02 \tag{7.50}$$

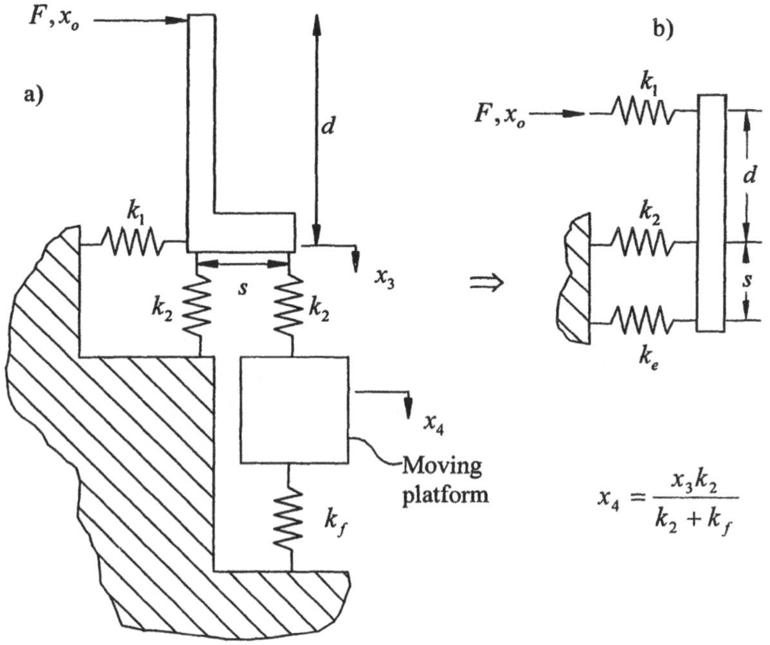

Figure 7.19: Mathematical model of lever mechanism a) complete model, b) reduced form to be equivalent to that in figure 7.13

Having a value of greater than 1, this design is uncomfortably near to the buckling load and may be unstable at the extreme extension of the platform. Ignoring this potential error, the axial stiffness of each lever cantilever is

$$k_{axiallever} = \frac{EA}{l} = 7.8 \times 10^7 \qquad (\text{N m}^{-1}) \qquad (7.51)$$

Although this is significantly higher than that of the linear translation spring, there are other errors due to the compliance of the lever acting as a linear spring, $k_{linearlever}$, and the cantilever displacement of the lever itself, $k_{levercant}$. For these deformation modes the stiffness values are respectively

$$\begin{aligned} k_{linearlever} &= \frac{24EI}{l^3} = 390000 \\ k_{levercant} &= \frac{3EI}{(d - 2l/3)^3} = 3.0 \times 10^6 \end{aligned} \qquad (\text{N m}^{-1}) \qquad (7.52)$$

Based on these three stiffness values, it is possible to develop a model of the lever as shown in figure 7.19. Rather surprisingly, it can be seen that *compliance in the lever can be added to that of the coupling.*

In fact, distortion of the base also adds to the linear distortion of the lever, this is ignored in the following analysis. These compliant elements are connected in series for which the equivalent stiffness is given by

$$k_1 = \frac{1}{\dfrac{1}{k_{linearlever}} + \dfrac{1}{k_{levercant}}} = 345112 \qquad (\text{N m}^{-1}) \qquad (7.53)$$

For computation of lost motion, the above stiffness values can be used to compute the ratios β = 4.13x10^{-2} and α = 9.33. Assuming an infinitely stiff coupling between the drive and the lever, the fractional lost motion is given by (7.29)

$$\frac{x_3}{x_0} = \frac{-n}{\left(n^2\alpha + n^2\beta\left(1+\dfrac{1}{n}\right)^2 + 1\right)} = 0.122 \qquad (7.54)$$

From this, it is expected that the lever ratio will change from 1:6.4 to 1:8.2 corresponding to an *increased* attenuation. It is also necessary to estimate the stresses in the two flexures at the maximum deflection of 40 μm. For the linear flexure, this will occur at the root of the flexures and is approximated by (2.69) and (4.78)

$$\sigma = \frac{Mt}{I2} = \frac{6Etx_4}{2L^2} = 200 \qquad (\text{MPa}) \qquad (7.55)$$

This is reasonably far from the yield stress for this particular material although some care should be exercised to reduce the effects of stress concentrations at the roots of the flexure by incorporating a fillet radius equal to the thickness of the flexure. For such dimensions the stress concentration factor is likely to add a further 10% or less to the stress computed from the bending equation.

For the lever flexure, stresses will be comprised from axial forces and bending of the cantilever. Bending stresses may reasonably be computed by considering each flexure to distort as the sum of a simple cantilever with a bending moment applied at the free end plus linear spring mode deformation due to the applied force. The maximum angular deflection of the lever, θ_{max}, is computed from

$$\theta \approx \frac{x_3}{s} = \frac{x_o n'}{s} \qquad \text{(rad)} \qquad (7.56)$$

$$n' = \frac{x_3}{x_o}$$

Table 7.5.1: Material property and geometric parameters for the levered linear spring mechanism

MATERIAL PROPERTIES	
E (GPa)	130
σ_Y (MPa)	600
General	
Depth b (mm)	12
Lever parameters	
Thickness t (mm)	0.1
Length l (mm)	2
Separation s (mm)	0.9
Force offset d (mm)	6.4
Lever ratio n	1:6.37 =0.157
Translation spring parameters	
Length L (mm)	8
Thickness t (mm)	0.82

The axial stress turns out to be 112 MPa while the bending stress in the lever is approximated by the equation

$$\sigma = \frac{Mt}{I2} = \frac{Et}{2l}\theta = 145 \qquad \text{(MPa)} \qquad (7.57)$$

Assuming the lost motion is almost entirely due to 's' shaped distortion of the lever springs, the bending stress associated with this deformation mode can be approximated from

$$\sigma = \frac{3Et}{l^2} x_3 \left(\frac{1}{n} - \frac{1}{n'} \right) = 725 \qquad \text{(MPa)} \qquad (7.58)$$

Because the axial stress may be assumed uniformly distributed, the worst case stress is simply the sum of axial and bending stresses and has a value of 982 MPa which is above the yield stress of the material. Clearly the lever flexures are likely to yield before buckling, a disaster either way. A number of design

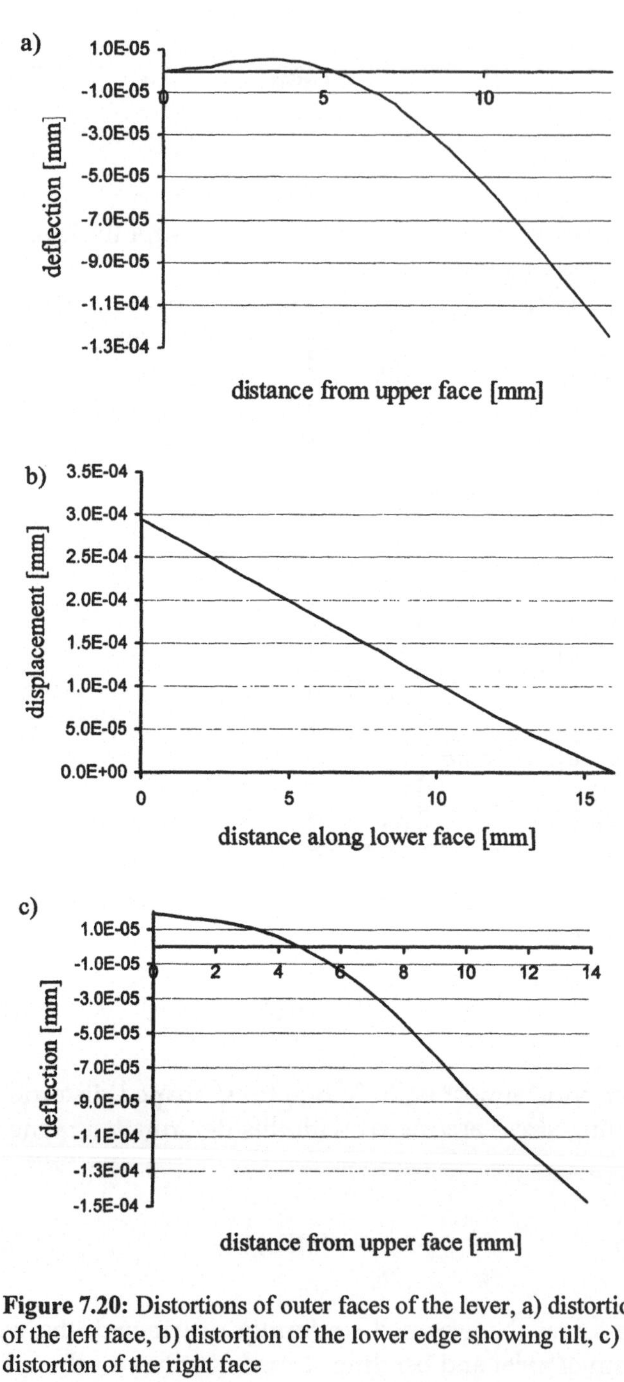

Figure 7.20: Distortions of outer faces of the lever, a) distortion of the left face, b) distortion of the lower edge showing tilt, c) distortion of the right face

modifications are possible to reduce this stress. Because the axial stiffness of the lever flexures are adequate, one solution is to reduce the thickness of the legs of the linear translation flexure to 0.4 mm for which a stress of 400 MPa is predicted.

To visualize the deflections and assess the fidelity of the above calculations finite element methods have been used to analyze this mechanism. Additionally, these flexures have been manufactured and measured using a vision-based, coordinate measuring machine to assess the accuracy of a commercial wire-electro-discharge-machine. This will be discussed in the following section.

From the finite element analysis it is possible to predict the distortions around the mechanism. To illustrate some of the previous discussions presented in this and the preceding chapters, distortions along three edges are shown. These correspond to; horizontal distortion of the left hand side of the vertical edge on the base, the parasitic displacement of the lower face and distortion of the vertical edge of the outer face of the of the moving platform in the direction perpendicular to the motion. These plots correspond to figures 7.20(a-c) respectively. The first and last of these figures show similar distortion corresponding to a net rotation primarily caused by a net rotation of the base support. Such a conclusion was made from a series of tests, not presented here, in which tilt was measured while the thickness value of both the left and right hand side supports were varied. This bending of the base support results in a displacement at the lower face of approximately 0.1 μm corresponding to around 4% of the output motion. Subtracting this tilt from these plots reveals distortion of the members of a shape similar to those analyzed in section 5.2. A simple calculation indicates a net rotation of around 8.7 μrad for 2.5 μm output motion. Over 50 μm this would correspond to 170 μrad. This rotation induces an additional displacement of the lever relative to the base. Consequently, this will result in additional lost motion thereby increasing the predicted attenuation from the previous value of 1:8.16. Figure 7.20(b) shows the parasitic motion of the lower surface. This has been computed from the vertical displacement relative to the point B that is used as the output displacement, which has a value of 2.5 μm in this analysis. Tilt of this member is approximately 2 ¼ times worse and at the extreme end contributes around 12% of the total motion. In fact, for this mechanism, the lever ratio between the displacement of the point *B* in figure 7.17(a) and the lever was found to be 1:9.54. Clearly, for this rather poor design there are a number of significant factors influencing the mechanism behavior that can be summarized as follows

1. Bending of the base will result in additional lost motion and unwanted rotations
2. Distortions of the supports also result in lost motion and parasitic rotations
3. Relatively large axial forces experienced by the lever flexures could result in buckling
4. Large lateral forces induce significant linear spring mode distortions that result in lost motion plus some additional stresses not accounted for in the above models
5. These large forces produce combined loads in the flexures that are likely to result in significant departure from behavior predicted from simple beam theories.

All of the above problems are significantly influenced by the relatively high stiffness of the linear spring. Reducing the thickness of the linear translation flexure will correspondingly reduce the stresses and forces in the mechanism. To assess this, flexures of thickness 0.6 and 0.4 mm have been modeled using finite elements. The lost motion predicted by these models is shown in figure 7.21. This plots theoretical predictions of the lever ratio are computed from the displacement of the midpoint of the lower edge of the moving platform to that at the lever with results normalized to an ideal lever of ratio 1:6.37 and the results from three different finite element models. Because the effect of finite stiffness of the base support is not included in the theoretical model, an increased lost motion, as observed with the finite element analysis, is expected. Additionally, the effects of stress concentration are not included in the model. For the relatively small lost-motion, when the thickness of the linear translation spring is 0.4 mm, the theoretical and finite element models correspond reasonably well.

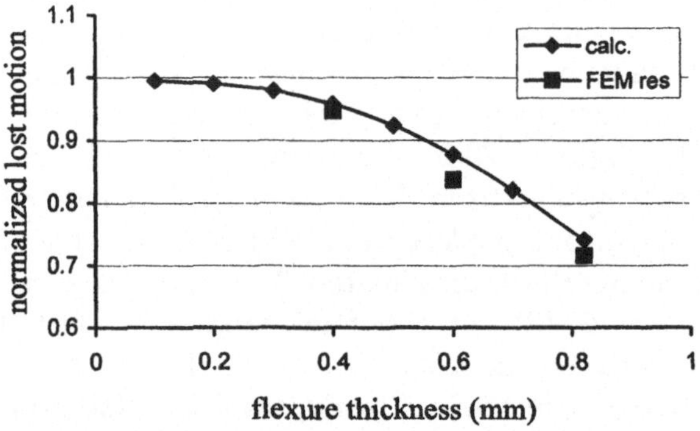

Figure 7.21: Lost motion of the lever mechanism as a function of the thickness of the linear translation flexures

It is also interesting to compare the theoretical stresses in the lever with results from the finite element analysis for these three designs. Maximum stresses occur at the ends of the lever and linear translation flexures. These have been analyzed using finite elements and the results compared with theoretical predictions. Because of the complexity of this mechanism and the finite mesh size, FEM results are probably accurate to within 5%. Based on results for the three different thickness of linear translation flexure, the error between the calculated stress and that predicted by the FEM analysis are tabulated below.

Table 7.5.2: Comparison of the differences between theoretical stress predictions and those produced from finite element analysis

t (mm)	% difference linear spring	% difference lever flexure
0.82	-10.00	-24.42
0.60	-8.10	-22.75
0.40	-2.50	4.90

a)

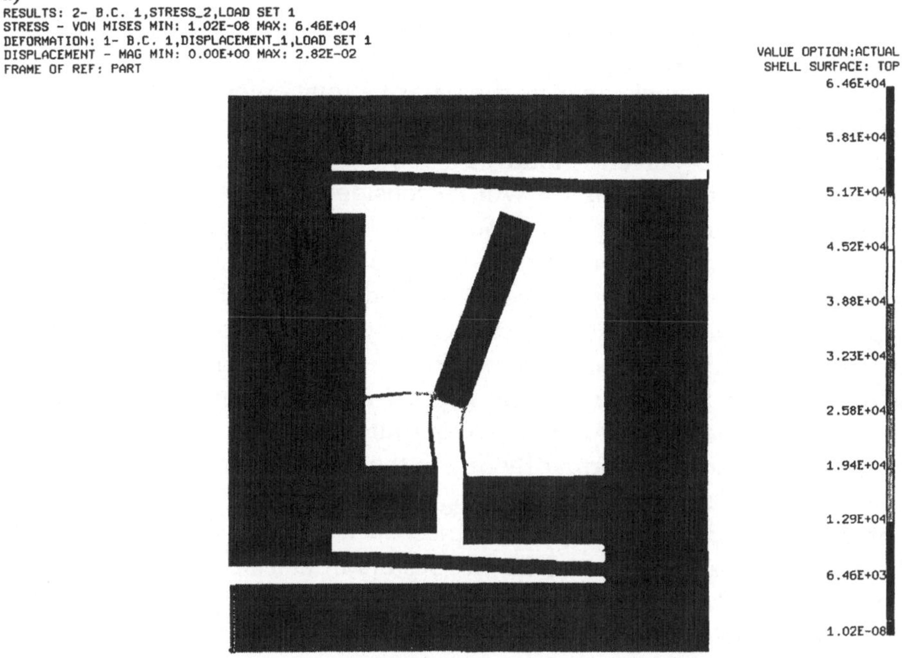

b)

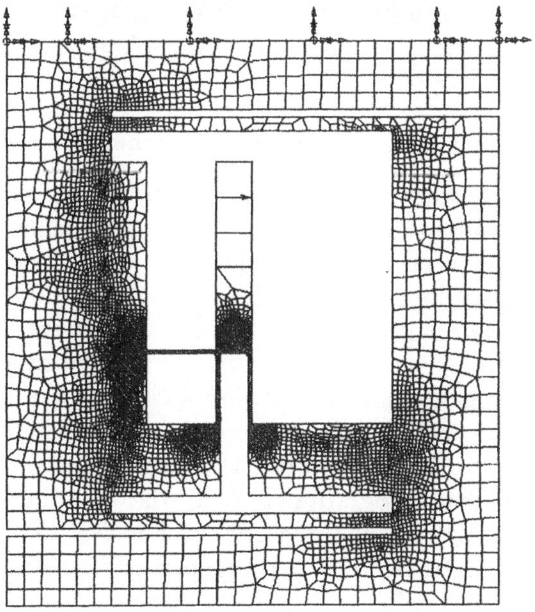

Figure 7.22: Alternative lever design in which horizontal motion of the lever is constrained, a) Solid model showing distortion, b) typical finite element mesh used in lever analyses

At both locations it can be seen that the difference between the two values increases with the thickness of the linear translation flexure. In this case a negative error corresponds to the calculated stress being higher than that produced by the finite element analysis.

Finally, it is informative to analyze the parasitic rotation of the lower edge of the moving platform as the linear flexures reduce in thickness. To illustrate this, the parasitic tilt as the percent motion of point *A* contributed by the tilt in comparison to the motion at point *B* will be considered. For the 0.82 mm thick linear flexures, this corresponded to about 12%. This value reduced to 9% at a thickness of 0.6 mm and 3% for 0.4 mm. For this last design, a parasitic tilt of 0.03*0.05/15 mm = 100 μrad is predicted for a displacement of 50 μm of the moving platform. As for the lost motion and analysis of stresses, the trend with thickness shows an improved performance for the linear flexure thickness of 0.4 mm and this is considered to represent a good design in terms of maximum stress and minimum parasitic motion of the output platform.

Because the lost motion is primarily due to the linear mode translation of the lever flexures, constraining this motion using a further leaf spring might result in a more predictable and consistent performance. Such a design is shown in figure 7.22. The addition of the horizontal flexure results in half of a cartwheel hinge producing a pivot at the lower left corner of the lever. Assuming an ideal pivot at this point, a lever ratio of 0.9/4.4 = 0.205 would be expected. Based on finite element analysis, figure 7.22(b), a measured value of 0.194 corresponds to within 5%. In this analysis, a maximum stress of 64.6 MPa occurred in the cartwheel hinge for a 43 μm displacement of the moving platform. The reader is left to ponder this design.

7.5.2 Manufacture of the lever using wire electro discharge machining

Both of the above designs have been manufactured with a wire electro-discharge machine. Typical machining times were around 2 ½ hours for a flexure of 10 mm depth. Once produced, accuracy of manufacture was assessed using a vision based coordinate measuring machine. Dimensions were obtained by measuring the positions of a number of points along an edge and computing a best-fit line to the data. Mean separation between these lines is then used as a measure of the relative dimension. In general, from measurements on five nominally identical flexures, it was found that the dimensions were consistently in error by an amount that would occur if the machining radius of the wire were approximately 6-7 μm larger than that programmed into the path-generation algorithm. However, it is not clear that this apparent error is that of the geometry of the flexure or a systematic effect due to the edge detection method used to measure this part. Either way, in principle, programming the cutting path for a larger wire easily compensates this. Typically, the standard deviation of the measured dimensions was of the order 2-3 μm which was only a little in excess of the measurement standard deviation of around 1.5 – 2 μm. In general, it may be concluded that the machine is capable of producing components with accuracy's of better than a few to 5 μm as is currently claimed by manufacturers of electro-discharge machines.

There were complications when trying to produce the thin flexures. First attempts resulted in the complete removal of these thin beams. It was speculated that this could be a consequence of combined forces from the high-pressure water jet used to both cool and flush the debris from the machining zone and thermal damage during the 'roughing' cuts. Reducing the water pressure from a few to less than 1 bar and leaving a larger amount of material for fine finishing resulted in acceptable flexure devices.

7.6 Case study II: Optical levers, galvanometers and the filar suspension (a long, happy marriage)

Towards the latter part of the 19th century, the quest to connect the fundamental electrical units with mechanical force led such esteemed scientists as Helmholtz, Joule, Kelvin, Maxwell and Weber into flexure design. Many of the instruments developed for this purpose consisted of null balances in which two equal and opposite forces are created. For example, currents between two adjacent wires will induce a force that can be balanced by that of a known mass in a known gravity field (which can be computed from equation 1.1). In many cases, because of the relatively small forces it is necessary that the disturbed mechanism be brought back to the null position will minimal residual forces. Requiring precise force control of small motions, this is an ideal application for a well designed flexure. Probably the most common design for such applications is to suspend the moving element from a filar (long thin rod) suspension. Based on the remarkably accurate results from these initial experiments this mechanism has continued to enjoy widespread use to the present day, particular examples being the galvanometer, torsion magnetometer and optical lever (called the 'mirror method' in Maxwell's treatise). A relatively low cost competitor to the filar suspension is the jeweled pivot, which provides a more robust rotary bearing with a surprising low torsion resistance. Figure 7.23 shows a typical moving coil mechanism originally designed by William Thompson (later Lord Kelvin). In this, the coil is suspended between two filar suspensions arranged so that the common axis is coincident with the center of gravity of the coil. Surrounding the coil are the poles of a magnet (either permanent or electro-magnetically generated) arranged so that the magnetic flux cuts the wires of the coil perpendicular to the current. Within this coil, but not touching it, is a soft iron core to direct the magnetic flux. Passing a current down the filar suspension to the wires results in a Lorentz force acting in equal an opposite directions either side of the coil. Under the action of this pure torque, the coil will twist through an angle θ. The torque can be computed from the torsion equation of chapter 2.

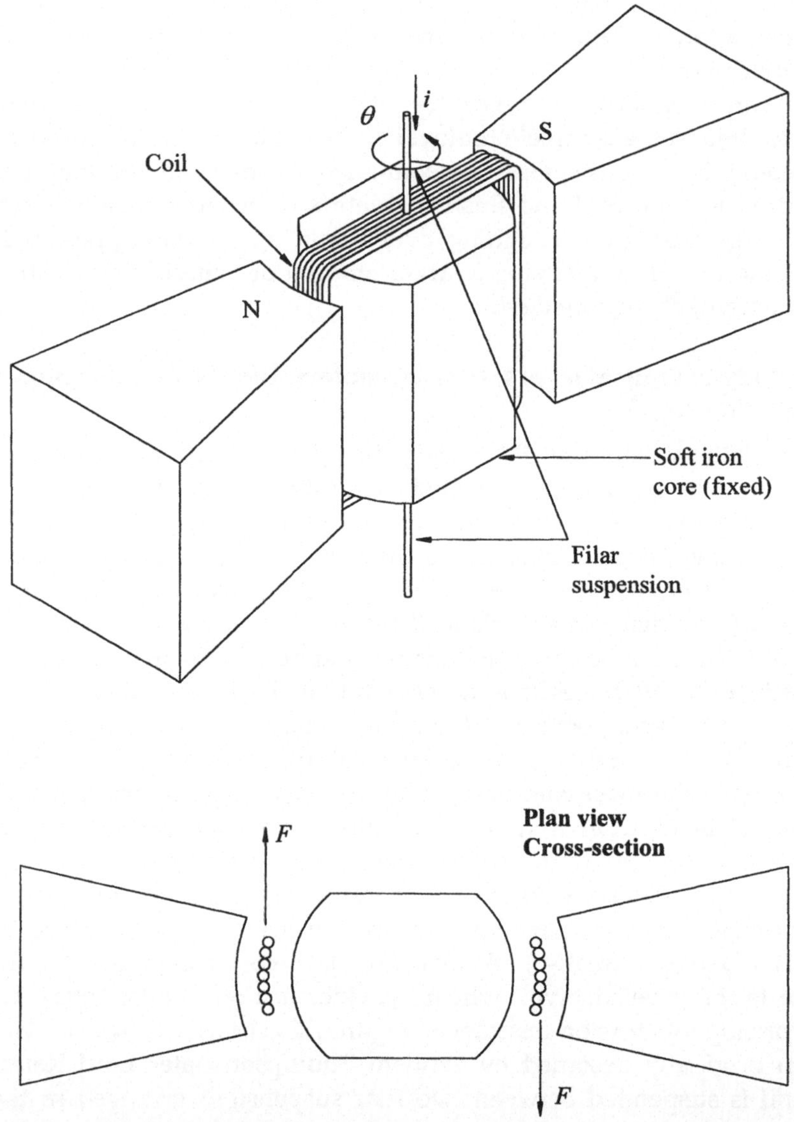

Figure 7.23: Schematic diagram of a moving coil galvanometer in which the coil is suspended, and pivots, by a filar suspension

A carefully wound coil and a magnetically soft iron core ensures the fidelity of this instrument, for a full discussion see Maxwell, 1891.

Three parameters determine the operation of this actuator. These are the angular twist due to torque applied by the coil, the current through the coil and cross product of the magnetic field and current. Instead of an electromagnet, the magnetic field can be produced using a second coil and, in this configuration, the direct force between the two conductors can be measured. Although systems involving two, three and four suspended coils were developed to null the error due to 'terrestrial magnetism', the principle of operation is similar. If either the two of the above are known, the third can be deduced. For the multiple coil measurements, the current is passed through all of the coils and subsequent deflections or forces measured. Two methods were commonly used. One is to suspend the coils on long wires and measure the pendulum rotation. Based on a knowledge of the mass of the coil and length of the suspension, the electromagnetic force can be deduced. Alternatively, to avoid complications associated with the 'lost' forces in the filar suspension, the twisting or displacements of the coil could be nulled with masses of 'known' value appropriately placed. The former techniques employs the so-called pendulum lever, Howard and Fu, 1997, the latter being referred to as a 'force balance'. In both cases, it is necessary to have an accurate knowledge of the geometry of the instrument and mass of the moving parts. Problems associated with a definition of mass have plagued researchers to this day and the current standard relies on an artifact. For large masses (1 g to 1 kg), uncertainties of better than 1 part in a million are possible. Bellow this, influences due to environmental contamination and instrument limitations can result in unacceptably large uncertainty, Howard and Fu, 1997. Because the electrical units can now be derived and experimentally realized from fundamental quantum phenomena, it is being proposed that the kilogram be replaced by the electrical definition of the Watt thereby linking the artifact kilogram, the meter and the second to the ohm and the volt. Recent experiments are reported in the paper of Williams *et al.*, 1998. To this day, a realizable definition of mass still hangs on a wire (in this case 50 strands).

In instruments such as the torsion magnetometer, the optical lever continues to be the main method for the measurement of angular deflection of the coil, see for example Hadfield, 1962. Typically, the coil of the galvanometer is suspended from one wire while a long rod hangs from bellow with the 'free' end hanging between two poles of an electromagnet. Specimens can then be placed on a platform attached to this end of the rod and subject to known magnetic fields. If the specimen has magnetic properties this will result in a torque on the rod and subsequently on the coil. Deflections are measured using the optical lever and fed back through an integrating controller to a current amplifier, which acts to null the deflection. If the instrument is calibrated, subsequent measurement of the current through the coil and knowledge of the applied field and orientation of the specimen enables magnetic properties of the specimen to be measured.

Applying a current to the galvanometer will produce a known angular deflection. A mirror placed on the moving element can then be used to direct a beam of light. Such devices are often used in printing devices, 'light shows' and may have a future as color displays (the electromagnet may be replaced by an electro-static actuator). Whether or not this latter application transpires, the marriage of the filar suspension to moving mirror looks like it will see a second diamond anniversary.

Another, more exacting, application of the torsion balance is that used for the measurement of the Newtonian gravitational constant. Initially used by Cavendish and Boys towards the end of the 20th century, new torsion balances are being proposed in an effort to reduce the uncertainty from the current value of 0.1% to better than a few parts in 100,000, Quinn *et al.*, 1997.

References

Furukawa E. and Mizuno M., 1990, Displacement amplification and reduction by means of linkage, *Bull. Japan Soc. Prec. Engg*, **24**(4), 285 - 290

Furukawa E., Mizuno M. and Terada K., 1991, A magnifying mechanism for use on piezo-driven mechanisms, *Bull. Japan Soc. Prec. Engg*, **25**(4), 315 - 320

Furukawa E., Mizuno M. and Hojo T., 1994, A twin-type piezo-driven translation mechanism, *Bull. Japan Soc. Prec. Engg*, **28**(1), 70 -75

Hadfield D., 1962, *Permanent Magnets and Magnetism*, J. Wiley and Sons, London.

Hart M., 1965, An Ångstrom ruler, *Brit. J. Appl. Phys.* (*J. Phys. D.*), **1**, 1405 – 1408.

Howard L.P. and Fu J., 1997, Accurate force measurements for miniature mechanical systems; a review of progress, *Proc. SPIE*, 3225, 2 – 11.

Johnson K.L., 1985, *Contact Mechanics*, Cambridge University Press, chpt. 9.

Marsh D.M., 1961, Micro-tensile testing machine, *J. Sci. Instrum.*, **38**, 229 - 234 (also discussed in Smith and Chetwynd, 1991, chapter 6)

Maxwell J.C., 1891, *A Treatise on Electricity and Magnetism*, Vol. II, 3rd Ed., Dover Publications Inc., NY. The reader will be well rewarded for perusing part III, chapters VII and VIII and part IV, chapter XV which includes a first hand account of these early developments will close attention being paid to the mechanical construction of a variety of coil suspension apparatus. The optical lever was well known and of great utility to these workers.

Mindlin R.D. and Deresiewicz H., 1953, Elastic spheres in contact under varying oblique forces, *Trans. ASME, Series E, J. Appl. Mech.*, **16**, 327 - 344

Quinn T.J., Speake C.C. and Davis R.S., 1997, Novel torsion balance for the measurement of the Newtonian gravitational constant, *Metrologia*, **34**, 245-249.

Smith S.T. and Chetwynd D.G., 1991, *Foundations of Ultraprecision Mechanism* Design, Gordon and Breach, NY.

Snyder J.J., 1993, Accurate, inexpensive, thermal expansion microtranslator, *Rev. Sci. Instrum.*, **64**(5), 1351 - 1354.

Williams E.R., Steiner R.L., Newell D.B. and Olsen P.T., 1998, Accurate measurement of the Planck constant, *Phys. Rev. Letts.*, **81**(12), 2404-2407.

Xu W. and King T., 1997, Flexure hinges for piezo-actuator displacement amplifiers: Flexibility, accuracy and stress considerations, *Precision Engineering*, **19**(1), 4 - 10

8 Manufacturing and assembly considerations

8.0 Overview

Having selected the appropriate material and geometry for a flexure there remains the task of building and assembling the mechanism. At this stage it is necessary to consider the methods of manufacture and fabrication. Often, conventional machining methods will adequately produce the component to the required tolerances. In many cases the flexure will consist of a thin-walled section. Some care is necessary to avoid imposing large forces during machining. Two approaches can be adopted. Either the tool geometry and cutting path strategy can be chosen to direct the cutting forces towards the bulk material or temporary supports in the form of conforming materials can be applied to the opposite face. For this latter purpose, a range of reusable waxes and low-melting temperature alloys are discussed. However, for some designs, conventional methods are either incapable of producing the complex shapes required or may introduce unacceptable surface damage in the form of residual stresses. Other times, there may exist more economic techniques for large-scale manufacture. In these circumstances, it is necessary to utilize less conventional methods of manufacture and it is these that occupy the first part of this chapter.

Flexures may be wrought from a single monolith of material or fabricated from individual components. While the former of these eliminates a lot of the problems associated with assembly, it is often impractical due to the flexure geometry or because the mechanism must integrate with other components for which a monolithic element would not be suitable. Other times, it might be desirable to change the stiffness of the flexure by interchanging flexible elements that, for example, might simply comprise flat plates. Consequently, the fabricated flexure is commonly found in many instruments and machines. For reliable performance, it is necessary that the flexure is designed for assembly and a few tips and tricks comprise the second part of this chapter. By definition, elastic mechanisms do not provide energy dissipation. Although in practice some losses will be intrinsic to all elastic mechanisms, it is often necessary to add viscous or visco-elastic damping. Some considerations are also covered in this section.

For some materials selected for these applications, it may be necessary to machine the material in its annealed state after which it is then necessary to heat treat the component. In some cases a post heat-treatment finishing process such as grinding or polishing are required. For conventional steels, such processes are common and procedures can be determined from any of the many handbooks on machining available in any good library. Beryllium copper is often used because of its high yield strain. The machining and subsequent heat treatment of this material is briefly described.

8.1 Manufacture

In most workshops, available machine tools usually comprise lathes, milling machines, grinders and boring machines. For large flexures, these machine tools, in combination with standard fastening techniques (including welding and gluing), probably represent the optimal manufacturing method. Commonly, either separate components are made and subsequently assembled or the flexure may be directly machined into a solid to form a monolithic mechanism. For the production of small, instrument flexures, a broad range of alternative processing techniques may better achieve a particular design. Also, in some cases, it may be necessary to produce the flexure from hard or brittle materials that are not amenable to single point cutting or grinding processes. Under these circumstances, it is necessary to utilize less conventional production methods. Techniques commonly used for the manufacture of flexures include

- Conventional machining
- Electro-discharge machining
- Lithographic etching
- Electroplating (or electro-forming)
- Diamond grinding

These are discussed, in turn, below

8.1.1 Conventional machining

The objective of this brief section is to provide a few tips on the use of conventional machine tools to produce leaf or notch-type flexure geometry's. Detailed machining parameters for most materials can be found in available literature and therefore are not covered here.

Consider a flexure mechanism produced from a single monolithic solid. By definition, toward the end of the machining cycle, the monolith will consist of two or more rigid bodies connected by relatively compliant elements. Clearly, it will be necessary to ensure that these rigid bodies are suitably constrained on the machine tool. With care, such constraints can often be supplied using standard fixtures. Additionally, during manufacture, it will be necessary to machine the leaf and notch hinges. This is problematic as the limited strength of the thin webs of the hinge render it susceptible to deformation if large cutting forces are applied in the wrong direction. If possible, machining using small, sharp, cutters running at relatively high speeds will tend to minimize these undesirable forces.

Additionally, it may be possible that the flexure can be machined in a way that ensures that the cutter is always removing material from a relatively stiff region on the monolith. When machining, the general rule is to direct the cutting forces towards stiff or supported solid material.

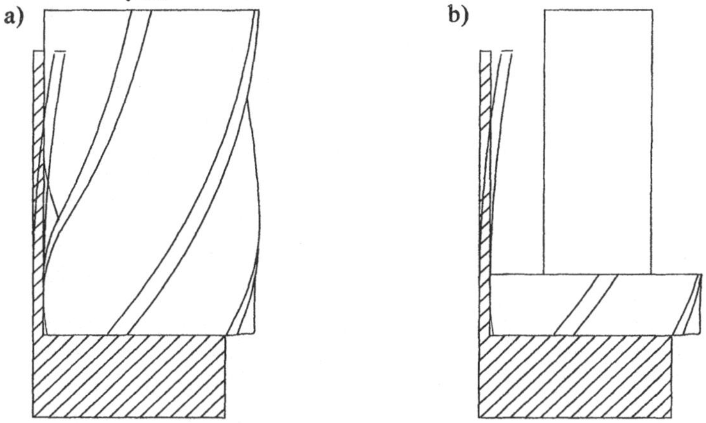

Figure 8.1: Using the side face of an end mill to produce a thin walled section, a) using a conventional cutter, b) use of a modified cutter to reduce surface damage in the presence of chatter

During the cutting of thin walled structures using a milling machine, it is important to keep in mind the machining forces at the point of cutting. Cutting with an end mill will introduce both axial and radial forces. Figure 8.1(a) illustrates the cutting of a thin walled section using a standard end mill. In this figure the cutter is moving towards or away from the reader (i.e. it is moving in a line that is perpendicular to the plane of the paper) and the lower face of the workpiece is considered to be rigidly fixed to the machine tool frame. Compliance with the principle of machining from solid or high stiffness it is obvious that the most desirable cutting planes are either on the upper face of the solid block or, near to the bottom of the thin web. Clearly, the worst place to attempt machining is at the surface of the thin web towards the top. Also illustrated in this figure is an exaggerated distortion of the thin web when it is vibrating. Predominantly, it is the periodic cutting forces introduced by the teeth of the cutter the source of excitation. When the rotational speed of the tool produces periodic forces in harmony with the natural frequency of the workpiece then the forced vibration amplitudes increase dramatically. Often, after excitation of the workpiece begins the interaction between the moving surface and the tool will introduce additional forces. Because these forces will be at a similar (or 'sympathetic') frequency as the vibration, they will further amplify the distortion leading to a form of self-excited vibration commonly called 'chatter'. Inevitably this will diminish surface finish and geometry of the web. It might appear that choosing a different spindle speed or irregularly spacing the cutting teeth around the tool periphery could solve this. However, the natural frequency of the web will change as machining progresses. Although it might not be possible to eliminate vibration of the web, the effect of this on the geometry of the final component can be reduced by changing to the cutter geometry shown in figure 8.1(b), for more detailed discussions see Smith and Dvorak, 1997. In this case, the flutes of the end mill extend a short axial length from the free end of the cutter after which they are removed. Above these flutes, a clearance provides a space in which vibrations of the web will have little effect. A drawback with this approach is that the web must be produced by multiple step-and-repeat machining cycles.

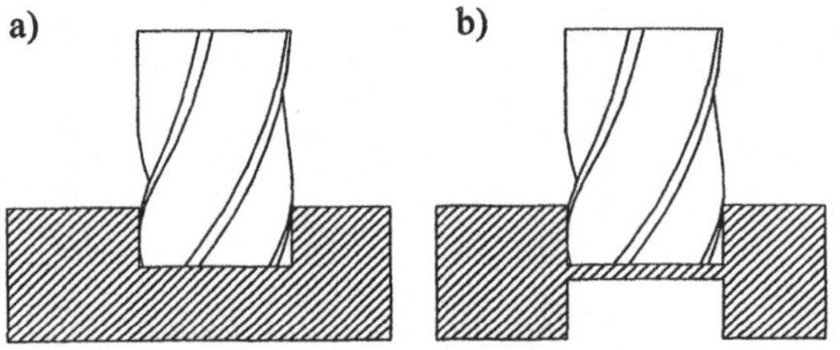

Figure 8.2: Machining of a monolithic flexure using an end mill, a) a slot is produced in a solid block, b) the block is turned upside down and another slot produced. The flexure hinge consists of the surfaces generated by the end face of the cutter during the two machining operations.

An alternative method for producing thin sections is represented in figure 8.2. In this case, the machining direction is the same as for the previous figure but the flexure is being produced from the horizontal surfaces machined by the end face of the cutter. While some vertical forces will be produced during this machining operation, these are relatively small compared to those imparted on the rigid blocks by the side face of the cutter. In this illustration it is assumed that the flexure is to be produced by first of all machining the upper slot after which the block is turned upside down. The flexure will emerge after the second slot is complete (this is when the cutter, which is moving away from the reader, is clear of the workpiece). In practice, to reduce stress concentrations at the roots of this flexure, a radius end mill should be used.

For flexures of higher slenderness it may be necessary to use smaller cutters operating at higher speeds in an effort to reduce forces in the direction of the thin web. Again, it may be possible to generate the thin web using end milling using the side face of the cutter to machine primarily on the side face from solid material. Such a scenario is shown in figure 8.3. The slot is machined in a similar manner to the previous example. However, in this case, it is also necessary to translate the cutter along the length of the flexure as shown. More detailed discussions and examples giving all machining parameters for thin web manufacture from an aluminum workpiece can be found in the papers of Tlusty *et al.*, 1996, and Smith and Dvorak, 1998. Under suitable machining conditions manufacture of webs of thickness 0.5 mm and lengths of 50 mm or more (i.e. slenderness ratio's of 100) can be manufactured to relatively high precisions.

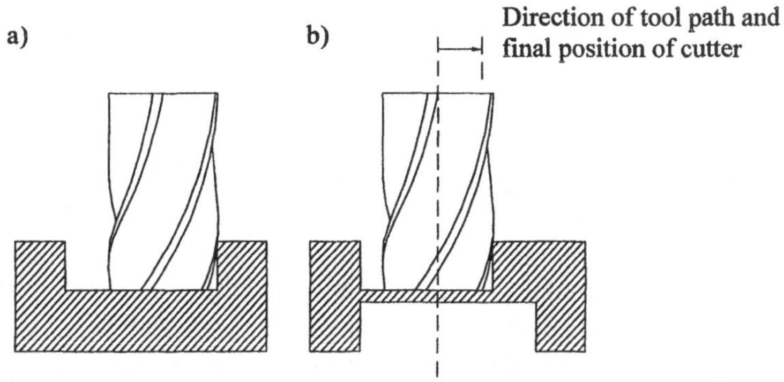

Figure 8.3: Machining a slender, monolithic flexure, a) a slot is produced in a solid block, b) the block is turned upside down and the process repeated

An important advantage of notch hinges is that they are easier to produce using conventional machine tools. In principle, it is only necessary to drill the holes for the webs with any precision. Because this can be performed on the solid blank, provided the webs are of sufficient thickness, these webs will have tolerances near to the precision of the machine tool used to produce them. After the holes for the notches have been produced, removal of the remaining material will have little influence on the performance of the mechanism. Often, in the production of the webs, the drilling of the second hole will significantly distort the web of the notch. This is particularly troublesome because the most distortion will occur at the thinnest portion of the web that, in turn, imposes the most significant influence on its performance. In many cases, a remedy to this problem is to drill (and, preferably, ream) the first hole and insert a closely fitting dowel pin. This will then provide the necessary support of the thin web while the second hole is being drilled or reamed.

For non-circular webs, manufacture of fixtures to support the rear face during final machining on the other side becomes time consuming and expensive. To ensure stiff support, it is also necessary that the support closely conforms to the surface and hence the fixture must be either precisely manufactured or some fine adjustment must be incorporated. Alternatively, if a pocket can be formed that incorporates one flexure surface, support can be added by temporarily filling this with liquid filler that can be solidified and removed after machining is complete. Choosing the appropriate liquid is still a matter of debate. Contenders include

- Waxes
- Low temperature metals that do not contract upon cooling
- Two part epoxy adhesives
- Solders

In principle, application of these temporary fillers is readily implemented. While in the liquid state the filler material is poured into the plenum and solidified after which it functions as a support on the opposite face of the web during machining. After machining, the filler is removed, dissolved or transformed back to its liquid form and drained from the flexure. With this latter option, the filler can be reused.

Waxes possessing a very wide range of thermal and mechanical properties can be readily obtained. Most commonly these are supplied for the purposes of optical bonding, candle making or as reusable machine tool prototype material.

The so-called 'optical waxes' have been in use for centuries for the temporary bonding of optical components during lapping, polishing etc. These can be obtained with a large variety of hardness values and softening temperatures. Primarily these are made from varying mixtures of beeswax and rosin sometimes with additives such as turpentine or rosin oil. In general, the higher the proportion of rosin the harder the wax. A soft optical wax can be readily produced from a mix of equal parts refined bee's wax and optical rosin. A typical medium hardness wax is produced from 16 parts rosin, 2 parts bee's wax, 1 part Carnauba wax and a small addition of Venetian turpentine or rosin oil. A large variety of waxes for different applications are commercially available. Generally, softening temperatures tend to correlate with hardness of the wax. Typically the softer waxes will soften at temperatures of around 40 to 60 °C while the harder waxes tend to soften at temperatures in the region 70 to 140 °C. Usually, increasing the temperature above the softening point will result in a reduction of the viscosity. For example, a hard wax with a softening temperature of 80 °C typically will not flow 'freely' to form a thin surface film until heated to 140 °C. In the softened state these waxes have good wetting properties and will adhere to almost any clean metallic or ceramic surface. Upon cooling this will provide a relatively stiff bond or, in bulk, a strongly adhered solid surface. Although the wax contracts upon cooling, this does not cause many problems. Because of its low conductivity it is likely to cool slowly from the outer surfaces and, therefore, shrinkage tends to result in a slight 'coring' in the center. Most waxes will form a strong bond that will resist moderate machining forces. A lot of the commercial waxes will slowly dissolve in detergents and many coolants used in machining processes. However, at room temperature, dissolution usually occurs slowly with reduction in strengths of bonded components only becoming noticeable after having been immersed in coolant for many days.

When machining is complete, the wax and any temporary backing materials can be removed by re-heating the assembly above the softening point. When the wax has been removed or scraped away, there will often be a residual film. These films can be removed using a suitable solvent. Typically, warm acetone or petroleum provide relatively rapid action while alcohol, which is most commonly used, is rather slow. In all cases, solubility increases with temperature. Information of other, more efficient, solvents can often be obtained from the supplier. Excessive heating can result in the wax film becoming burnt or fused to the surface. Invariably, this will be very difficult to remove.

Waxes provided for the purpose of prototyping in machine tools have been used to provide support during the machining of thin flexures.

Some reusable, low-melting point metal alloys are supplied specifically to provide a temporary supporting material for machining thin-walled structures. Because these alloys can expand upon solidification, components held in the frozen matrix will experience a clamping force, and so, they are also used for the securing of awkwardly shaped components. For this reason they are also referred to as 'fixturing alloys'. One such alloy has a melting point of around 70 °C and can therefore be removed by placing it in a container and putting it in a hot bath[1].

A disadvantage with both of the above techniques is that it is necessary to heat up the workpiece. Often, to do so, it is necessary to remove it from the machine tool, after which, precise relocation of the workpiece is problematic. Heating the workpiece in the machine tool may cause unacceptable time delays because it is often necessary to wait for the system to return to an acceptable thermal equilibrium. Room temperature adhesives appear to provide a solution.

Current efforts to utilize solders as a filling material have not met with success. There are three reasons for this; they shrink upon solidification, they do not wet many of the materials used for flexures and their high conductivity causes problems due to rapid solidification and, therefore, inadequate flow during pouring.

8.1.2 Electro-discharge machining

Two types of electric discharge machines are commonly found in engineering workshops. These are the plunge and wire electro discharge machines outlined in the following sections

8.1.2.1 Plunge electro-discharge machining

Electro-discharge machining, as the name suggests, utilizes electric discharges to erode local areas of a surface as a shaped electrode is brought into close proximity to a grounded workpiece surface. Over the last four decades, this technology has undergone significant development and such machines are now considered to be capable of producing components to a relatively high precision.

The principle of operation is relatively simple. A shaped electrode is produced using conventional machining, sintering or forming operations. The work-piece is immersed in a dielectric fluid (mineral oil, kerosene or purified water). A potential is then applied to the electrode and it is lowered towards the workpiece surface until electric discharges are produced. A servo-control maintains a constant separation to enable pulsed discharges at rates ranging

[1] Supplied by MSC Industrial Supply Co., PA.

from tens to hundreds of kiloHertz. Continuously plunging and flushing debris from the work zone enables the surface to be eroded to any desired depth. After machining the workpiece will become a replica of the electrode. To produce the discharge, it is necessary to apply potentials ranging from tens to hundreds of volts resulting in currents of 0.1 to 500 amps. At the highest removal rates, the surface will melt and re-solidify leaving a relatively thick damage layer with poor surface finish. After rough machining, a lower power finishing process is required to remove this re-solidified, or recast, layer (see below).

Electrodes are usually made from materials with a high melting temperature and high latent heat of melting. Graphite is often used for shaped electrodes, while tungsten wires are commonly used to produce small holes of high aspect ratio. Alternatively, copper tools can be used in a process called 'no wear' electric discharge machining, Kalpakjian, 1995.

8.1.2.2 Wire electrodischarge machining

Although the material removal mechanism is similar to the above, the principle of operation for this process is entirely different. In this case, the electrode is a thin wire that is continuously fed through the work-piece. This is then electrically energized and used to erode a thin line through the workpiece material. Any wood worker will recognize this as an equivalent to jig sawing, or grocers, the cheese cutter. Moving the wire in a controlled path through the specimen results in the desired geometry. In practice, wires ranging in size from 0.025 to 1 mm are common. Presently, capital cost of these machines is high and the continued costs of wire and filtering of the flushing fluid limits this process to the production of precise, high value flexures. It is not clear that this will always be so. Some of the advantages of wire electro-discharge machining for flexures are

- Components of almost arbitrary complexity can be produced.
- Any reasonable conductor can be machined enabling choice from the hardest materials
- For thin wires, the gaps between surfaces can be correspondingly small, enabling the design of compact mechanisms.
- Being capable of machining to depths of 100 - 200 mm, geometry's of high aspect ratio can be produced making possible designs with high ratios of stiffness between the free and constrained directions.
- Highly automated machines capable of re-threading accidentally broken wires can be left unattended, therefore reducing labor costs.
- Work-pieces can be stacked to produce multiple components with a single machining cycle.

The economics of this process are not straightforward and the reader is left to assess their own particular needs and applications. In passing, it is noted that the erosion process requires the volumetric removal of material in contrast to machining in which only new surfaces are produced through extensive plastic deformation. However, unlike machining using conventional cutting tools, wire EDM can be used for coring. Because the core remains in one piece, it can be further used as stock for other, smaller, components. For removing large volumes of material it is sometimes faster and requires less energy to use wire electro-discharge machining. For comparative purposes, some typical machining parameters are outlined below.

Cutting speeds through the work-piece typically range from less than one up to a few millimeters per minute for thin wires of diameter 0.25 mm or less. In general, the feed-rate will slow down with increasing thickness of the work-piece. At the faster cutting speeds or higher removal rates, surface finish tends to become rougher with R_a values typically in the region of a few micrometers. Applying 'finishing' cuts can produce better surfaces with R_a values of a small fraction of a micrometer. However, this order of magnitude gain in surface finish is obtained at the expense of discharge power. Consequently, machining times are likely to increase with each finishing cut (sometimes the feed rate for the finishing cut is one third of that for the 'fast burn'). For low surface damage and surface finish values, it is not unusual to perform four finish cuts. Consequently, the machining time may increase dramatically. Typically, the electrode wire feeds through the specimen at a rate of 100 - 300 mm s^{-1}.

Under controlled conditions, it is not unreasonable to produce components holding tolerances within a few micrometers and with sub-micrometer surface finish. Flexures measuring 0.1 mm in the thinnest region are routinely manufactured. Although some surface alteration must occur, no significant problems caused by this have been observed in flexures of these dimensions. Current machines can produce components to accuracies of better than a few micrometers, see for example the case study 1 of the previous chapter.

As an example, a semicircular notch hinge of radius 6.4 mm, $\beta = 0.172$ and a depth of 15 mm could be produced with four finishing cuts in a total time of approximately 30 minutes.

8.1.3 Lithographic etching

Manufacture of flexures using lithographic etching processes is a widely used technique for the mass production of complex planar mechanisms. One does not have to look far to find one. Magnetic read heads on almost all hard disk storage devices are supported on thin stainless steel flexures, accelerometers commonly used in automobiles as crash detectors contain flexures integrated into micro-electronic sensing circuits, and thin beryllium copper flexures are commonly used in fine scientific instruments. In fact, the former two devices are produced in quantities measured in tens of millions per year. The first uses a proprietary etching process.

8.1.3.1 Flexures produced using microelectronic processing techniques (MEMS)

Production of flexures using microelectronic manufacturing techniques is well known. Laboratory devices manufactured more for scientific interest than market potential were being made in the late 1970's. It was not until these devices were presented in a seminal paper by Petersen, 1982, that the scientific community became alerted to the potential of micro-electro-mechanical devices (now often referred to as MEMS). Since this time, the number of applications has grown enormously although the main markets appear to be for displacement sensors, accelerometers or resonators for mechanical frequency selection.

In fact a simple dimensional analysis of flexures indicates the advantageous dynamic characteristics of miniature devices. Consider a simple cantilever beam for which the natural frequency is given by (3.232)

$$\omega_n = 1.875\sqrt{\frac{EI}{ml^3}} = 1.875\sqrt{\frac{Et^2}{12\rho l^4}} \tag{8.1}$$

If we now change *all* of the dimensions by a factor k, equation (8.1) above becomes

$$\omega_n = \frac{1.875}{k}\sqrt{\frac{Et^2}{12\rho l^4}} \tag{8.2}$$

For a reduction in the dimensions there is a corresponding linear increase in the dynamic response of the system. Such a scaling is, at least intuitively, obvious when observing relative heights that can be jumped from a standing start by elephants, dogs and their fleas.

Manufacture of the micro-mechanical devices traditionally starts with a flat silicon wafer. Patterned masks are transferred to the surface using photographic techniques and thin layers are either grown or deposited. Removal of the mask leaves a thin film on the surface that is a replica of the original pattern. Common film deposits are silicon dioxide (often used as a spacing layer to be removed during a later processing stage), polycrystalline silicon (structural components), silicon nitride (insulator) and aluminum (conductor usually on the upper face). After multi-layered structures have been deposited, selective etches can then be used to remove underlying layers leaving thin films 'free'. Components deposited on these layers can be left completely free to form rotors or more complex shapes or may remain attached at certain points so that the substrate becomes the fixed link of a planar flexure mechanism.

Currently, the thickness of the films and, therefore, these devices are limited from a few to tens of micrometers. Consequently, devices with planar dimensions considerably larger than this tend to be rather slender. Another technique for producing components that have a higher aspect ratio (ratio of long dimension to thickness) is called LIGA, see Guckel *et al.*, 1991. This utilizes thick films that can be decomposed using high energy x-rays. Screening the thick film with a mask and exposing it to high intensity x-rays can produce high aspect ratio holes. These can then be filled with, for example, nickel, after which, removing the surrounding film will leave correspondingly high aspect parts.

Fortunately, the analytic techniques outlined in the previous chapters are equally applicable to these devices. It is clear that the mass production of flexures using micro-scale lithographic processing will provide inexpensive, fast and repeatable flexure systems with integrated sensing and computing capacity.

8.1.3.2 Lithographic etching of copper sheet

For many decades, workers in electrical industries have been producing complex shapes from thin copper films. Printed circuit manufacture can be readily adapted for the production of planar flexures in thin beryllium copper sheet. Advantages of this process are

- Arbitrary planar shapes can be produced from computer generated masks
- Beryllium copper can be processed in worked and hardened state as supplied by the manufacturer[2] thereby avoiding post process heat treatments.

During the production of a flexure, the following steps represent a typical manufacturing procedure

1. The thin sheet is first of all cleaned using a solvent such as acetone or isopropyl alcohol.
2. A thin layer of photo-resist is deposited onto the surface preferably spinning the sheet to produce a thin and relatively uniform deposit. Note: Positive acting photo resists tend to be easier to process.
3. Baking to around 100° Celsius for approximately one minute then hardens the photo resist.
4. An opaque mask is placed over the surface, which is subsequently exposed to ultra-violet light. A 150 W light source requires an exposure of around 10 seconds. If this is not available, exposure in strong sunlight for ten to twenty minutes is often adequate.

[2] Brush Wellman Inc., Ohio 44110 will supply detailed booklets relating the properties and heat treatment of a range of beryllium copper alloys

5. The photo resist is then developed (sodium hydroxide developer) leaving the original mask pattern on the surface and this is then baked for around twenty minutes at 80 –100 Celsius.

Finally, the rear face of the sheet is painted with etch resist (often referred to as 'dope' by printed circuit manufacturers) and the sheet then etched in ferric chloride. To accelerate etching, the ferric chloride solution is often heated to around 40 - 60 Celsius. Etching continues until all unexposed copper has been dissolved. Removal of the flexures as soon at this occurs ensures minimal undercutting.

In general, if the flexures are removed as soon as etching is complete, undercutting by the etchant into the pattern will only extend a distance roughly equal to the thickness of the film. For large thin sheets, such an effect will be insignificant. In practice, sheets of up to 200 μm thick can be readily etched at minimal expense.

8.1.4 Electroplating (or electro-forming)

Bellows produced by electro-deposition are commercially available[3]. If the film deposited is removed to become the component, this process is called electro-forming. Typically, these are produced by forming a mandrel, depositing a desired thickness of metal (nickel), trimming the metal and then dissolving out the mandrel. All metals can be electro-deposited and, because precise control of deposition is possible, films ranging from near atomic dimensions to tens of micrometer thickness can be readily produced. Other metals commonly used for electro-forming include copper, gold and silver.

For coating onto insulating materials such as plastics, the surfaces must be pre-coated using electro-less processes. Nickel is amenable to a relatively straightforward electro-less deposition.

The advantages of this process are

I. Complex three dimensional geometry's can be readily produced

II. Because the film can be made very thin, relatively large distortions can be achieved. This is immediately apparent from the bending equation, which, for a flat plate of thickness t bent to a radius R, there will be a stress at the surface of magnitude

$$\sigma = \frac{Et}{2R} \tag{8.3}$$

As the thickness t reduces the curvature $(1/R)$ of the film can increase.

III. Seamless, non-porous geometry's can be produced.

[3] Servometer Corporation, Cedar-Grove, NJ 07009. This company supplies informative booklets outlining applications and detailed design calculations.

IV. Components of almost arbitrary size can be produced ranging from bellows of less than 1 mm outside diameter to components weighing hundreds of kilograms

It should also be noted that thin film devices often exhibit excellent fatigue resistance.

8.1.5 Diamond grinding

In some applications, it might be desired to use ceramics or other brittle materials for the flexure. The advantages offered by these materials include

- Being brittle, they will exhibit linear behavior only. In contrast, flexures produced from metals and other ductile materials will only exhibit linear behavior if the maximum stresses are below the elastic limit. In brittle materials, the elastic limit and catastrophic failure occur at the same stress.
- Materials can have excellent combinations of properties. For example, the glass ceramic Zerodur™ has a thermal expansion coefficient that can be a factor of 200 or more lower than steels at room temperature; silicon can be obtained in very pure, single crystal form, titanium boride has a very high hardness and speed of sound (however, it is difficult to machine with any cutter); diamond has superior mechanical and thermal properties to *any* other material, etc. (for a more complete discussion see Smith and Chetwynd, 1992).
- Brittle materials tend to maintain dimensional stability over extended time periods.
- Many ceramics and insulators provide excellent immunity to hostile environments.

Many ceramics and non-metallic solids can be readily machined by grinding with either diamond, cubic boron nitride or other hard grits impregnated in a matrix. Because of the broad range of materials that can be processed using diamond cutters, these shall be the focus of this section. However, in many aspects, similar process parameters (i.e. speeds, feeds, etc.) apply to grinding wheels comprising the alternative grit types.

Diamond wheels can be purchased in a variety of shapes and sizes. Invariably, the diamond grits are attached to a metal wheel or to the ends of spindles (often called 'pins'). These, in turn, mount into standard grinding spindles that can be intrinsic to the machine or retrofit to milling machines and lathes. Grit sizes ranges from very fine 400 grit (this number indicates the number of grits on average that would form a straight line one inch long) up to coarse sizes of around 50 (approximately 0.5 mm). For super fine finishing, 800 grit wheels are sometimes used. The graph of figure 8.4 provides a rough guide to the expected surface finish as a function of grit size. Although it would appear that the finer grits are desirable for all applications, the maximum depth of cut

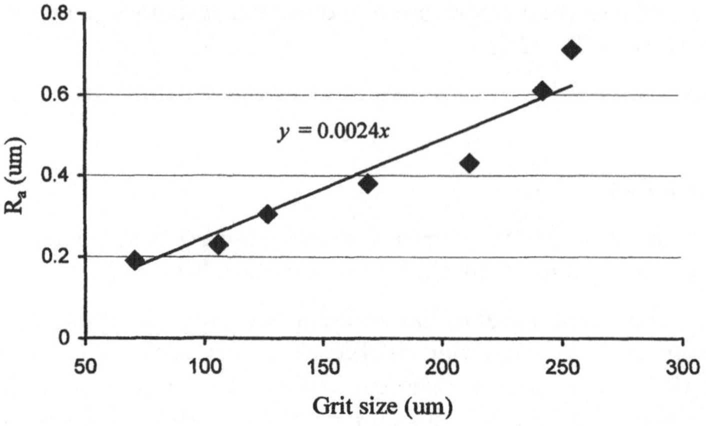

Figure 8.4: Expected surface finish as a function of the grit size used for grinding

reduces at nearly the same rate. As a consequence, the volumetric removal rates drops off even more rapidly as the surface finish improves.

Typically the abrasive part of the wheel is produced using one of two techniques. Possibly the least expensive method is to coat the metal shank with diamond grits and then deposit over these a thin nickel plating for retention. Although inexpensive, it is difficult to true such a wheel and the grits tend to have a relatively low packing density. Consequently, these are usually used for rough machining where surface finish and dimensional tolerances are not important.

A better quality wheel can be produced by mixing the diamond grits in a softer matrix (any matrix will be softer!) and bonding this to the wheel or pin. Common matrix materials are either resin or metal (brass or bronze) based. The latter matrix proving to be more difficult to produce and, therefore, more expensive to purchase. However, both wheels have the advantage that they can be trued for optimal cutting conditions. Because of their excellent durability, it is often possible to dynamically balance the metal bonded wheels after which they can be used for many hours without further attention. It is the longevity of the metal bonded wheels that makes them desirable for precision machining applications. In the author's experience, the quality of all types of wheel can be very variable with wheels appearing to wear out almost immediately or some, seemingly, last forever. As a consequence, many machinists can be very possessive of their 'favorite' wheels.

When machining ceramics, there are some general rules that can help the novice to get started. These are

- Keep the workpiece surface as cool as possible. Flush the grinding process with a fast flow of coolant taking care to direct the flow directly into the cutting region.
- Make sure that there is plenty of opportunity for the debris to be removed from the cutting interface. Clogging of ceramics at high speed can lead to rapid heating and solidifying of aggregates, often with catastrophic result. Unlike metals, which will tend to make a lot of noise and glow before failing, ceramics give no warning signs before shattering.
- Use safety guards.
- Do not machine steel, it will wear out the diamond wheels very rapidly.
- For efficient machining rates, use surface speeds of the grinding wheel of the order 5 to 15 m s^{-1}.

This last parameter is important for the machining of complex shapes using small cutters similar to those used for end milling. Under these conditions, the cutter consists of a small end mill often having a diameter of a few millimeters or less. To achieve a requisite surface speed of 10 m s^{-1}, a cutter of diameter 4 mm must rotate at an angular speed of 47,750 rpm. Fortunately, air turbine and electric motor driven spindles that can readily achieve such speeds are commercially available. Although it may not have sufficient rigidity for precision applications, a common dentist drill will typically be rotating at speeds higher than 100,000 rpm.

Usually, combinations of slicing and end milling are sufficient to produce most flexures. Provided that the debris can be removed from the cutting zone, arbitrary depths of cut are possible. In general the feed rate will depend on the size of the grits. Typically the depth of cut per pass of a grit through the work piece should not exceed one hundredth of the grit size and for fast cutter speeds and deep cuts this should be reduced by an order of magnitude. However, this is only a guide and can vary considerably depending on the machining configuration. For example, with small cutters rotating at speeds of 1000 revolutions per second, a grit size of 100 μm corresponds to a feed rate of 0.1- 1.0 mm s^{-1}. As an example, single crystal silicon is readily machined with a 100 grit, 2 mm diameter wheel rotating at 50,000 rpm with a feed rate of 2 mm min^{-1} or 0.3 mm s^{-1}. Machining with slow feeds and large depths of cut is usually referred to as creep-feed grinding. This has the advantage of utilizing more of the cutter surface therefore increasing the life of a tool.

Holding a brittle workpiece in a machine tool is a problem. Most materials are non-(ferro)magnetic and, therefore, cannot be held directly using a magnetic chuck. Being brittle, clamping between the hard jaws of a chuck can cause the component to crack if there are any local stress concentrations. High stresses can be caused either by hard dust particles or if the faces of the vise are not coincident with the work piece so that there is a clamping force along an edge. Both scenarios are likely. A common method for mounting the work piece in a machine tool is to first stick them to the surface of a metal block using optical waxes, see section 8.1.1. This metal block can then be mounted in the machine tool in the usual manner. If the component is to be sliced through, it is usual to place an intermediate glass or ceramic plate between the work piece and metal mounting block. Float glass, being inexpensive and readily available in most hardware stores, is often the material of choice for this.

Again, the economics of creep feed grinding are not straightforward. Typically, it is feasible to finish the machining process with a single cut. When manufacturing metal components, it is common to saw the blank from larger stock, remove bulk material using roughing cutters and complete the finishing process with a relatively slow, fine finishing (usually grinding) operation. Creep-feed grinding, although requiring a slow feed rate, can combine roughing and finishing in one process. Consequently, it is likely to become economic for the cutting of large slices. Another consideration is that of mounting these specimens. Melting and cooling of optical waxes is time consuming. This may only be economic for components with large profit margins or in cases in which multiple flexures can be produced from a single machining cycle.

Drilling represents a more difficult challenge. In principle, it should be simple to use a diamond-coated pin as a 'drill'. The key problems appear to be the removal of debris from the hole and the fact that no matter how fast the cutter is rotating, the surface speed is always zero at the bottom of the hole in the center. Although it is feasible to drill using pin type cutters rotating at high speeds, it is difficult to prevent the occasional clogging. Invariably this results in the drill 'binding', often with catastrophic result. A more conservative approach is to rotate the cutter slowly (the author has used 70 rpm with a 3 mm cutter for 'drilling' a 10 mm deep hole) and apply constant force to the cutter. To avoid clogging, the cutter is continuously retracted and the hole flushed with coolant. Such an operation corresponds to a pecking cycle in which the cutter is fed at constant force instead of constant feed rate.

8.2 Assembly

Invariably, whether it be fabricated from many components or machined from a single monolith, a flexure will eventually be integrated into more complex instrumentation or machinery. Not only will it be necessary to clamp the flexure into a larger frame, but actuators for translation and sensors to monitor position must be added. Fortunately, most sensors consist of electromagnetic coupling (capacitance, inductance, optical interferometry) which adds little mass or stiffness to the flexure. Parallel plate capacitance electrodes can add significant damping due to shearing or squeezing of the air in the gap between them. There remain some issues associated with coupling of actuators to the flexure and subsequently attaching the completed assembly to a rigid frame. Finally, for optimal dynamic response, it is often necessary to add damping (or energy dissipation) mechanisms. Consequently, this section addresses

Assembly of flexures

Coupling the actuator to the flexure

Flexure mounts

Adding damping to flexure systems (extracting energy)

8.2.1 Assembly of flexures

Consider the exploded view of a single degree of freedom parallelogram flexure shown in figure 8.5. It is apparent that all components are relatively simple to manufacture and assemble. Important features of this assembly are the clamping blocks, leaf type flexures and the upper and lower platforms.

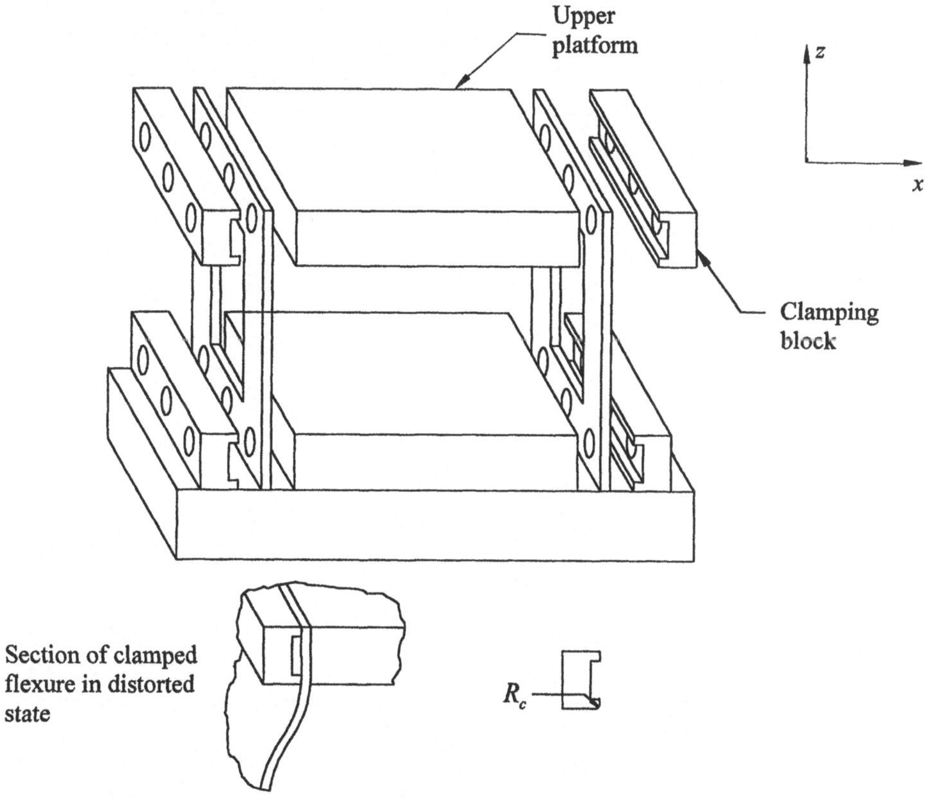

Figure 8.5: Exploded view of a simple leaf type linear flexure assembly

It can be seen that the clamping blocks have a recess to provide clamping forces along two lines. There are two reasons for this. Firstly, the function of the clamp is to provide a bending moment to the flexure. Resolving this into two, equal and opposite forces acting at an equivalent distance either side of the bolt, it is clear where the forces act in this case. Obviously, for a given bending moment these forces reduce with increasing separation. There are obvious limits to this separation. To ensure that the clamping forces applied to the leaf are a minimum, the contacts will be lines at the extreme edge of the clamp. However, this will result in large stresses. Consequently, it is necessary to extend this to a rectangular area contact preferably having a minimum thickness much greater than that of the leaf spring. Another consideration is that of stresses imposed by the clamp screws. In practice, screws are usually tightened until near to the yield stress. Most screws, being made from steel, have a yield stress, at least, near to that for a typical leaf. Consequently, to reduce the stress applied to the leaf, it is necessary that the clamping area of the contact is larger than the combined cross-sectional area of the screws. This is less important if the hardness of the leaf spring is considerably higher than that of the screws and materials of the clamp. Typically, it is better to design the recess to be a little wider than the thread of the clamping screw. Consideration of the two clamps shown in figure 8.6 should reveal the difference between good and bad design. Always desirable are spring washers that help to regulate the rate at which forces are applied with rotation of the bolt.

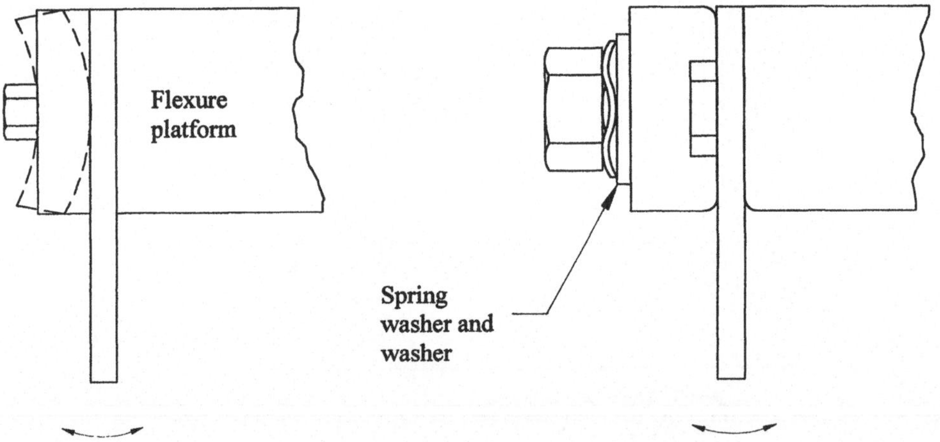

Figure 8.6: A typical, leaf spring clamp, a) clamp element too thin, of incorrect geometry and fastened with undersized bolt (dashed line shows exaggerated distortions), b) correct clamping method

Another consideration is the maximum stresses in the leaf which coincide with the edge of the clamp. Firstly, there will be a compressive stress due to the clamps. Applying superposition, the total stress will be the sum of this plus the stress due to bending. For a given displacement, δ, the radius of curvature due to bending for the leaf spring shown above is

$$R_f = \frac{EI}{M} = \frac{L^2}{6\delta} \tag{8.4}$$

If the corners of the clamp surfaces were sharp at this point there would be a local stress concentration of infinite magnitude. Clearly, something would yield. Choosing clamp materials of lower hardness will cause the yielding of these sharp corners and not of the flexures. Consequently, if the flexure is deflected to its maximum position after assembly, there will be an automatic plastic deformation of the clamp edge, after which, the flexure should behave in a reasonably linear fashion. A more reliable design would apply a small radius at this edge of the clamps and platforms. Often, a machined deburr will suffice.

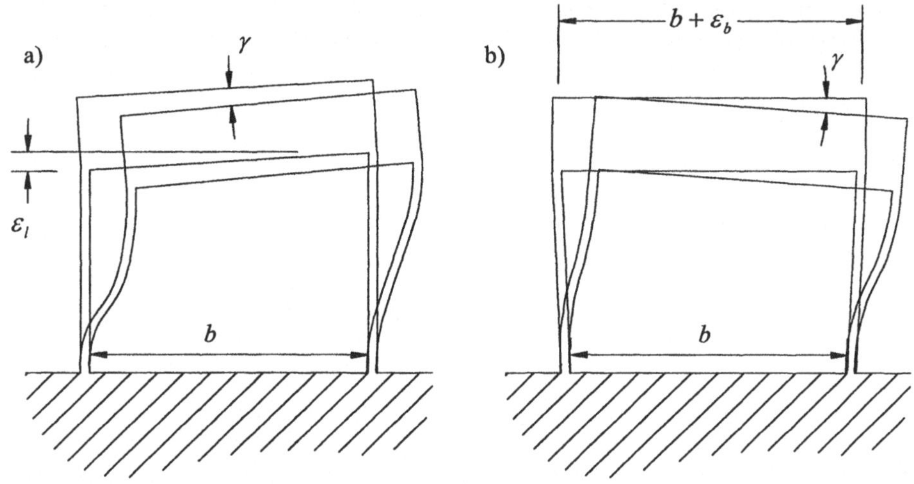

Figure 8.7: Potential parasitic errors of a simple linear spring due to manufacture and assembly tolerances

Manufacturing and assembly tolerances present another source of error. Two sources are those due to variations between separation of the springs at platform and base, ε_b, and those due to difference in length of the flexures, ε_l, see figure 8.7. For a linear translation δ, the parasitic rotations due to these errors can be approximated from the equations

$$\gamma = \frac{\delta^2 \varepsilon_b}{2bL^2} \tag{8.5}$$

$$\gamma = \frac{\delta \varepsilon_l}{2bL} \tag{8.6}$$

Tolerances on assembly are tighter for equal length of the flexures than they are for lengths of the platforms. To minimize the latter, both moving and stationary platforms should be machined as pairs. The former error is usually introduced during assembly and can be minimized by placing a parallel separator (i.e. gage blocks) between the platforms during assembly.

Often, a long, leaf type spring has insufficient torsion or buckling resistance. Splitting the leaf springs and separating them as far as possible from the center of the platform can enhance the torsion resistance. Buckling resistance can sometimes be enhanced, at the expense of maximum displacement range, by applying a stiffening element in the middle portion of the flexure. The bending moment along the axis of the flexure is zero in the center and increases linearly to a maximum value at each end. Consequently, applying constraints at the middle portion does not have a large effect on bending stresses for a given deflection but can dramatically increase its ability to withstand buckling loads. Applying central constraint results in a linear flexure comprising four, leaf-type hinges. This results in more precise location of hinge action and, therefore, assessment of parallelogram platform motion.

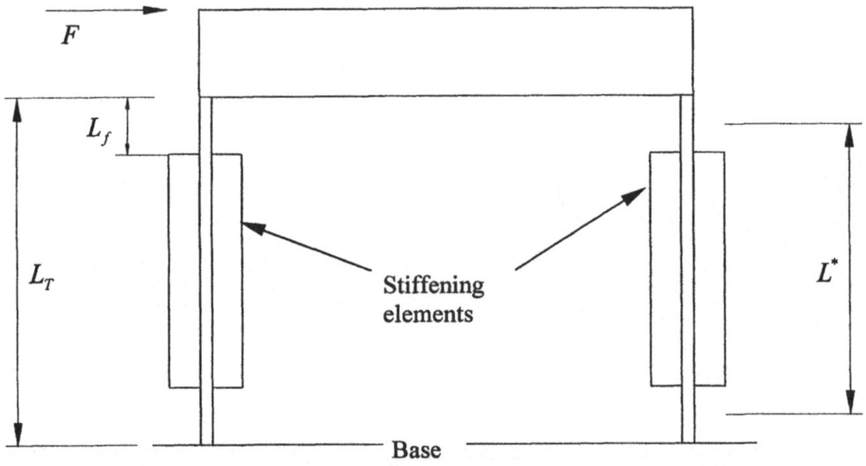

Figure 8.8: A simple linear spring in which four, leaf type hinges are produced by clamping the central portion of two leaf springs

To assess the effect of applying stiffening to the central portion of the flexures, consider the simple spring system shown in figure 8.8. For a simple leaf type spring, the maximum stress for a platform displacement δ is

$$\sigma_{L_T} = \frac{3E\delta t}{L_T^2} \tag{8.7}$$

For the same mechanism of four leaf type hinges this is

$$\sigma_{L^*} = \frac{E\delta t}{2L_f L^*} \tag{8.8}$$

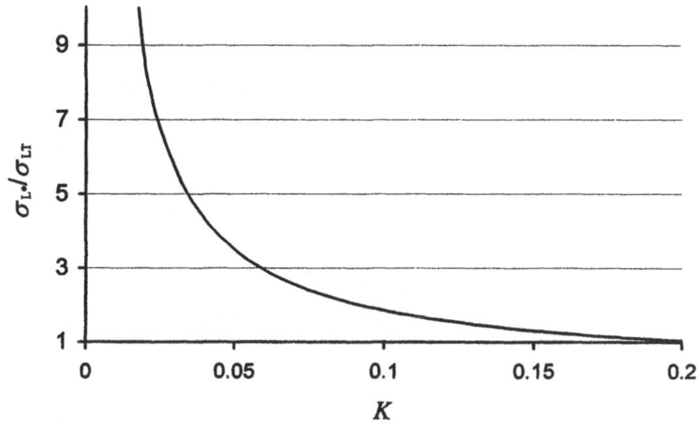

Figure 8.9: Graph of the ratio σ_{L*}/σ_{L_T} as a function of the dimensionless ratio K

Assuming that the length of the leaf hinges is small in comparison to the total length between platforms, the ratio of stresses in the two flexures for a given displacement is given by

$$\frac{\sigma_{L_T}}{\sigma_{L^*}} = \frac{6L_f L^*}{L_T^2} = \frac{6(L_f L_T - L_f^2)}{L_T^2} = 6K(1-K) \tag{8.9}$$

where $K = L_f/L_T$

Plotting this in figure 8.9 indicates that there will be little difference in the stresses for the two designs if the hinges are around one sixth of the length between two platforms (or with stiffeners of 2/3 the length). For shorter hinges, the displacement range drops off rapidly.

8.2.2 Coupling the actuator to the flexure

Actuator couplings represent a classic 'Catch 22' design problem. On the one hand, it is desired to rigidly connect the actuator to the flexure to avoid, as far as possible, 'lost motion'. Denoting the equivalent stiffness values of the flexure in the desired and undesired axes k_d and k_u and of the actuator and coupling k_a and k_c respectively gives an approximate model for the lost motion (for a more detailed discussion of this problem see chapters 5 and 7)

$$\frac{x_o}{x_i} = \frac{k_a k_c k_u}{k_a k_c k_u + k_d(k_u k_a + k_c k_a + k_u k_c)} \tag{8.10}$$

where x_i and x_o are the actuator and flexure platform displacements respectively, see figure 8.10.

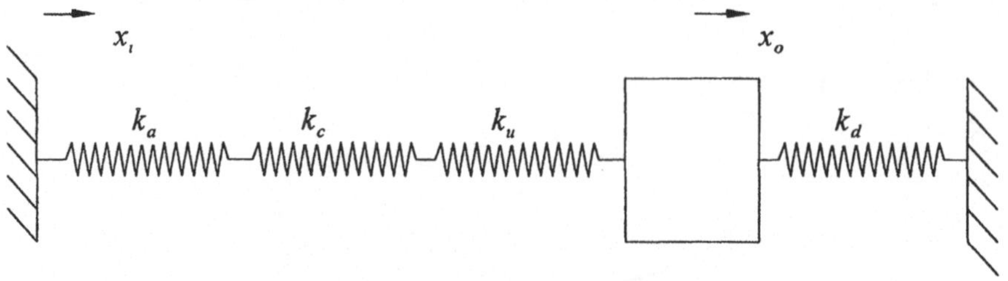

Figure 8.10: Generalized 'static' model representing displacements within a single degree of freedom actuator and flexure mechanism

For an ideal system, the ratio in (8.10) should be unity. Unless elements having negative stiffness are added (i.e. elements in a buckled state, see chapter 4), it is only possible to approach this ideal as all of the actuator, coupling and stiffness values in undesired freedoms are large *and* the flexure stiffness becomes vanishingly small. Unfortunately, as the stiffness of the flexure reduces in the desired freedoms, so too will its resistance to off-axis (or parasitic) forces. Consequently, the flexure will be become increasingly susceptible to parasitic errors in the drive mechanism. In the limit, the motion of the flexure will correspond exactly to that at the drive. The only reason for using a flexure in the first place is to provide a guided motion and so this latter option is not available. A stiff coupling that provides a high compliance in the off-axis freedoms always presents a design challenge.

Potentially, a near perfect coupling could be achieved using either an electromagnetic or hydrostatic drive. Sensing and maintaining a specified gap between driver and platform often produces the effect of an extremely stiff actuator with high compliance in the off-axis freedoms. For relative large devices, such drive couplings have been successfully used to provide precisely controlled motion, Peirce *et al.*, 1994, Chen *et al.*, 1995. However, for small flexures, such drives tend to contribute a significant cost while introducing problems of heat dissipation. For most instrument and machine flexures, the favored drive is either a direct motor driven screw or, more popularly, the piezoelectric actuator. Curiously, for much smaller scales in which the flexures are measured in micrometers, electrostatic drives which in principle dissipate little to no heat, are commonly used. A variety of couplings between a rigid actuator and moving flexure stage for intermediate sized mechanisms are shown in figure 8.11. Generally, these consist of a mechanism that is required to be stiff in the drive axis while freedom of motion to accommodate errors in the drive. As might be expected, flexures can be readily designed for this purpose and three variations are shown in the figure. A simple wobble-pin or sphere on flat is also commonly used. Jeweled pivots for this purpose are commercially available. A low friction equivalent to the sphere on a flat is provided by contacting the outer

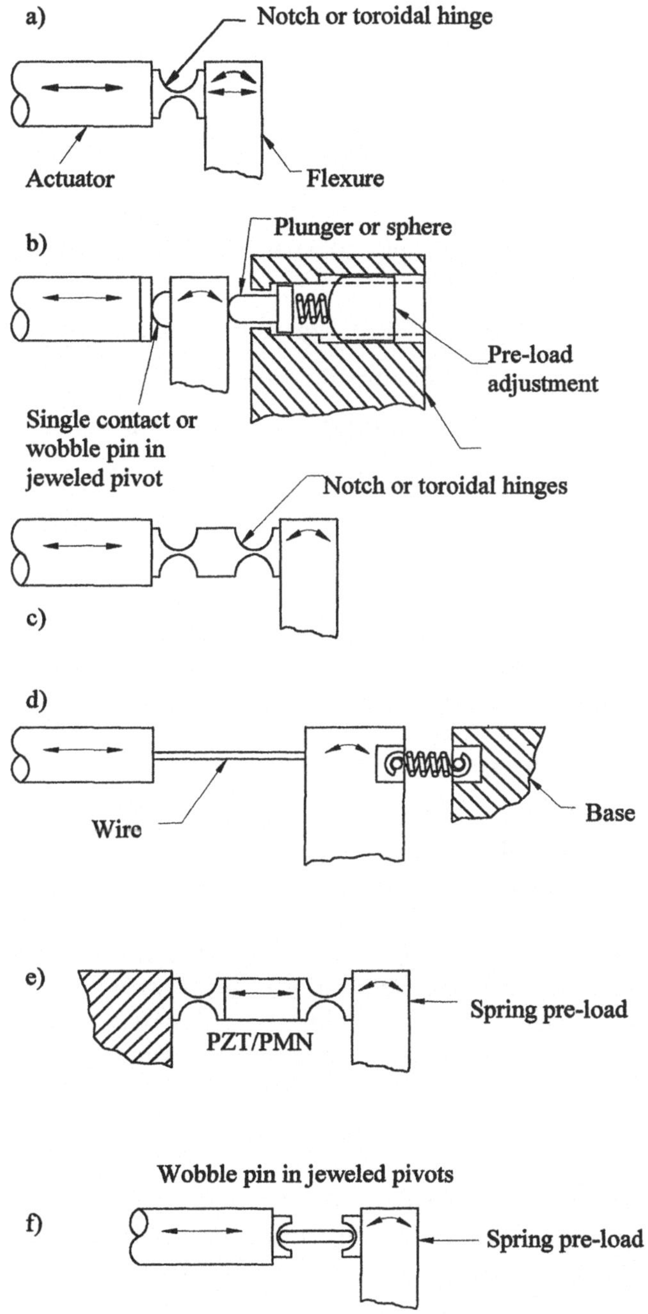

Figure 8.11: A variety of couplings for connecting a rigid drive to a moving flexure element

races of two rolling element bearings with their axes at right angles, see Smith and Chetwynd, 1992. It is important to note that some form of pre-load is necessary if a contact based coupling is chosen.

8.2.3 Flexure mounts

Having produced a flexure mechanism and integrated it with drivers and other auxiliary components, it is often necessary to secure this into some larger instrument mechanism. Generally, this must be secured using bolts and clamps. Again, good design principles will help to ensure minimal influence of the fastening mechanism. Useful rules are

Use kinematic or pseudo-kinematic design of the mounting points.

Match symmetry between the flexure and the mount.

Choose mounting points to be symmetric about the centroid of the flexure.

Try to match thermal expansion effects and rate dependent thermal responses.

Tighten clamps incrementally to ensure equal clamping forces at each fastening point.

Introduce compliance at the clamp. This can be achieved by either locally weakening one of the clamping members where it connects with the bolt or introducing a spring washer (for the same purpose).

8.2.4 Adding damping to flexure systems (extracting energy)

For optimal response, it has already been suggested in chapter 3 that, for many applications, it is desired that the critical damping ratio, ξ, lays in the range 0.4 - 0.8. Almost by definition, if a material has been chosen for its favorable characteristics as a flexure, it is unlikely to provide such a large value of energy dissipation due to work of distortion. In fact, energy dissipation can only occur with an accompanied hysteresis in the stress-strain characteristic thereby directly effecting performance. In this section, internal friction of materials will be briefly discussed. This will be followed by a discussion of mechanisms for increasing the energy dissipation in a flexure.

8.2.4.1 Internal friction in solids

All materials subject to cyclic loads will dissipate energy, Wert, 1986. Unfortunately the physical mechanisms whereby this occurs are not easy to determine or discriminate. For metals, a number of phenomena have been identified. These are

Internal friction caused by thermal currents and atomic diffusion

The Snoek effect

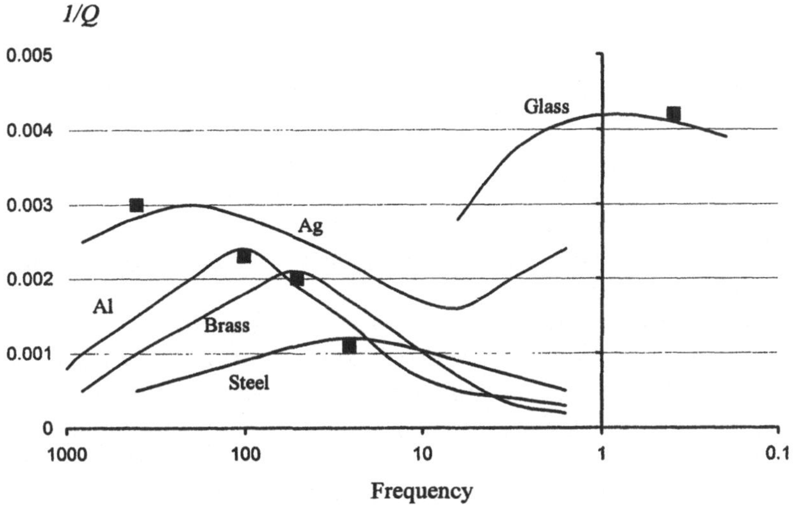

Figure 8.12: Internal friction of thermoelastic origin in vibrating reeds, the dark squares represent the frequency at which maxima are predicted by the theoretical model of Zener (reproduced from Wert, 1986)

Zener relaxation

Dislocation damping

Thermodynamic considerations led Carl Zener to the analysis of energy losses due to temperature gradients developed across the face of a thin beam subject to cyclic bending. Compression on one face leads to a temperature increase and vice versa on the tensile side. Subsequent irreversible heat flow then provides the mechanism for energy dissipation. Depending on the thickness of the beam, the rate of heat transfer leads to energy loss that will be a function of the frequency. In his original work Zener showed that the maximum energy loss for a thin cantilever beam occurs at a frequency

$$f = \frac{A\kappa}{C_v t^2} \qquad \text{[Hz]} \qquad (8.11)$$

where C_v is the specific heat per unit volume, κ is the thermal conductivity, t is the thickness of the cantilever and A is constant that depends on the cross-section.

Subsequent experimental results by Bennewitz and Rötger, 1936, validated this mechanism, see figure 8.12. Extension of this analysis to include the effects of heat transfer across crystal interfaces in polycrystalline materials also showed considerable promise, see figure 8.13. For our purposes, it is informative to consider the scale in figure 8.12. Being considerably less than unity, this can be expressed using (3.90) as 2ξ (it will be shown shortly that this can be directly related to hysteresis as a material property). Being less than a few parts in 1,000 the damping ratio is far too low for most flexure applications. It may also be reasonably assumed that for a limited range of frequencies the damping ratio remains reasonably constant.

Two other loss mechanisms, due to either interstitial atoms or substitutional alloys within a crystal, have been extensively studied over the years. The former was first studied by Snoek for determination of macroscopic strains in the region of interstitial carbons in bcc iron directly relating to the martensite response. Based on this analysis, it was determined that the internal friction could be related to the relaxation time. Experimental studies over temperatures ranging from near zero to around 700 K indicate a good correlation between relaxation time and diffusion of carbon, with the diffusion coefficient varying by up to twelve orders of magnitude over this range! Zener studied the effect of diffusion in substitutional alloys. In particular, it was proposed that, in the unstrained state, these alloys would form in local clusters. For example , zinc in 30 - 60 brass will tend to form into pairs imposing a local tetragonal strain. Upon application of stresses, these will move to a more favorable configuration by diffusion. Again, this will result in a relaxation of the material that can, in turn, be used as a measure of the diffusion coefficient.

Another dissipation mechanism, for which verifiable theoretical models have

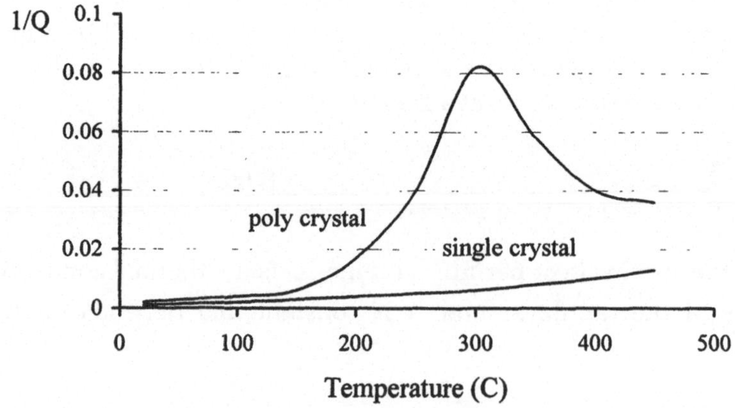

Figure 8.13: Grain boundary relaxation as a function of temperature for single and poly-crystalline aluminum

yet to be realized, is that due to motion of dislocations. In general, significant dissipation is often experienced only at high frequencies although a lesser effect covering a broad range of frequencies is likely in metals. From the above, the general rules for the avoidance of damping might be

- Use single crystals or poly-crystals of optimal size to minimize thermal dissipation
- Avoid alloys with high diffusion rates or single elements
- Avoid interstitial additives

Unfortunately, most of the above are not possible for the production of flexures from common engineering metals. An alternative compromise is

- Reduce the mobility of dislocations, interstitial elements and alloy compounds

This is a more realistic option and can often be readily achieved after processing by a suitable heat treatment to increase hardness.

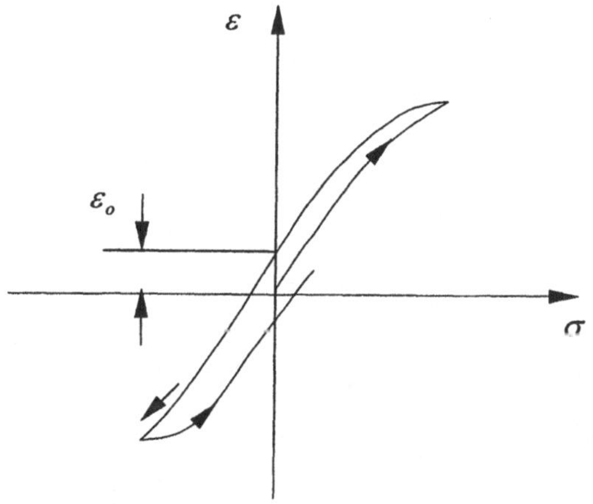

Figure 8.14: Stress-strain characteristic with hysteresis damping. Arrows on the graph indicate progression of time.

Modeling of the energy dissipation due to internal friction turns out to be relatively straightforward. Because this is due to hysteresis in the stress-strain characteristic, it may be assumed that such a mechanism is independent of how fast the stresses and strains are being applied. Within a lumped model, this can be inserted in the form of a damping element that is independent of frequency but provides a force that lags displacement by 90°. There are two ways to introduce this effect into a mathematical model. Before proceeding, it is informative to speculate the response of a spring/mass system for which hysteretic energy losses are significant. If the spring mass system is deflected slowly and released, it is expected that the deflection will follow that of the stress strain characteristic of the spring material, figure 8.14. Consequently, at low frequencies of cyclic load, the displacement due to an applied force will not be the same as that for a 'perfect' spring. At excitation frequencies near to resonance, it would be expected that, for small dissipation, similar resonant phenomena to that for viscous damping should be observed. To derive a dissipation function that is independent of frequency, consider the models of figure 8.15. In these, the hysteretic damping is represented respectively as

1. A damping factor that varies as the velocity of distortion divided by the frequency
2. An additional complex term added to the spring.

For both systems the equation governing motion for a periodic excitation force may be expressed in either form

$$m\ddot{x} + \frac{h}{\omega}\dot{x} + kx = \mathrm{Re}\{F_o e^{i\omega t}\}$$

or (8.12)

$$m\ddot{x} + (k + ih)x = \mathrm{Re}\{F_o e^{i\omega t}\}$$

As usual, assuming a linear solution of the form

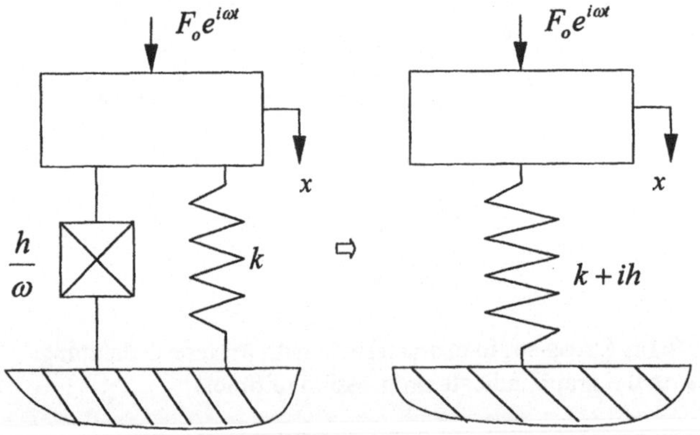

Figure 8.15: Mathematical model of a single degree of freedom spring mass systems with hysteretic damping, a) using a hysteretic damping element, b) equivalent model using complex stiffness

$$x(t) = \mathrm{Re}\{H(i\omega)F_o e^{i\omega t}\} \tag{8.13}$$

The frequency response of this single degree of freedom system derived from either of equations (8.12) is

$$
\begin{aligned}
H(i\omega) &= \frac{1/m}{\omega_n^2 - \omega^2 + ih/m} \\
&= \frac{1/k}{1 - \dfrac{\omega^2}{\omega_n^2} + i\dfrac{h}{k}}
\end{aligned}
\tag{8.14}
$$

Expressing the above 'normalized' to the displacement for a static load applied to a perfect spring yields

$$
\left|\frac{X}{X_{static}}\right| = \left|\frac{X}{F_o/k}\right| = \left|\frac{1}{1 - \dfrac{\omega^2}{\omega_n^2} + i\dfrac{h}{k}}\right| = \frac{1}{\left(\left(1 - \dfrac{\omega^2}{\omega_n^2}\right)^2 + \left(\dfrac{h}{k}\right)^2\right)^{½}}
\tag{8.15}
$$

The phase shift as a function of input frequency can be readily obtained as

$$
\phi = -\tan^{-1}\left(\frac{\dfrac{h}{k}}{1 - \dfrac{\omega^2}{\omega_n^2}}\right)
\tag{8.16}
$$

The subsequent frequency response is plotted in figure 8.16. From equation (8.15) it is apparent that the value for *1/Q*, which occurs at the undamped natural frequency, is simply *k/h* which is the ratio of real to complex stiffness. For solid materials, this ratio will also be that of the real to imaginary elastic modulus and is therefore a material property. Consequently, to assess this value it is common to use a rod of the material as part of a torsional pendulum and quote the material property in the form *1/Q* as is the case in figures 8.12 and 8.13. Extensive tables of complex elastic moduli have been developed over the years, references to which can be found in Lazan, 1968, and Snowden, 1968.

Often, the compound elastic moduli in linear and shear loading are denoted by

$$
\begin{aligned}
E^* &= E' + iE'' \\
G^* &= G' + iG''
\end{aligned}
\tag{8.17}
$$

Also the phase shift at zero frequency is commonly quoted as the loss angle that, for small values, is given by

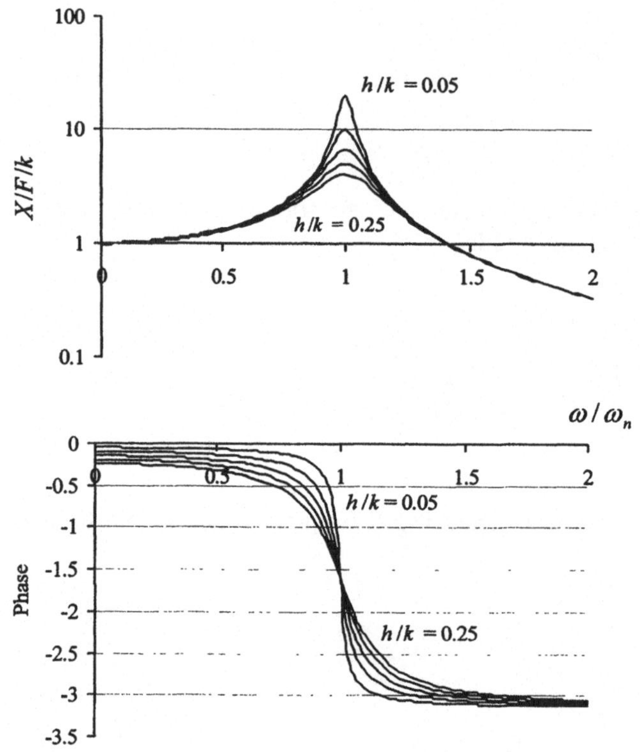

Figure 8.16: Frequency response for a spring mass system with hysteresis damping, a) magnitude response, b) phase lag

$$\begin{aligned}\phi &= \tan^{-1}\left(G''/G'\right) = \tan^{-1}\left(E''/E'\right) \approx E''/E' \\ &\approx h/k\end{aligned} \qquad (8.18)$$

A more extensive discussion of this field of study plus some tabulated values for materials having high loss angles can be found in Goodman, 1996.

Value of Q for resilient materials are difficult to measure. In practice, measurements are hampered by the energy losses introduced by the specimen mounts. Even for non-metal springs of optimal geometry, it is difficult to achieve Q values of greater than 6,000, Schindel *et al.*, 1997. With great care, and for some low loss materials such as fused silica, Q values of greater than 10^7 have been measured, Startin *et al.*, 1998. However, in the experiments of this reference, it is considered that the Q value was primarily limited by losses through the mounting interfaces.

8.2.4.2 Adding damping to a flexure

In much of the above, it has been implicitly assumed that the damping coefficients are inadequate. Simple analysis of systems for applications such as vibration isolation, highly linear springs or optimal dynamic response reveals that structural damping falls in the intermediate region of being either too high or too low. For systems that are required to follow dynamically varying demands on position it has already been stated that optimal values are likely to be in the range $\xi = 0.4$ to 0.8. To achieve this value it is usually to provide some additional energy dissipation. There are two techniques commonly used to achieve such a goal. These being

Active dissipation

Passive mechanisms

Dissipation occurs if there is a force that is out of phase with the position and acceleration. An obvious method is to derive a signal based on the velocity.

Active dissipation

To provide active dissipation, it is necessary to monitor the velocity of the moving members and apply a force in direct proportion. Commonly, there are two means of doing this

Exploiting the velocity directly to produce a feedback force.

Measuring the position and differentiating with respect to time prior to feeding this back as a force to the moving member

The first of these is often referred to as velocity feedback or, for motor controllers, tachometer control. Exploiting the Faraday induction is a common method for both deriving a signal and creating a velocity dependent force. For example, it might be possible to attach a magnet and coil to the moving and stationary components of the flexure. Motion of the magnet will produce a Lorentz force on the electric field in the coil and this can be used to produce a current or voltage proportional to the velocity. In the former case, it is only necessary to provide a resistor across the coil, which will then dissipate energy in the form of i^2R losses. In reality, the added mass of either the magnet or coil attached to the moving platform of the flexure will outweigh the benefits of such a simple, passive isolator. Leaving the coil open circuit, there will be an induced voltage proportional to the products of magnet flux, length of coil and velocity. This velocity signal may then be amplified and returned to a force actuator acting directly on the platform. Liu et al., 1993, were able to used such a technique on a flexure based stylus mechanism to adjust the critical damping ratio from an intrinsic value of around 0.02 to values greater than one.

The second approach is to measure the position of the platform, differentiate it with respect to time and then feed this back to a force actuator (usually electromagnetic or, less often, piezoelectric in origin). For a perfect system, this will have the effect of improving the transient response of the flexure. Often, the derivative is combined with an integrator and proportional feedback for providing a fast response with zero steady state errors. Known as a PID (proportional-integral-derivative) controller these individual terms effect the general response, steady state errors and transient response respectively. A disadvantage to this approach is the adverse effects of noise on the derivative term. As frequency increases, so too does the derivative with respect to time. Consequently, high frequency noise, albeit mechanical or electronic, will tend to be amplified and fed back to the system, often resulting in reduced performance in terms of both stability and precision. For a more detailed discussion of closed loop control strategies see for example, Dorf, 1980, Richards, 1979.

Passive damping mechanisms

Passive damping mechanisms have the advantage that, in comparison to active mechanisms, they are relatively 'noise free'. By passive mechanisms, it is implied that some form of mechanical device is to be added, the physical nature of which tends to dissipate energy due to relative motion between two links of a flexure. Invariably, this will be achieved by applying forces (thereby inducing relative velocities) through damping elements connecting these links.

Three techniques are commonly employed, these being

Addition of viscous dampers

Introduction of interfaces

Coating with high loss materials

These are discussed in turn below.

Viscous dampers

Shear of any fluid or gas at low Reynolds number (defined as the ratio of inertial to viscous force) will result in a force that is proportional to both velocity and viscosity. For macroscopic structures involving large forces, it is possible to utilize viscous damping by sealing a fluid in a hydraulic or pneumatic piston and cylinder arrangement. Arranging for controlled leakage either through or around the piston results in a force that is a definite function of the relative velocity between piston rod and cylinder. For smaller instrument mechanisms, problems due to frictional forces at the seals often results in unacceptable hysteresis. As a consequence, it is more usual to utilize some form of 'free paddle' arrangement. Three common types are

The paddle in a bucket

Shear film damping

Squeeze film damping

The principle of operation of paddles is relatively simple. A fluid is placed in a container that is both stationary and upright. A paddle immersed in the fluid and rigidly connected to the moving platform provides the damping forces. Such mechanisms are common in torsion magnetometers, galvanometers and other instrument mechanisms with masses relatively low in comparison with the stiffness. Some care is necessary to arrange that the center of force of the damping element does not introduce unwanted bending moments into the mechanism.

Shear damping relies on the relative motion of two parallel, solid surfaces with a fluid between them. For two parallel surfaces of area *A* and separation *d*, the shear force as a function of the relative velocity *v* is given by

$$F = \frac{\eta A v}{d} \tag{8.19}$$

where η is the viscosity of the fluid between the two and has units of Ns m^{-2}. Examples of such mechanisms applied to flexures are given in the papers of Chen *et al.*, 1995, and Holmes, *et al.*, 1997.

Interestingly, for relative small separations, it is possible that surface tension forces, independent of orientation will retain the fluid. Consequently, for flexures it is sometimes possible to place a surface near to the moving flexure and deposit a small amount of fluid in the gap. Some mechanisms have been known to retain damping fluid between two vertical, parallel surfaces for tens of years.

Sometimes, it might be arranged that the parallel surfaces change separation, or, in other words, move in a direction towards or away from each other. Under these circumstances, any intervening fluids will be physically squeezed out from the gap. This will require a force that is a function of the ratio of the minimum dimension of the plates to their separation i.e. the slenderness of the gap. In general, it is found that the squeeze film force is related by

$$F \propto \frac{\eta}{d^n}\frac{\partial d}{\partial t} \tag{8.20}$$

where *d* is the instantaneous separation of the plates.

For two parallel plates moving in a direction normal to their surfaces, *n* is 3 while for a sphere and flat this reduces to 2, Vidic et al., 1997. Alternatively, squeeze film damping might be utilized for the damping of flexures undergoing distortion. Under these conditions, it is necessary to solve Reynold equation to determine the relationship between the squeeze film forces and separation velocity, see for example Gross, 1962. Even when the intervening fluid is air, shear and squeeze film effects can produce significant forces, particularly in MEMS, Muller and Howe, 1990, and some small instrument mechanisms,

Howard and Smith, 1992. Because of this near inverse cube dependence on the nominal separation, this can dramatically increase the damping ratio in small mechanisms, often to values considerably greater than unity. The effects of squeeze film damping between a cantilever with a small mass attached to the free end and a solid base parallel and near to one face has been assessed by Xu and Smith, 1995. Significant squeeze films can also develop in larger structures if the viscosity of the intervening fluid is high or surfaces conform closely. For example, laying one large glass sheet on another can result in a squeeze film of air that, under self-weight loads, is maintained for several seconds. Because the lateral friction is very low, such an effect can be hazardous if the plate is unconstrained.

8.2.5 Effects of manufacturing tolerance on the stiffness of flexure elements

Intrinsic to all manufacturing processes is the unpredictable deviation between the desired and actual geometry of parts. Commonly, this is simply accounted for in the tolerance. More often than not, tolerances represent worst case values encompassing the complete scatter of differences over all possible geometry's for a given process (i.e. this might be ±0.05 mm for a milling machine on all dimensions for any component that might be produced in one operation on the machine). Quite often, the tolerance arises because of hysteresis in a machine tool drive and slideway causing 'dead' bands during reversal. If there are significant Abbé errors between the line of action of the tool and the measuring scale used to monitor slideway position, it is not possible to detect, and therefore compensate, these errors, see for example Slocum, 1992. Consequently, this tends to remain relatively constant and is independent of the size of components to be produced. Other possible errors such as undetectable parasitic rotations of the slideway, thermal expansion effects and tool wear tend to be relatively small in comparison. It is possible to reduce hysteresis using software compensation. However, because the direction of the backlash depends on the history of the motion and instantaneous friction coefficients, it is not possible to accurately predict, and therefore eliminate, this error (i.e. reduce it to a value considerably lower than other error sources). In this section, the effects of manufacturing tolerances on the stiffness of flexure elements are discussed.

Predominantly, flexure action is provided by deflection of a beam element or a locally notched region in a solid. For small deflections and applied loads, it is often possible to derive an approximate expression for the stiffness, k, of such an element by

$$k \propto \frac{Eb^{u}t^{v}}{L^{w}} \tag{8.21}$$

To determine the influence of tolerances in the above, it is possible to use the chain rule of differentiation to estimate the change in stiffness from

$$\begin{aligned}\delta k &= \left(\frac{\partial k}{\partial E}\right)\delta E + \left(\frac{\partial k}{\partial b}\right)\delta b + \left(\frac{\partial k}{\partial t}\right)\delta t + \left(\frac{\partial k}{\partial L}\right)\delta L \\ &= \frac{b^u t^v}{L^w}\delta E + u\frac{Eb^{u-1}t^v}{L^w}\delta b + v\frac{Eb^u t^{v-1}}{L^w}\delta t - w\frac{Eb^u t^v}{L^{w+1}}\delta L \\ &= \frac{Eb^u t^v}{L^w}\frac{\delta E}{E} + u\frac{Eb^u t^v}{L^w}\frac{\delta b}{b} + v\frac{Eb^u t^v}{L^w}\frac{\delta t}{t} - w\frac{Eb^u t^v}{L^w}\frac{\delta L}{L}\end{aligned} \tag{8.22}$$

The common multiplier in the last of equations (8.22) is simply the original stiffness value for a perfect flexure. Consequently (8.22) can be rearranged to produce the fractional uncertainty in stiffness

$$\frac{\delta k}{k} = \frac{\delta E}{E} + u\frac{\delta b}{b} + v\frac{\delta t}{t} - w\frac{\delta L}{L} \tag{8.23}$$

Each of the numerators in the fractions on the right hand side of this equation represents the uncertainty of that particular parameter. To assess the influence of manufacturing tolerance on the predicted stiffness of the flexure, it is only necessary to determine the exponents *u, v, w* in (8.21). For example, the stiffness of a leaf spring subject to uniaxial tension is given as *Ebt/L*, from which it is readily apparent that $u = v = w = 1$. For a few flexure elements the appropriate exponents are given in Table 8.2.5.1 below

Table 8.2.5.1: Exponents of equation (8.21) for a variety of flexure elements undergoing small distortions (in the case of a notch hinge *R* replaces *L*)

Flexure type	Distortion mode	Applied load	u	v	w
Beam	Axial	Axial tension	1	1	1
	Lateral	Applied moment	1	3	2
	Lateral	Applied force	1	3	3
	Lateral	Uniformly distributed	1	3	4
	Angular	Bending moment	1	3	1
Notch	Angular	Bending moment	1	5/2	1/2
	Lateral	Lateral force	1	5/2	5/2

Equation (8.23) indicates that the fractional variability of the stiffness is directly proportional to the fractional uncertainty of the individual parameters multiplied by the exponent in the original equation for predicting the stiffness. It is worth spending a little time considering the implications of this in view of the errors most likely to dominate during manufacturing.

Taking, for example, a leaf-type hinge element it is worth speculating the relative contribution of the errors that are likely to be encountered. Surprisingly, the first of these, elastic modulus, is difficult to know accurately. Usually, the error is of the order 1 % or more with higher accuracy often requiring considerable effort and expense. However, for material from the same batch the value of the elastic modulus is likely to be constant to considerably better than this. Similarly, the depth of the flexure is often dependent upon the parallelism between the front and back surfaces of the material from which the flexure elements are produced. Alternatively, this will depend upon the thickness of the strips from which the hinge is fabricated. In most cases this dimension is relatively large and is likely to be both accurate and constant. In view of the low value of the exponent, this parameter is unlikely to be a source of significant errors for commonly used manufacturing processes. In contrast, the last two parameters can have relatively small dimensions and large exponents. Consequently, if manufacturing tolerances are constant and independent of dimensions, the ratio of tolerance to dimension can be relatively high. For example, it is not uncommon for leaf or notch hinges to require thickness values of the order of a few hundred micrometers. A manufacturing tolerance of 10 μm on a thickness of 100 μm represents an uncertainty ($\delta t / t$) of 10 %. In view of the exponents given in table I, this can result in stiffness variations of around 25 - 30 %.

As is clear from equation (8.23), the variation in stiffness is simply the sum of scaled variations for each parameter. The negative coefficient in *w* simply indicates that stiffness will reduce with increasing *L*. Of more importance for assessing the consistency of products is an estimation of the variance of the stiffness. It might be speculated that this may be obtained by taking the expected value (indicated by the quantities inside $\langle\rangle$) of the square of (8.23) given by

$$\begin{aligned}\left\langle\left(\frac{\delta k}{k}\right)^2\right\rangle &= \left\langle\left(\frac{\delta E}{E}+u\frac{\delta b}{b}+v\frac{\delta t}{t}-w\frac{\delta L}{L}\right)^2\right\rangle \\ &= \left\langle\left(\frac{\delta E}{E}\right)^2\right\rangle + u^2\left\langle\left(\frac{\delta b}{b}\right)^2\right\rangle + v^2\left\langle\left(\frac{\delta t}{t}\right)^2\right\rangle + w^2\left\langle\left(\frac{\delta L}{L}\right)^2\right\rangle \\ &\quad + 2\left\langle\frac{\delta E}{E}u\frac{\delta b}{b}\right\rangle + 2\left\langle\frac{\delta E}{E}v\frac{\delta t}{t}\right\rangle ...\end{aligned} \tag{8.24}$$

Adopting standard statistical notation and denoting the standard deviation by the symbol σ (not to be confused with the stress!), equation (8.24) can be expressed as

$$\sigma_k^2 = \sigma_E^2 + u^2\sigma_b^2 + v^2\sigma_t^2 + w^2\sigma_L^2 + 2\rho_{Eb}u\sigma_E\sigma_b ... \tag{8.25}$$

The subscripts in this equation refer to the *ratios* and therefore the variances, σ^2, are dimensionless. Choosing this notation, it is important to note that the tolerances must be specified in terms of the standard deviation measured about the mean. The symbol ρ represents the correlation coefficient between the appropriate variables indicated by the subscripts.

In many instances, each individual parameter is likely to be independent of the others. In this instance the correlation coefficients are zero and (8.25) reduces to

$$\sigma_k^2 = \sigma_E^2 + u^2\sigma_b^2 + v^2\sigma_t^2 + w^2\sigma_L^2 \tag{8.26}$$

One particularly troublesome consequence of variable stiffness hinges within a flexure is the disturbance of the symmetry of the structure. This asymmetry of forces within the mechanism can result in significant deviations in the reaction of the flexure to applied forces. In many cases, although they will still exhibit a linear response between displacement and applied force, this difference between the *direction* of applied force and motion might introduce significant errors in terms of the function of the mechanism. An analysis of these effects is presented in the paper of Ryu and Gweon, 1998.

The variability of the stiffness may also introduce uncertainty in the operation of the mechanism. In particular this may lead to considerable variability in the work required to produce a given displacement. For many flexure designs comprising many hinges a truly random variability of the hinge stiffness will tend to average towards the design value. For example, for the simple and compound rectilinear spring mechanisms discussed in section 5.4 for which each hinge is subject to nominally similar strains, the mean stiffness of hinges will tend towards the true value with the variance of the mean $\sigma_{\bar{k}}$ given by

$$\sigma_{\bar{k}}^2 = \frac{\sigma_k^2}{j} \tag{8.27}$$

where j is the number of hinges and the variance of the mean hinge value becoming normally distributed in accordance with the well known central tendency of statistical distributions.

In some cases, it may not be reasonable to assume that there is no correlation between geometric parameters of a hinge. As an example, for a notch hinge produced by making two nearly 'touching' holes variations in the radius will directly effect the thickness of the web. In this case there will be a unity correlation coefficient connecting the two parameters. In reality, the radius of the notches will tend to be relatively constant while the web thickness will be effected by more significant variations in the center location of the holes plus the undesirable effects of lobing of the holes common during drilling operations.

8.2.6 Typical flexure drives

Some form of drive must be included in the design of any flexure mechanism. Although outside of the scope of this book, drive techniques are intrinsically linked to the flexure design process. Consequently, to indicate possible design solutions, some of the more common drives and their relative merits and limitations are listed below, see also Smith and Chetwynd, 1992,

Feedscrews. These provide an inexpensive, relatively stiff and precise drive for long (potentially unlimited) range motions of larger scale flexures. Backlash and non-linear stiffness for small displacements can produce significant errors upon reversal of motion, thereby limiting performance.

Friction drives. A friction drive is achieved by gripping a prismatic rod between two rollers. Rotation of the rollers is converted to a linear translation of the prismatic bar. Such a drive can have stiffness comparable to a feedscrew with significantly reduced backlash. A drawback is the requirement for precise rotational control of the motor drive. Transmission of rotational motion through contact between the outer rims of smooth disks is also common.

Piezoelectric. These are commonly used for fine motion stages because; they are stiff, can be directly voltage controlled, produce a fixed displacement almost independent of size (and therefore are favorable for small devices), have a relatively fast dynamic response, have a low thermal expansion coefficient and can operate with a high preload. Disadvantages include hysteresis, creep, sensitivity to thermal variations and internal losses that generate heat. High (0-1500 V) and low (150 V) voltage materials as well as electro-strictive devices are commercially available. Longer motions are often obtained by incorporating flexure based lever mechanisms into the drive. Less commonly, translation range can be increased using 'bimorph' designs in which two piezoelectric beams are bonded to produce a 'bi-metallic' mode translation when they are actuated in equal and opposite directions.

Electromagnetic. When applied directly, electromagnetic actuators can be categorized into four groups, these being

The solenoid and soft magnetic target

Coil in a field produced by a permanent magnet or 'voice coil'

Magnetostatic

Electrostatic

In general, the above might be considered to be ordered from largest to smallest force generation mechanism and a very brief overview of the relative merits and limitations of each is presented below.

Usually the solenoid is mounted to the base of the flexure and will consist of a copper winding surrounding a soft magnetic material to provide a low reluctance path for the magnetic field. The flux path is then broken by an air gap with another magnetically permeable 'target' being on the other side of the gap and attached to the moving platform. To estimate the force, *F*, on the target it is necessary to compute the inductance, *L*, of the magnetic circuit. This can be computed from the equation

$$L = \frac{N\psi}{I} = \frac{1}{I^2} \iiint_{vol} \mathbf{B.H} \mathrm{d}vol$$

$$\approx \frac{N^2}{\dfrac{l_c}{\mu_c A_c} + \dfrac{l_g}{\mu_g A_g}}$$

where μ represents the permeability of the soft magnetic core/target and the gap respectively, *A* the respective cross-sectional area across the flux path, *N* the number of turns of the solenoid, **B** is the magnetic flux density, **H** is the applied magnetic field, *I* is the excitation current and *l* is the path length of the magnetic flux. When the permeability of the core material is significantly greater than that of air and for reasonably large air gaps, the above reduces to

$$L \approx \frac{N^2 \mu_g A_g}{l_g}$$

Force on the target can be computed from the gradient of the magnetic potential, *U*, given by

$$F = -\frac{\partial U}{\partial l_g} = \frac{1}{2}\frac{\partial (LI^2)}{\partial l_g} \approx \frac{N^2 \mu_g A_g I^2}{l_g^2}$$

Although relatively large forces can be generated using such a design, it is non-linear in the length of the gap and the permeability will vary with magnetic field and temperature and exhibits hysteresis. However, for systems under closed loop control, these effects may not be particularly troublesome, Holmes *et al.*, 1999.

Potentially, a more reproducible force can be generated with a voice coil design. Two designs are commonly used to drive flexure mechanisms. Probably the most common electromagnetic flexure drive exists in the loudspeaker. In this, An arrangement of permanent magnets is positioned around a coil so that its windings are subject to a uniform and constant magnetic flux of density B. Under these circumstances, when energized with a current, I, the coil will experience a force, F, given by

$$F \propto BIl$$

l is the length of the winding within the flux path and in this it has been assumed that winding, flux and force are mutually perpendicular. Of particular interest with such a design is the fact that, at least from this simple analysis, the force is linearly proportional to the coil current and is independent of the relative position between the coil and permanent magnets.

Often, to reduce mass, the winding is attached to the moving platform of the flexure. In some applications, this may produce an undesirable heat source. One solution, might be to make a coil consisting of two opposing windings and generate the coil magnetic field by differential currents operating at constant power. However, the extra winding will add mass. Alternatively, it is sometimes desirable to attached the permanent magnet to the moving platform and surround this by a winding that is rigidly attached to the base, the moving magnet design. For a circular cylindrical winding surrounding a permanent magnet, the maximum force on the magnet is experience when it is positioned with poles along the coil axis and these being equidistant about the ends of the coil, Smith and Chetwynd, 1990. In this case, the force on the magnet in the direction of the coil axis can be computed from

$$F = vol_{magnet} B_{rem} \frac{dH}{dx} \propto I$$

For suitably chosen coil and magnet geometry such an actuator can again be designed to provide a force that is linear with current and, to first order, is independent of relative coil/magnet position.

Performance characteristics of both of the 'voice coil' designs will vary with temperature and magnitude of the applied field. However, for 'hard' (usually rare earth) magnets subject to relatively small fields non-linearity and hysteresis may be negligible.

If the permanent magnets of the above design were replaced with a 'perfect' magnet, then the above voice coil designs would provide a correspondingly perfect linear relationship between coil current and force. A winding of fixed geometry through which constant current flows can produce an effectively perfect magnet. Such 'perfect' designs are currently being used to compare electrical to mechanical power, Olsen *et al.*, 1991..

Electrostatic drives utilize the force produced between two electrodes separated by a small gap (usually air) with a potential difference across them. Forces can be easily derived from the derivative of the electrostatic potential computed from the product of capacitance and potential difference divided by 2, see for example Smith and Chetwynd, 1992. For mechanisms of typical engineering dimensions, such forces are invariably small. However, at small scales such as those typically used for MEMS devices significant forces may be produced and, in many cases, this is the preferred drive mechanism.

Magneto-strictive. Manufactured from alloys based on terbium, iron and dysprosium, this material is often called Terfenol. Suitably pre-stressed, this material will undergo a strain when subject to an applied magnetic field. Such actuators appear favorable for relatively large, high-force drives.

Hydraulic. Direct drive mechanisms employing hydraulically driven pistons can produce very large amounts of work in reasonably small volumes. They differ from piezoelectric and electromagnetic systems in that the pressurized fluids can be generated remotely. Hence hydraulic actuators can produce these large amounts of work with relatively low heat generation at the point of actuation. A current drawback is the expense required for the generation and control of high pressures and sealing of the fluids is always problematic.

Poisson's ratio or 'ballooning'. For solid chambers, the problem of sealing is obviated. Application of a pressure change to a pressure vessel will result in corresponding strains and subsequent motion of points on the surface. Two methods for achieving fine motion over small ranges have been employed. One consists of a simple cavity that, upon application of internal pressure change, results in relative motion of walls on opposite sides of the cavity. The other method may be visualized by a device consisting of a cylindrical cavity with a solid cylinder running through the center and a small clearance in which the pressurized fluid can be fed. Application of pressure to this fluid will impose a uniform stress on the central solid. By Poisson's ratio this solid rod will extend in the axial direction. Both techniques can provide stiff actuators for short-range, fine motion control. Similar costs of pressure generation and control exist as for the aforementioned.

Pneumatic. Pneumatic (mainly air) driven actuators have the advantage that reasonably high-pressure supplies are readily available in most manufacturing environments in the form of a mains supply of in pressurized canisters. Although the pressures are relatively low compared with hydraulic systems, they can exhaust to the environment without pollution or mess. The compressibility of gases means that heat will be generated whenever irreversible work is done on the system (it is always irreversible).

Thermal. Changing the temperature of a system element, often by passing a current through it, will result in both thermal softening and expansion with changing temperature. Using such an element to pre-load a flexure so that it is in tension produces actuation, the effect of temperature resulting in a change of the pre-load. Such designs are relative slow for large-scale devices, are asymmetric in response (heat can be 'driven in' to the material at a different rate than it can be extracted) and, by definition, produce heat.

8.3 Machining and heat treatment of some common flexure materials

By definition, optimal materials for flexures should have a relatively high elastic limit. For metals, this corresponds to the material being in its hardened state. However, this often precludes machining of the material using single point cutting (turning and milling) processes. This is unfortunate because these machine tools represent some of the fastest and cheapest methods for material removal. In many cases, there are direct cost benefits if the material can be free machined in an annealed state and subsequently hardened. For steels, machining technologies have existed for well over a century and numerous texts adequately cover the machining characteristics of this material.

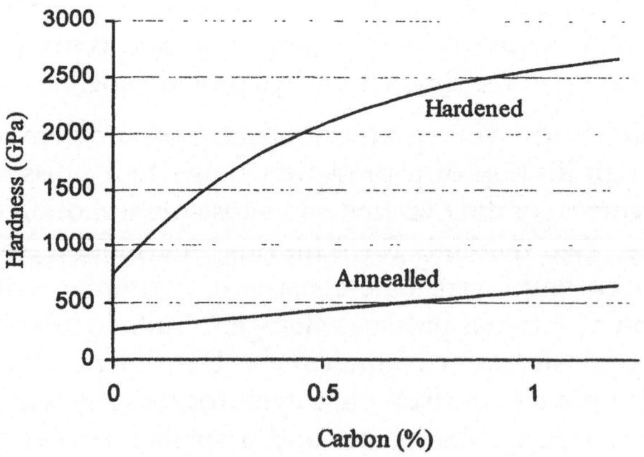

Figure 8.17: The hardness of steel as a function of carbon content

8.3.1 Steels

For the manufacture of precision flexures, it is desirable to use steels with the maximum difference in hardness before and after heat treatment. It is also necessary that the hardness of the material in its softened state is sufficiently hard so that it will machine relatively freely. In general this corresponds to steel having a carbon content greater than 0.3 % by weight and, where possible, close to the critical point at 0.83%, see figure 8.17. However, with such high carbon content this material tends to be both difficult to machine and may contain large residual stresses at the machined surfaces.

8.3.2 Beryllium copper

In its annealed state beryllium copper can be readily machined using either carbide or high-speed steel tools. For flexure applications, it is usual to select a high strength alloy with the addition of lead producing a 'free' machining alloy without effecting its mechanical properties. Precipitation hardening of this material is achieved by heating below a phase transition temperature of around 1100 F. for high strength applications it is important to use a material that has undergone significant cold working. This material can be readily annealed by heating above 1300 F, which results in a phase transition involving dissolution of the beryllium into the copper to form an α phase. Rapid quenching effectively freezes the material in its soft state. Heating in air will lead to discoloration from a copper color to dark brown.

After machining, the component can be hardened by heating to a temperature below the phase transition and leaving it at that temperature for a specified time period. As an example, a typical age hardening might require heating to a temperature of 700 F for 1 hour. Heating at too high temperatures or for excessive time periods can result in over-aging of the material resulting in a reduction in strength. For example, 800 F the optimal hardening time might reduce from 1 hour at 700 F down to 10 minutes[4]. High strength alloys have good fatigue resistance with 10^7 cycles endurance stresses of up to 600 MPa being possible.

As an aside, it is noted that clean beryllium copper readily glues to ceramic surfaces and can serve as a reasonably good conductor of electrical signals (high conductivity alloys are available but these have less desirable properties for flexure applications).

[4] A complete guide to beryllium copper can be obtained from Brush Wellman Inc., 17876 St. Clair Ave., Cleveland, Ohio 44110.

References

Bennewitz Von K. and Rötger H., 1936. Über die innere reibung fester körper; Absorptiosfrequenzen von metallen im akustischen gebiet, *Phys. Z*, **37**, 578 - 578-588.

Chen Kuo-Shen, Montiero A., Trumper D.L., Smith S.T., and Williams M.E., 1995, Spring dominated regime design of a high load capacity, electro-magnetically driven X-Y-θ Stage, *Proc. ASPE*, **12**, 199-202.

Dorf R.C., 1980, *Modern Control Systems*, Addison-Wesley, Ma.

Goodman L.E., 1996, Material damping and slip damping, in *Shock and Vibration Handbook*, 4th edition, ed. Harris C.M., McGraw-Hill, NY, chapter 36.

Gross W.A., 1964, *Gas Film Lubrication*, J. Wiley and Sons, Inc., NY.

Guckel H., Skrobis K.J., Christenson T.J., Klein J., Hans S., Choi B., Lovell E.G. and Chapman T.W., 1991, Fabrication and testing of the planar magnetic micromotor, *J. Micrmech. Microeng.*, **1**, 135-138.

Holmes M., Trumper D. and Hocken R.J., 1999, Long range scanning of an electro-magnetically suspended platform with shear damping, to be published.

Howard L.P. and Smith S.T., 1992, Long range constant force profiling for measurement of engineering surfaces, *Rev. Sci. Instrum.*, **63**(10), 4289-4295.

Kalpakjian S., 1995, *Manufacturing Engineering and Technology*, 3rd ed., Addison-Wesley Publishing Co. Inc., Ma.

Lazan B.J., 1968, *Damping of Materials and Members in Structural Mechanics*, Pergamon Press, London.

Liu X., Chetwynd D.G., Smith S.T. and W. Wang, 1993, Improvement of stylus measurement fidelity by active damping control, *Meas. Sci. Technol.*, **4**(12), 1330-1340.

Muller R.S. and Howe R.T., 1990, Technologies for microdynamic devices, *Nanotechnology*, **1**, 8 – 12.

Olsen P.T., Williams E.R. and Elmquist R.E., 1991, Monitoring the mass standard via the comparison of mechanical to electrical power, *IEEE Trans. on Inst. and Meas.*, **40**(2), 115-120.

Peirce S.B., Tran H., Wiedemann M. and DeBra D., 1994, Quiet hydraulics for ultraprecision actuation, *Proc. ASPE*, **9**, 20-25.

Petersen K.E., 1982, Silicon as a mechanical material, *Proc. IEEE*, **70**(5), 420-456.

Richards R.J., 1979, *An Introduction to Dynamics and Control*, Longman, London.

Ryu J.W. and Gweon D.G., 1998, Error analysis of a flexure hinge mechanism induced by machining imperfection, *Precision Engineering*, **21**(2/3), 83-89.

Schindel D.W., Hutchins D.A. and Smith S.T., 1997, A study of materials at high temperature using miniaturized resonant tuning forks and non-contact capacitance transducers, *J. Acoust. Soc. Am.*, **102**(3), 1296–1309.

Slocum A.H., 1992, *Precision Machine Design*, Prentice Hall, NJ.

Smith S. and Dvorak D., 1998, Tool path strategies for high-speed milling aluminum workpieces with thin webs, *Mechatronics*, **8**, 291-300.

Smith S.T. and Chetwynd D.G, 1990, Optimization of a magnet/coil force actuator and its application to linear spring mechanisms, *Proc. Inst. Mech. Engrs.*, **204**(C4), 243-253.

Smith S.T. and Chetwynd D.G., 1992, *Foundations of Ultraprecision Mechanism Design*, Gordon and Breach, London.

Snowden J.C., 1968, *Vibration and Shock in Damped Mechanical Systems*, J. Wiley and Sons Inc., NY.

Startin W.J., Beilby M.A. and Saulson P.R., 1998, Mechanical quality factors of fused silica resonators, *Rev. Sci. Instrum.*, **69**(10), 3681-3689.

Tlusty J., Smith S. and Winfough W.R., 1996, Techniques for the use of long slender end mills in high-speed milling, *Annals of the CIRP*, **45**(1), 393-396.

Vidic M., Harb S. M. and Smith S.T., 1997, Observations of contact measurements using a resonance based touch sensor, *Precision Engineering*, **22**(1), 19-36.

Wert C.A., 1986, Internal friction in solids, *J. Appl. Phys.*, **60**(6), 1888 - 1895.

Wilson F.W. and Cox R.W., 1965, *Machining the Space-Age Metals*, The American Society of Tool and Manufacturing Engineers, Michigan.

Xu Y. and Smith S.T., 1995, Squeeze film damping of a cantilever beam with a concentrated mass on the free end, *Precision Engineering*, **17**(2), 94-100.

Author index

Subject index

A

B

C

D

G

H

I

J

L

M

N

O

P

Q

R

S

T

U

V

W

Y

Z